CENTRIFUGAL PUMPS
AND
ALLIED MACHINERY

4th Edition

CENTRIFUGAL PUMPS
and
ALLIED MACHINERY

4th Edition

Harold Anderson

ELSEVIER
ADVANCED
TECHNOLOGY

ISBN 1 85617 231 7

Other books in this series include:
Hydraulic Handbook
Seals and Sealing Handbook
Handbook of Hose, Pipes, Couplings and Fittings
Handbook of Power Cylinders, Valves and Controls
Pneumatic Handbook
Pumping Manual
Pump User's Handbook
Submersible Pumps and their Applications
Centrifugal Pumps
Handbook of Valves, Piping and Pipelines
Handbook of Fluid Flowmetering
Handbook of Noise and Vibration Control
Handbook of Mechanical Power Drives
Industrial Fasteners Handbook
Handbook of Cosmetic Science and Technology
Geotextiles and Geomembranes Manual
Reinforced Plastics Handbook

Published by

Elsevier Advanced Technology
The Boulevard, Langford Lane, Kidlington, Oxford OX5 1GB, UK

Transferred to digital printing 2007

Preface

Harold Anderson's contribution to the world of pumping spanned more than 60 years and included many publications on the technology of pumps and allied machinery. This book provides updated information to many of his previous works and concludes his life in a manner so typical of his generous professional style; presenting his knowledge and experience for the benefit of others.

A Fellow of The Institution of Mechanical Engineers, a Life Fellow of The American Society of Mechanical Engineers and several prestigious international awards are indication of the regard in which Harold Anderson was held by his peers.

Throughout his career as an Engineer, a Pump Designer, an Author and Inventor, HHA loved to discuss engineering and to communicate his own ideas, innovations and his solutions to technical problems but he was also quick to acknowledge, compliment and support others on their work.

HHA's enthusiasm for engineering was matched in his encouragement and support for all who worked with him. His many published articles, papers and his books now provide a lasting source of help and inspiration to both academic and vocational workers in all aspects of pumping.

This last HHA book incorporates and updates a lot of his earlier work, including his unique "Area Ratio" method for pump hydraulic design and performance estimation. It also gives some very sound advice on pump mechanical design which is typical of his innovate and practical approach to engineering.

"Centrifugal Pumps and Allied Machinery" is a reference book for engineers at all levels who are concerned with pumps, their specification, design, behaviour or operation. The book is set out in a clear and convenient style; nine sections deal with the practices of pump engineering from its fundamentals, through pump and system design, pump applications, installation and operation. Information, pump design data and sound practical advice is presented on many pumping and related topics and most types of pump from the very smallest standard design to large high power machines.

HHA died before this book was published. As one of many pump engineers who was fortunate enough to be schooled by HHA in the art and science of pump design, I am privileged to have this opportunity to introduce his last publication.

DEREK J. BURGOYNE C.Eng., F.I.Mech.E.
Pump Engineering & Training Services
Cuckfield, West Sussex, England.

Acknowledgements

The publisher would like to thank all of the companies and individuals involved in the publication of this book.

A special debt of thanks is owed to Harold Anderson's family – especially his son Christopher – and also Thompsons, Kelly & Lewis of Victoria, Australia who supplied much needed information after Harold's death.

Harold had worked closely with Elsevier for a number of years and his sound advice and expertise will be sorely missed by all who knew him.

Harold H. Anderson (1906–1994)

Contents

SECTION 1

Fundamentals

INTRODUCTION

UNITS OF THE SYSTÈME INTERNATIONALE (SI)

GENERAL REVIEW OF PUMPS

GENERAL HYDRODYNAMIC PRINCIPLES

PUMP PERFORMANCE AND DIMENSIONS

THEORY OF CENTRIFUGAL PUMPS

INTRODUCTION

Historical

The raising of water has been one of man's earliest needs, and indeed, the first call for ingenuity in providing power arose from pumping duties. To this extent it may be suggested that the art of pumping is older than the art of power production.

The earliest devices for lifting water are still in operation in India, Egypt, etc. A typical example is the basket or scoop lowered into a river by means of a balance beam or shadouf, or in later development by a pulley, rope, and a beast of burden. Such devices would appear to be as old as man himself.

A definite step forward in the mechanization of pumping was the development of a lifting wheel carrying on the rim a number of buckets which dip into the river, and raise the water to an irrigation channel, the wheel being propelled by primitive gearing which, in turn, is driven by animal or human power.

A later development is to be found in the self-propelled wheels which are a combination of an outer water wheel driven by the river flow and an inner wheel having buckets which lift the water from the river to an irrigation channel. Typical examples of these wheels may be seen on the Yellow River in China. Other developments were the Archimedean screw, the hand operated reciprocating pump, and the chain pump.

The reciprocating steam engine was first applied to the duty of pumping water from mines. Reciprocating pumps, and engines, were later developed in parallel to large powers, culminating in the 1000 kW waterworks pumping engines of the 1890's, some of which are still in use today.

Centrifugal pumps driven by slow speed engines were applied to low head duties in the 19th century. As an illustration, two experimental gunboats, the Water Witch and the Viper, were fitted by the Admiralty in 1866 with jet propulsion. The relatively low efficiency of the pumps then available rendered this method of propulsion uneconomic, although present day efficiencies have justified the development of a small jet propelled boat which by avoiding the conventional propellers, can float in a few inches of water and can eliminate risks of propeller injury to water skiers.

In 1875 Osborne Reynolds produced the first turbine pump, ie diffuser pump, with considerably improved efficiency.

Modern Pumps

The invention of the steam turbine and the electric motor provided high speed power which was ideally suited to the centrifugal pump. As a result the rapid development of the centrifugal pump enabled it to replace reciprocating pumps on all but the highest pressure small quantity medium pressure duties, particularly when handling viscous liquids.

Application of Pumping Machines

Pumps have now been developed for every possible duty of lifting or raising pressure of liquids; quantities, pressures, temperatures and viscosities varying within the widest boundaries, and the powers ranging almost to the size of the largest prime movers yet made. The duties can be sub-divided as follows:

(a) Transport of valuable water etc, to where it is required. This section includes the provision of a clean water supply to domestic and industrial consumers. The problem of irrigation of land, the pumping of water from boreholes, the movement of water for fire extinguishing and the transportation of oil are important duties in this section.

(b) The removal of objectionable water, etc. This group includes the problems of mine drainage, dry dock drainage, sewage pumping and land drainage. Mines often sell water to water works.

(c) The circulation and process group. Here we are concerned with processes which demand a circulation of water to provide or withdraw heat, to feed steam boilers, to transfer various liquids in chemical and process industries, and in the refining of oil. This group also includes the pumping of pulp for the paper industries and the scrubbing of gases, for example in the nitrogen fixation industry.

(d) The transmission of power by the pumping of liquids. Here we have a large group of pumps providing high pressure water to operate cranes, presses, shears, lifts, etc in docks, steel works and similar environments. In a smaller degree oil is pumped to operate brakes, servo mechanisms, aeroplane controls, etc., where the value of the hydraulic transmission lies in the remote and sensitive control obtained and in weight saving.

(e) The use of a jet of water to cut, peel or dig. This group includes the descaling of steel bars, the removal of bark from trees and the sluicing of china clay and gold ore by water jets.

(f) The storage of energy by pumping water to an elevated reservoir. such duties demand very large pumps in order to justify the civil engineering costs, but such storage of energy is a material factor in improving the overall efficiency and the life of thermal and nuclear power stations.

(g) The jet propulsion of ships and space vehicles.

(h) The use of water for the purpose of transporting solids. In the food industry, potatoes, coffee beans, sugar beat, fruit and fish; in civil engineering work, dredging and sand stowage; in power stations, the handling of ashes, coal, etc, and

Figure 1.1 – Vertical borehole pumps in a South American water works each handling 1.4 m³/s against a head of 55 m.

in a multitude of industries the transporting of the various slurries is readily effected by the use of water.

(j) Pumping of food liquids, chemical and process liquids.

The Design and Manufacture of Pumps — The Types of Pump

The simplest type of pump is the reciprocating pump, of which the village pump is the earliest example. Being a displacement machine it will force a given quantity, corresponding to its speed and swept volume of cylinder, through the piping of a pressure system. The pressure developed is determined by the pressure differences, static lift and the friction losses in the system. The volume pumped is determined by the speed. If, however, the discharge valve of a reciprocating pump is closed we have the problem of the irresistible force and the immovable object. If the inertia of the rotating parts is sufficiently great the pump cylinder will burst. If the cylinder has sufficient strength the reciprocating and rotating parts will be brought to rest and the drive stalled since liquids are almost incompressible. It is interesting to note that this feature may in certain cases be used to develop very high pressures by taking advantage of the inertia of rotating masses. The reciprocating pump is therefore of value for developing very high pressures and has the advantage that an approximately constant quantity can be handled against varying pressure.

A centrifugal pump impeller throws the water outwards with sufficient force to generate a head, thus overcoming the static pressure and friction head in the system to which the pump is connected. When a centrifugal pump operates against a closed valve or against a system having a greater pressure than the pump can generate, no harm results, the impeller merely churning the liquid. (It may be necessary to remove the resulting excess heat.) The rotational speed determines the head of a centrifugal pump but determines the quantity of a reciprocating pump.

Rotary pumps are true displacement pumps. They do not rely upon the operating speeds to produce the pressure but obtain pumping effect by their elements passing in turn through the same space or by sliding vanes; they include vane pumps, screw pumps, gear pumps, etc.

Reciprocating Pumps

A reciprocating pump comprises: piston, cylinder, inlet and outlet values, and a driving mechanism. In the simplest case the pump has a single cylinder, the piston operating in one direction only, ie single acting. The inlet water is raised by vacuum from the sump or well, atmospheric pressure serving to force the water through the suction piping and the inlet valve. When the piston reverses, the inlet valve closes and the outlet valve is forced open by the pressure imposed upon the water by the piston, water being delivered through the outlet pipe.

It will be appreciated that if the speed of a reciprocating pump is increased, a limit will arise beyond which atmospheric pressure is insufficient to force water into the pump at the same speed as the piston. This causes a break in the continuity of the water in the suction pipe, referred to as cavitation, and gives rise to vibration, noise, and chemical attack by any dissolved gases which may be released from the water owing to the high vacuum and

Figure 1.2 – Large double entry split casing pump

the breaking of the water column. When the water passes into the pump and reaches a higher pressure as a result of the piston movement, the vapour recondenses with a severe shock and causes an impact on the metal walls of the pump. The damage to the metal is known as cavitation pitting, and is a combination of chemical attack and mechanical impact. It causes the metal to become spongy and, if allowed to persist, ultimately penetrates the metal wall.

In order to avoid this contingency the amount of suction lift and the operating speed must be carefully chosen using the experience gained on previous machines, so as to ensure that the plant is operating within safety limits.

Centrifugal Pumps (including Axial Flow Pumps)

A centrifugal pump comprises an impeller rotating at a high speed within a stationary casing. Centrifugal force is best explained by the following examples. If tea in a cup is stirred vigorously it will rise at the end of the cup to a slightly higher level than at the centre. This is due to centrifugal force and is further explained by another example. A train travelling at high speed along a straight line will try to continue along that straight line by virtue of its inertia. When a curve is approached the speed must be reduced to avoid the risk of a train leaving the rails in its endeavour to run along a straight line. In a similar manner the tea in the cup trying to travel in a straight line is constrained by the curvature of the cup with the result that the liquid piles up at the restraining wall.

This phenomenon is taken advantage of in a centrifugal pump where the liquid is spun round at sufficient speed to raise the level or pressure by an amount ranging up to 1000 or 2000 m in extreme cases.

The function of the casing is, firstly to restrain the water into an approximately circular or spiral path, and secondly to collect the water as it is delivered from the impeller. As the water passes through the pump it is accelerated in a tangential direction. Work is expended in raising the speed of the water. The pump head is therefore made up of centrifugal head due to rotation and impact head due to the increase of speed as the water passes through the impeller. The casing picks up the water and converts part of its kinetic energy or energy due to velocity into energy of pressure times mass.

In designing centrifugal pumps it is found that, up to a certain point, the performance can be calculated by known hydraulic and mechanical laws but beyond this point experience of previous machines must be called into account. In this respect it is perhaps true to say that pump manufacture is to some extent an art as well as science. Geometric scaling from a known small machine to a new large one is a convenient method of designing but the phemonena attending such change of size must be thoroughly investigated.

Head and Quantity

The head generated by a pump will obviously have some reference to the velocity of rotation since the head is primarily due to the kinetic energy imparted to the water. Kinetic energy varies as the square of the velocity.

As a first approximation, therefore, it can be said that the head generated varies as the square of the peripheral velocity of the impeller. Similarly, the quantity pumped will depend upon the area of the pump passages and upon the speed of flow which in turn is determined by the peripheral speed of the impeller.

Tests are analyzed by plotting the variation of head and quantity from the above relationships against a suitable variable in the proportions of the pump itself.

Efficiency

The efficiency of the pump is primarily determined by the size, speed and proportions of the impeller and casing. The head generated by the pump suffers from losses due to the friction within the walls of the impeller and casing. This friction is proportionately greater in the case of a smaller pump. The reason for this may be explained as follows: if we consider a one cm square passage, the capacity of the passage for handling water is 1 cm²; the resistance to flow is represented by the perimeter of the walls, that is 4 cm. If however the dimensions of the walls are increased to 2 cm square, giving an area of 4 cm², it will be seen that the area has increased four-fold, but the perimeter has only creased two-fold to 8 cm. The wall friction, therefore, which is represented by the length of perimeter, has a relatively smaller effect on larger areas.

A thorough investigation of friction was made by Professor Osborne Reynolds, who provided a relationship between the friction loss due to the rubbing against the walls and the friction loss due to churning eddies of water within the passage itself. The relationship between these two losses was expressed by a 'Reynolds' number.

Professor Reynolds in one of his experiments passed a quantity of water through a glass tube. In the centre of the glass tube he introduced a thin thread of dye. On low water velocities the dye passed through the pipes as a straight clean thread. When the speed of the water was increased, eddies started and, as a result, the thread was completely broken up. The point at which the break-up occurred is referred to as the critical velocity. Below this critical velocity, pipe friction was found to vary as the velocity, being almost entirely wall friction. Above the critical velocity the pipe friction varied as a power of the velocity approximately in the neighbourhood of 1^2. The exact value of this power depended upon the Reynolds number, larger numbers representing mainly eddy losses, giving a higher power to the loss. Kinetic energy varies as velocity square. Friction losses vary as some other power of velocity. A change of velocity and size, therefore, will alter the friction relative to the total head and consequently alter the efficiency of the pump.

It is, therefore, possible to use the Reynolds number as a determinant of friction and from this, it is a simple step to use a form of Reynolds number to determine by change of relative friction losses in the pump how efficiency will change with size and speed.

A very simple and every-day example of Reynolds' experiment is seen in the smoke rising from a cigarette. The smoke is lighter than the surrounding air and the difference in weight will give it an upward thrust. The thrust acts continuously, thus causing a progressive increase in the speed of the smoke as it rises.

The smoke near the cigarette will have a slow speed and the further the smoke is from the cigarette the higher its speed will be. At some point it will reach its critical speed above which flow is unsteady. In still air it is particularly noticeable that the smoke rises from the cigarette in a straight thread for a foot or so then very definitely breaks up in eddies and waves.

Figure 1.3 – Multistage drainage pumps in a South African mine. 1 000 m head.

Cavitation of Centrifugal Pumps

As mentioned above in the case of reciprocating pumps the suction of the pump is capable of lifting water to a certain height, the movement of the water being due to the atmospheric pressure on the surface of the well or sump being greater than the partial vacuum in the impeller entrance. In similar manner to the case of reciprocating pumps it is possible to increase the speed of a centrifugal pump to a point where the suction stream can break up so that actual cavitation occurs. This is dependent upon the operating speed and the friction in the inlet passage. The friction loss can be determined at the cavitation point by the use of Reynolds number in similar manner to that described in the paragraph on Efficiency above. It is obvious that the cavitation will be dependent upon the velocity of water entering the impeller and on the relative velocity of the impeller blades where the water is picked up. A combination of these two speeds is the main reference of cavitation studies.

The Use of Experience

In·practice, a pump is designed, manufactured, tested and put to work on site. Experience of this and other pumps will permit the building of larger and higher speed machines. After several pumps have been manufactured it is possible to survey the whole range of experience in respect of efficiency, cavitation and head quantity characteristics so that larger machines can be designed. It is, of course, that any step outside the realm of known experience brings unknown hazards and it is in this respect that design is to some extent an art as well as science.

A step towards a machine of higher speed, pressure or size is a step in the dark in respect of the performance, efficiency, cavitation, and wear and tear. The experience range permits us to lay down rules of design and performance which will to some extent hold good for steps into the darkness. Uncertainties are of course inevitable but this is true of all engineering work and indeed the successful overcoming of such difficulties provides its main interest.

Variables in Manufacture

The variations of performance of pumps are primarily due to the fact that water appears to have a will of its own and may travel in any direction according to the roughness of the passage — an uncertain factor with cast surfaces — and the amount of eddies that may be incurred. The friction losses, efficiency, head and cavitation characteristics are therefore liable to vary from pump to pump.

Empirical Designing

The most satisfactory method of investigation of performance is to plot all available test results against some variable in the dimensions of the pump. A considerable degree of patience is involved in the investigating the effect of all possible factors on the performance of a pump. The correct choice of a factor will be indicated by the closest grouping of test points on the plotted chart. A central line can then be drawn through the mean of the points.

It is therefore possible to say that all the results agree with the centre line of the curve, within an error of plus or minus a percentage determined by the spread of the test points. It is a reasonable but not absolutely certain forecast that other machines will behave similarly in respect of the variations from the central line of points. As an example, the plotting of efficiency against Reynolds number gives a fairly close curve, thus permitting forecast of other sizes.

A similar analysis of the dependence of the head and quantity of a pump on the internal proportions gives a reasonably consistent curve, though not however quite so consistent as the efficiency curve.

In addition to the friction loss mentioned above, centrifugal pumps suffer from leakage, since rotating and stationary parts must have a clearance to prevent actual contact and through this clearance liquid can leak away, thus wasting power. There is also the friction loss due to rotating an impeller in a chamber of water and, in addition, gland and bearing power losses.

Mechanical Problems

The first problem concerns the design of the casing as a pressure vessel capable of withstanding all operating pressures without bursting or distortion. These problems become very complicated when the pumps must be capable of operating with sudden changes of temperature. The casing must also be able to resist chemical attack by whatever liquid is being pumped.

The rotating impeller must be strong enough to transmit all the turning forces, resist hydraulic forces due to the pressure distribution within the casing and must resist bursting stresses due to high speed.

The shaft is required to transmit the turning forces and provide a sufficient margin of strength to avoid any critical vibrations. It must also withstand axial and radial forces due to uneven distribution of pressure around the impeller, especially at part loads.

UNITS OF THE SYSTEME INTERNATIONAL (S.I.)

This book uses SI units since these units are embodied in the pump specifications of the International Electrotechnical Commission, the International Organization for Standardization, the British Standards Institution and the DIN. The units mentioned are those approved at the time of writing.

Rotational speeds are in rev/min. Pump heads are given in metres, based upon normal gravity acceleration of $g = 9.81$ m/s^2, since all empirical pump data are in this form. Pump pressures are given in bars, so that with $g = 9.81$ a head of 10.2 m of cold water is equivalent to one bar.

The use of bars facilitates determination of pump input power and determination of stresses of pump casings etc in hectobars. One bar equals 100 kilonewtons per square metre (14.504 lb/in^2).

Pump flows are given as cubic metres per second or litres per second. Temperatures are in degrees Celsius, kinematic viscosities in square metres per second.

Reason for Use of SI Units

No system other than SI is capable of explaining pump behaviours in the universal sense which must now include space travel.

Effect of Earth's Gravity

In general, pumps have been designed, manufactured, tested and set to work in areas where variation of gravity is negligible and a record of empirical design data has been based upon this gravity field of $g = 9.81$ m/s^2. The charts in this book are based on this gravity and pump relationships are now expressed as

$$gh = uw \text{ and } gh = v^2/2$$

Now, however, consideration must be given to pumps operating under different gravities; for example, in a spacecraft where gravity is almost zero, the pump head in the above equations becomes almost infinite. The head is described as the height of a column of liquid supported by the pump against gravity. It is here that the SI system of units is valuable, since the pressure generated by a pump can be expressed as specific energy in joules per kilogram. J/kg equals Nm/kg equals gh, of which the first two terms are independent of gravity.

Shape number

Shape number (previously referred to as specific speed) is a shape number differentiating (with respect to efficiency and to head and quantity coefficients) between narrow impellers, small volute throat areas (low shape number) and wide impellers, large volute throat areas (high shape number).

As a pure number, the actual value is of little importance, provided it is used consistently in empirical charts of efficiency and of head and quantity coefficients.

The simple expression

$$\text{Shape number equals } N^2 \text{ equals}$$

$$\frac{\text{rev/min l/s}}{M^2}$$

(avoiding unnecessary coefficients and assuming the constant gravity of these charts) is in line with standard specifications and is used in this book.

The strictly correct and dimensionless shape number

$$N^2 \text{ equals } \frac{\text{rad/s } \sqrt{m^2/s}}{\left[\dfrac{Nm}{kg}\right]^2}$$

involves in every case gravity conversion from m to Nm/kg and conversion from rev/min to rad/s, which would waste the time of designers who make shape number calculations several times a day. A simple l/s shape number of 1675 equals a dimensionless shape number of one.

Example

Boiler feed pump to deliver 1000 tonnes per hour at 150°C against a differential pressure of 200 bar. Density at 150°C is 920 kg/m², hence

$$1\,000 \text{ tonnes/h equals } 1\,087 \text{ m}^3/\text{h } 0.3 \text{ m}^3/\text{s}$$
$$200 \text{ bar equals } 20 \text{ million N/m}^2 = 20 \text{ M Pa.}$$

Liquid Power equals 0.3m³/s x 20 MPa equals 6 MW at 150°C 6.5 MW at 20°C

Input Power at 80% pump efficiency equals 6/0.8 equals 7.5 MW at 150°C, 8.1 MW at 20°C

Liquid power and input power have previously been expressed as water horsepower and brake horsepower.

The generated pressure must be expressed in metres, where g = 9.81 m/s², to determine the combination of impeller diameter/speed and the efficiency from the charts in the book.

Figure 2.1 – Large circulating pump

Previous Measures

To engineers accustomed to previous measures it is useful to know the follow approximate conversions for pumps.

Specific Speed in gal/min and feet multiplied by $^2/_3$ gives shape number in litres/sec and metres.

Gal/min multiplied by $^3/_4$ gives correct digits for litres/sec. Note: Divide answer by 10 for actual quantity.

Newtons per mm^2 equals megapascals (MPa).

One Hectobar equals $^2/_3$ tonnes per sq inch, equals 10 megapascals.

Abbreviations of SI Units

Length	metre	m
	millimetre	mm
Flow	litres/second	l/s
	cubic metres per second	m^3/s
Head	metres	m

Pressures	newtons per sq metre equals pascals	Pa
	bar equals 100 kN/m^2	bar
Revolutions per minute		rev/min
Radians per second		rad/s
Power	kilowatts	kW
	megawatts	MW
Stress	hectobar	hbar
Energy	joule	J
Time	hour	h
	minutes	min
	second	s

GENERAL REVIEW OF PUMPS

Technical Aspects of Centrifugal Pumps

A pump receives a flow of liquid at a certain inlet pressure, raises this pressure to a higher value and discharges the liquid through the outlet. Mechanical power in the form of shaft torque and speed is, therefore, converted into hydraulic power in the form of a flow rate raised in pressure level. Centrifugal and axial flow pumps are entirely dynamic in action, that is to say, they depend upon rotational speed to generate a head which is manifest as a difference of pressure between inlet and outlet branches. The quantity of liquid pumped and the power involved depend upon the system of liquid containers, static pressures and pipeline, etc, to which the pump is attached.

In contrast, a displacement pump, reciprocating or rotary, when driven at a certain speed, will force a given volume of liquid through the system to which the pump is attached, and the pressure obtaining and the power involved will depend upon that system.

Centrifugal and axial flow pumps are used up to the largest flow quantities for pressures varying from a few mm head to 10 000 m head in extreme cases. Considerations of efficiency demand that the centrifugal or axial flow pump should have a combination of quantity, head and rotational speed to give a favourable geometric shape to the pump. This is described in detail below under the heading of Shape number. It is sufficient to say here that the smallest flow quantities cannot be associated with the highest pressures, nor, because of power limitation, can the highest quantities be associated with the highest pressures.

Standard Pump Types

Small, general purpose, single stage pumps for minimum cost embody single entry overhung impellers in a volute casing with tangential discharge, the shaft being carried in two ball/roller bearings. Impeller, casing and bearing housing are made of cast iron. Such pumps are produced in very large numbers and have universal use in chemical, metallurgical and process industries, etc.

The more expensive type of pump with bronze impeller and renewable wearing parts has a double entry split casing, permitting examination and removal of the rotating element without disturbing pipe joints. These pumps will be batch produced in the smaller sizes and tailor made in the larger sizes for waterworks' duties and other such general purposes. The above single stage pumps are available for heads up to 100 m on the small sizes and up to 200 m on the larger sizes. Two stage pumps giving up to 200 m on the smaller sizes and up to 400 m on the larger sizes are useful for duties which are not sufficiently high in head to justify the more expensive multistage pump. The limitation of head on the smaller sizes is due to the aspect of shape number discussed below and to the difficulty of casting large diameter, narrow impellers.

The split casing pumps have bearings and glands with water seal on either side of the casing, ball/roller bearings being used on the smaller sizes with sleeve bearings on the larger sizes. Renewable neckrings, sleeves and gland bushes are provided.

Figure 3.1
Diagrammatic view of single stage split casing pump.

Combined motor pumps, mass produced in small sizes and tailor made in large sizes, have come into favour during the last 30 years. For very large duties and relatively high heads, a single entry impeller will permit a considerable increase of shaft diameter, thereby giving greater rigidity and simplification of layout. This is directly following water turbine practice, where double exit turbines are obsolete. Small high speed gear driven pumps are used for low flow high pressure duties.

For the lowest heads, 1–20 m, axial flow pumps and cone flow pumps are used. An axial flow pump impeller resembles a ship's propeller, usually, however, provided the inlet guides and outlet diffusers. On the larger sizes, axial and cone flow pumps can be provided with movable blades similar to a Kaplan Turbine in order to improve versatility and efficiency. The very high generated head and power demand of an axial flow pump at zero

Figure 3.2 – Single stage water supply pumps.

flow is usually avoided by providing a flap valve on the discharge so that the pump never runs at zero flow.

Multistage Pumps

In order to operate against higher pressures, several impellers are built on to one shaft in a multiple casing. Each stage comprises impeller, diffusers and return passages within a chamber which discharges the liquid into the next stage. This construction has the advantage of minimum cost, since the stages can be manufactured in batches and be available for building into pumps as required.

The stages are assembled between end covers containing the inlet and outlet branches, glands and bearings, the covers being held together by long bolts. The delivery cover generally contains a hydraulic balancing disc or drum to carry the end thrust of all the single entry impellers and to reduce the delivery pressure on the gland.

The above cellular construction is normal in Britain and on the Continent and is used in America for the highest pressures. For lower pressures up to about 70 bar, American manufacturers adopt split casing design with two, four, six or eight stages, in the end thrust being balanced by a suitable arrangement of flow passages. This design however is used only to a limited extent in Britain and on the Continent (Fig 3.3).

Technical Aspects of Rotary, Reciprocating and Other Pumps

Reciprocating pumps have positive action and will deliver a certain volume of water against a pressure determined by the system to which the pump is attached.

Figure 3.3 – Multistage water works pumps.

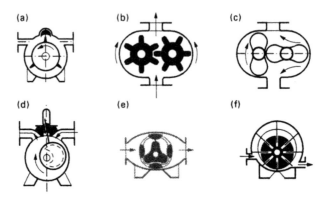

Figure 3.4 – Typical rotary pumps.

Rotary pumps (Fig 3.4) are also positive but generally have greater clearance leakage than reciprocating pumps and have lower inertia forces. Reciprocating pumps are therefore used for the highest pressures and the smallest quantities, rotary pumps being used for small quantities and medium pressures, especially oil duties, since lubrication of the internal parts is essential. Helical screw pumps are used for large powers at 200 bar, as shown at the end of this chapter.

Steam pumps, comprising a power cylinder and a liquid pumping cylinder, generally operate without rotary motion, steam valves being moved by linkage from the rams. Certain large steam pumps may have connecting rod, crank shaft and flywheel to provide means of governing and to give smoother operation. Only a small percentage of the total

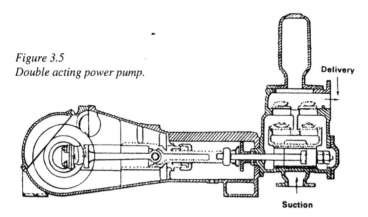

Figure 3.5
Double acting power pump.

energy, however, will be passed to and from the flywheel. Prior to the development of electric motors and steam turbines, large steam pump engines up to 1000 kW were installed for waterworks' duties, but very few of these are still in operation.

Reciprocating pumps (Fig 3.5) are used for small quantity high pressure duties where their efficiency can exceed that of a centrifugal pump. For example, 10 l/s 100 bar would be an entirely practicable duty for a centrifugal pump as far as mechanical operation is concerned and advantages of low costs, smooth delivery, easy regulation, small space and foundations would accrue, but a centrifugal pump efficiency of about 45% would take also twice the power of a reciprocating pump at 80% efficiency. This difference of power maintains the position of the reciprocating pump within its sphere.

The power pumps, with various combinations of single or double action and one, two or three cylinders, are generally driven electrically via gears. The multiple cylinders give smoother flow, for example, triplex pumps with cranks at 120°. Such pumps are often used for mine drainage duties and general industrial purposes handling water and other liquids.

Small reciprocating pumps for duties of a few litres per second against pressure up to 1000 bar are built with multi-plungers operated in line from one shaft with cams or from a swash plate (Fig 3.6). The latter pump is generally fitted with a gear type of booster to

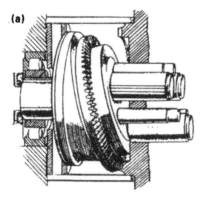

Figure 3.6
Axial plunger pump.

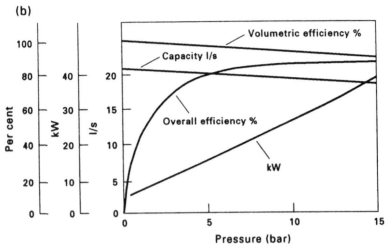

Figure 3.7 – Gear pump characteristics

keep the plungers pressed to swash plate, thus rendering the plate inactive but protected, in case of accidental oil starvation. The operating speed is 1500 rev/min.

Rotary pumps have positive displacement which is obtained in two ways.

(a) A rotor carries radially adjustable vanes, the outer tips of which are constrained by a circular casing, whose centre is remote from that of the rotor.

(b) Various combinations of gears, lobes, helixes, etc, within a casing arranged so that lobes or teeth attached to each rotor pass in sequence through the same pumping space.

Either of the above conditions can give direct displacement with smoothness and absence of pulsation, but since internal contact of rotating parts is inevitable, rotary pumps are normally used for liquids with some degree of lubricating properties.

Water-ring pumps are a combination of the rotary vane pump mentioned above and a centrifugal pump. Although the vanes do not extend to the periphery, the inertia of the water causes it to behave in a similar manner to the case of vane pumps so that a small water piston appears between each pair of vanes, thus giving a pumping effect. Water-ring pumps are used for priming duties and for extraction of water and air from the screen or wire of a paper mill.

Peristaltic pumps use a tube which is compressed to cause flow.

Reciprocating or Centrifugal Pump

In general, the choice between centrifugal and reciprocating types is determined by physical limitations of peripheral speeds, erosion, metal strengths, and the physical size of units from the aspect of manufacture, transport to site, and erection.

Choice of type is also largely determined by economic aspects and by the range of the duty and the cost of power. For example, a modest duty of 10 l/s at 30 m can be produced by a cheap single stage pump with an efficiency of 52%. Where power is cheap or where the operating hours represent a small proportion of the day, such a pump may be entirely economic.

On the other hand, where power is expensive and where a pump is running for a large proportion of the day, the multistage design, working on a more favourable shape number, that is with improved impeller proportions as a result of a lower head per stage, would perform the duty mentioned above with efficiency of 70%. The greater cost of the multistage pump would be justified on reduced power consumption. A reciprocating pump on this duty would give an even higher efficiency, but would be uneconomic on cost of installation.

As already mentioned, in the nineteenth century, reciprocating steam driven pumps were built for waterworks' duties in sizes as large as 1000 kW.

Such pumps comprising a vertical triple expansion steam engine with pump cylinders directly in line with the steam cylinders, would occupy a space of roughly 350 m² and give a pumping efficiency of the order of 90%.

At a time when the centrifugal pump efficiency was no greater than 75% and the water demand of cities was relatively modest, such pumping engines were a practical proposition. Since that time, however, centrifugal pump efficiencies have been developed up to the 90% efficiency level, and the steam turbine and electric motor have also been developed to provide adequate power and speed for today's demands.

An electrically driven centrifugal pump for the duty mentioned above would only occupy a space of 25 m³ and would have a pump efficiency of about 90%.

When we consider that turbines, ie reversed pumps, are now in operation with a power input up to 700 MW (one million hp), it will readily be seen that physical considerations alone would rule out the reciprocating pump for large quantity duties.

For the highest pressure duties, the reciprocating pump has a very much stronger appeal, but even there the disadvantages of vibration, physical dimensions and uneven flow severely limit its applications. For example, only one major power station in Britain has so far been equipped with reciprocating boiler feed pumps, and it is considered very unlikely that further reciprocating boiler feed pumps will be put forward even for super-critical steam boilers requiring feed pumps of 400 bar.

These points are further amplified by the study of the respective advantages of centrifugal and reciprocating pumps below.

Advantages of Centrifugal Pumps

When compared with a reciprocating pump, the centrifugal pump offers the following advantages:

1. Higher speed resulting in lower size and costs.
2. Continuous delivery free from pressure fluctuations.
3. Having known characteristics a centrifugal pump can
   (a) operate on minimum flow without exceeding a predetermined pressure;
   (b) operate on maximum flow without exceeding a predetermined power demand;
   (c) be designed to meet several duties, thus in many (but not all) cases matching the pump characteristic to the site requirement.
4. Applicable to direct drive in almost every case.
5. Absence of vibration and simpler foundation.

6. When handling liquids containing solids, eg paper pulp, a centrifugal pump tends to deliver a constant weight of solid matter irrespective of variation of consistency. The heavier consistencies cause a greater friction loss, thus reducing flow and compensating, in some measure, for consistency changes.

Advantages of Reciprocating Pumps

1. Within their field, higher efficiencies than centrifugal pumps.
2. Applicable to variable pressures without adjustment of speed – a useful factor where the head is uncertain.
3. Self-priming and, therefore, able to handle a certain amount of air or other vapour without failing to pump.
4. Capable of utilizing kinetic energy or rotation to give peak pressures and, as a result, often able to clear accidental obstructions in the delivery main.

Reciprocating pumps are, therefore, confined to duties where quantity is low and pressure high, for example, quantities up to 10 l/s and pressures up to 2 000 bar.

Field of Operation of Centrifugal Pumps

Centrifugal pumps are used for duties varying in quantity from the smallest to the largest, the upper limit being dictated only by the ability to manufacture and to transport. At the present time, the maximum flow achieved is of the order of a hundred cubic metres per second or ten million tonnes a day, the maximum stage head is 2000 m and the maximum pressure ifs 600 bar. Site welding of casings are splitting of impellers for transport has permitted the installation of water turbines of 700 MW and pumps of 200 MW each. In short, the centrifugal pump is second only to the electric motor in frequency of use in the world, and in this application has virtually no limits other than reasonable shape of impeller, and with it a reasonable efficiency. This latter aspect is discussed later under the heading of Shape Number.

General Construction

A centrifugal pump comprises, basically, an impeller and a casing together with driving shaft. The point at which the shaft enters the casing require a stuffing-box with gland or a seal. Sleeves, neckrings, wearing rings, etc. are provided at points where stationary and rotating parts are close together. The shaft will require journal and thrust bearings and a coupling from the driving unit.

The impeller may be of the single entry type (with hydraulic or mechanical balancing devices) or the double entry type. Single entry impellers are generally fitted to casings which are split on a plane normal to the shaft axis. Double entry impellers are generally fitted to casings split on the horizontal centre line, ie in the plane containing the shaft axis. The latter arrangement permits the placing of the branches below the centre line so that, by removing the upper half casing, the pump can be examined without disturbing pipes. The discharge from the impeller is collected in the volute or spiral portion of the casing and led to the delivery branch.

Impellers may be radial flow, mixed flow, or axial flow according to the direction of the

(c)

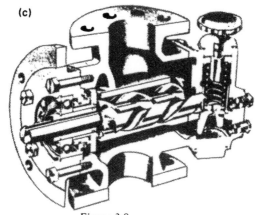

Figure 3.8
Helical screw pump.

water within them. The type of flow is determined primarily by the shape number. Backward curved vanes are universally used. Multistage pumps comprise several impellers on one shaft in order to generate a greater head than can efficiently be obtained from one stage. The construction of the pump involves some modification from the single stage type. For example, in order to save space and length, the volute is replaced by a series of diffusers and passages to return the water to the centre where it may enter the next stage. It is usual to fit a multistage pump comprising single entry impellers with a hydraulic balancing device.

GENERAL HYDRODYNAMIC PRINCIPLES

Before proceeding with the detail of centrifugal pump design, it is necessary to formulate certain general considerations of geometric design, of duty and of the liquid to be handled.

Geometric Designing

Since we are concerned with designing not just one pump but many complete ranges of pumps of all shapes and sizes, it follows that the basic calculation procedure for one pump could be applied by laws of similarity to other pumps.

Figure 4.1 – Small combined motor pump.

Since a centrifugal pump is a dynamic device for raising the pressure of a liquid, the increase of pressure, expressed as head of the liquid and determined according to Newton's laws, will vary as the square of the rotational speed.

Similarly, the quantity pumped, expressed as a volume of the liquid passing through the pump in unit time, will vary as the first power of rotational speed. In consequence, the power taken to drive the pump will vary as the product of the above, ie as the cube of the rotational speed subject to any small correction that may be made for change of efficiency with size or with speed.

The above relationships refer to the alteration of rotational speed of the same pump.

When we consider a geometric increase of size of pump whilst maintaining the same rotational speed, it follows that the head will vary as the square of this size, since the peripheral speed is only increased by the diameter change.

The flow will increase in proportion to the size, since the peripheral velocity has so increased, but, in addition, the flow will increase in proportion to the square of the size since the areas have so increased.

From this, it follows that the flow through a pump will increase as the cube of the size for a constant rotational speed and, furthermore, that the power, being the product of head and flow, will increase as the fifth power of the size, subject to any small change due to efficiency increase from a small pump to a larger one.

Hydraulic and Mechanical Limitations

The speed of a given pump can be increased until the limitations of mechanical strength and/or the capacity of the impeller to induce liquid are reached. The most convenient determinant of these limits is the operation head at the best efficiency point of the pump.

For example, if one particular pump under specified inlet conditions is capable, mechanically and hydraulically, of running at 200 m head at best efficiency point, if follows that all geometric machines when running at speeds to give 200 m at best

Figure 4.2 – Combined motor pump

efficiency point will have similar stresses and similar margins above cavitation. Small corrections, however, occur on efficiency and cavitation margins and on generated head, and these will be considered later.

Composite Similarity Rule

It is very convenient when determining the design data for one machine from those of another, to combine the above relationship in such a manner that the head is kept constant. This is achieved by reducing the operating speed inversely as the size or principal diameter is increased. By combining the above relationships, we see that a geometric change of size with inverse change of speed will result in the flow and power varying as the square of the size.

Model Tests

A model of moderate size can, therefore, be made and tested for hydraulic and for mechanical performance (using strain gauges), so as to forecast the performance of a prototype too large to test.

Models are used in the water turbine industry and in the manufacture of very large pumps, eg on pumped storage. The model power would be of the order of 2 to 4 MW.

Pumps up to about 15 MW can sometimes be tested to the maker's works, so that a record of behaviour of all sizes progressively from the smallest to the largest can be obtained, every pump being regarded as a model for every other pump of the same geometric shape, thus extending the field of data and giving more accurate laws of behaviour than a single pair of model and prototype would provide.

Summary of Similarity Formulas

Change of speed for the same size of pump

For a given operating point on the characteristic of the pump, flow varies as rev/min, head varies as (rev/min)^2. Since power is the product of flow and head, divided by efficiency, it follows that power is proportional to (rev/min)^2. A small change of impeller diameter gives similar variation.

Geometric change of size for the same rev/min

Flow varies as diameter^2, head varies as diameter^2, power varies as diameter^2 subject to efficiency changes.

Geometric change of size with inverse change of speed

Flow varies as diameter^2, power varies as diameter^2, rev/min varies inversely as the diameter. The above is subject to minor changes consequent upon change of efficiency with size and speed.

The following are constant: head; margin of safety above cavitation conditions; stresses in rotating and stationary parts; and margins of safety above critical speed.

Bearing loads due to rotor dead weight will vary.

Figure 4.3 – Two stage pumps

Liquid or Water

Here a purist would say that we should always use the term 'liquid' when discussing pumps.

However, the vast majority of pumps handle water, and water is the best known liquid to all of us in respect of its properties and behaviour when temperatures and pressures change.

The investigations on cavitation are more readily understood if we consider water, after which we can apply the general principles to other liquids, when needed.

The word 'water' will, therefore, be used in the general sense in describing pumps, and in the particular sense in cavitation calculations, unless otherwise specified.

Duty of the Pump

In design calculations and discussions, the duty of the pump will refer to its performance, litres/sec, metres head and revolutions per minute at the best efficiency point, unless otherwise specified, when describing the complete characteristics.

PUMP PERFORMANCE AND DIMENSIONS

Hydraulics has been defined as a subject having only one mathematical principle; that is to say, Bernoulli's theorem, and a mass of empirical formulas.

The design of centrifugal pumps naturally suffers in some measure from the uncertainties of cast surfaces and from the vagaries of water flow.

Nevertheless, considerable progress in improvement of efficiencies is resulting from a combination of careful statistical analyses of all tests and the application of Newton's laws and Bernoulli's theorem. Recent work on turbulence and boundary layer has been invaluable in giving that last bit of improvement so essential today.

Determination of Dimensions of Pump for a Given Duty

Since the head of a pump is proportional to the square of the peripheral velocity, that is to say the square of the product of speed and impeller diameter, it follows that the first approximation to the head of the pump is given by the expression

$$\frac{u^2}{g}$$

where u is the peripheral velocity of the impeller and g is the acceleration due to gravity.

The flow through a pump is determined by the peripheral speed of the impeller to which all velocities are related, and the flow area.

Tests are, therefore, analyzed by plotting the head generated by the pump as a ratio of $u^2/2g$ and plotting the flow velocity within the diffuser or casing throat, the beginning of the cone preceding the delivery branch, as a ratio of u.

These ratios are plotted against the ratio of the discharge vane flow area of the impeller to the casing throat area (see Fig 5.1).

A further refinement is necessary in respect of the efficiency of the pump.

A 50 mm pump with an efficiency of approximately 60% will have higher frictional losses than a 500 mm pump with an efficiency of 90%. The 50 mm pump will, therefore, have a lower head than a 500 mm pump of the same peripheral speed.

This variable of efficiency is eliminated in respect of head analysis by dividing the ratio of the head to $u^2/2g$ by the hydraulic efficiency of the pump.

The hydraulic efficiency is determined by a loss analysis — described later — wherein the mechanical and hydraulic losses in the pump are separated.

This analysis brings all pumps to the same level on a dynamic basis before any plotting is made of the variation of pump head with impeller to casing proportion and peripheral speed.

Impeller to Casing Area Ratio (Figs 5.1 and 5.2)

The importance of this ratio lies in the fact that with a universal use of backward curve vanes, the whirl velocity of the water is dependent on the relation between the forward velocity of the impeller itself and the relative backward velocity of the flow within it.

From a purely empirical aspect, we can consider two pumps having the same speed and same impeller diameter. The first pump has a very narrow impeller in a throat of very large area.

The best efficiency flow of the pump is determined by the combination of the impeller outlet area and the casing throat area. With a very small impeller area, the velocity of flow will be correspondingly large.

The whirl velocity of the water leaving the impeller, that is to say, the tangential component of its absolute velocity, will be determined by velocity triangle, from the peripheral velocity of the impeller forward and the relative flow velocity of the water within the impeller which, owing to backward swept blades, will be in a backward relative direction. Since the relative backward velocity is high, the resulting whirl will be low and the head generated by the pump will be low. At the opposite extreme, we can consider a pump with a very wide impeller in an extremely small casing throat.

The flow through the pump at best efficiency point will again be determined by the combination of impeller and casing area. In this case, however, since the impeller area is very large, the backward relative flow through the impeller passages will be small. the

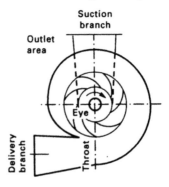

$$\text{Area ratio} = \frac{\text{Total outlet area of all vanes}}{\text{Throat area}}$$

Figure 5.1

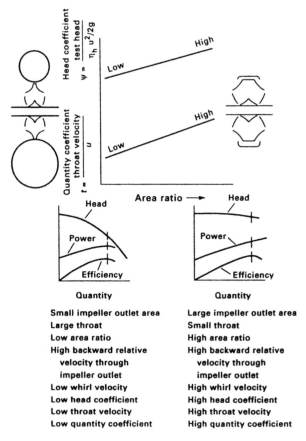

Figure 5.2

whirl velocity, which is determined by the impeller peripheral velocity forward and the relative impeller flow velocity backward, will, therefore, be considerably higher than in the preceding case, and the pump head will also be considerable higher.

Regarding the flow quantities in each of the above cases, it is obvious that with a very narrow impeller, the flow velocity through the large throat will be very small indeed.

Conversely, with the very wide impeller and small throat, the velocity within the throat will approach the peripheral velocity of the impeller.

Probable Shape of Performance Charge

Before plotting empirical data or going into deep theoretical considerations, we can, at this stage, by purely logical reasoning, form some idea of the general shape of the charge representing area ratio and pump performance.

With a low area ratio, that is a small impeller area with a large casing throat, we would expect to generate a low head and to have the best efficiency point corresponding to a flow velocity in the throat which is a very small proportion of the peripheral velocity.

At the other extreme, with a large impeller area and a small casing throat, we would expect to generate a head corresponding to a forced vortex, which, we will see later, approaches twice the velocity head, and to have a flow velocity in the throat at the best efficiency point which approaches the peripheral velocity of the impeller.

Between these two extremes, we can imagine a gradual change of performance, but the actual determination of the curve shape involves the plotting of every available pump test.

Influence of Water Turbine Design on Pump Design

In 1915, Professor Daugherty of Pasadena (Ref 1) showed how the velocities within the pump could be referred to the spouting velocity corresponding to the best efficiency head of the pump. This enabled a new pump to be designed with similar velocities to those in the existing one, but omitted to show how the varying head quantity characteristics could be obtained at a given shape number and introduced errors due to head measurement and to variation of hydraulic efficiency with size etc., into the design calculations.

This was directly derived from water turbine practice where efficiency variation was small, since turbines were physically large in contrast to pumps, wherein relatively small size introduced a large efficiency variation across the range.

Some fifty years ago, when the need arose to produce varying characteristics in a pump without changing its shape number, the above intuitive reasoning was used by the author to imagine the manner in which such varying characteristics could be produced. (Ref 2).

Subsequent Empirical Analysis

Over the years, empirical tests have been analyzed in the manner described above, which has gradually become the most accurate method of forecasting pump performance and is now in use all over the world.

This empirical analysis and the theoretical investigation that was carried out in parallel are described below.

References
1.   R.L. DAUGHERTY, 'Centrifugal Pumps', McGraw-Hill 1915.
2.   H.H. ANDERSON, 'Mine Pumps' Jl Min Soc Durham University. July 1938.

THEORY OF CENTRIFUGAL PUMPS

Pump Geometry

This section of a pump showing the basic hydraulic shape of impeller and casing appears in Fig 5.1, together with the relevant names for the salient parts.

The impeller blade in simplified form appears in Fig 6.1, together with inlet and outlet velocity triangles.

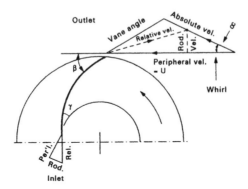

Figure 6.1
Velocity triangles shown diagrammatically and graphically.

Inlet Conditions

It is usual to design the pump on the assumption that the flow to the inlet or eye of the impeller will be normal to the line of motion of the relevant portion of the inlet vanes and in the plane of the streamline, that is to say, in the case of pure radial flow impeller, the inlet flow is assumed to be radial, or, in the case of an impeller having an axial flow inlet portion, the inlet flow would be assumed to be axial.

The above assumes that the liquid enters the impeller without any pre-whirl, although

in fact, such conditions probably only obtain at or about the maximum efficiency point of
the pump.

Outlet Conditions

The outlet velocity triangle is shown in Fig 6.1 and comprises, basically, the peripheral
velocity of the impeller, the absolute velocity of the water leaving the impeller and the
relative velocity of the water in the impeller at the point of leaving the impeller.

Many theoretical investigations, based upon the Euler equation (see below) assumed
that the water would leave the impeller with a relative velocity in the direction of the
impeller blade.

Daugherty 1915 (Ref 1), suggested that the relative velocity of the water at best
efficiency point should be at a smaller angle to the peripheral velocity than the vane angle
shown in Fig 6.1 since, otherwise, there would be no dynamic loading on the impeller
blade and therefore no torque.

This suggested that when the water passes through the pump at the vane angle of the
impeller, the resulting flow will correspond to the point of zero head because indeed any
increase of flow would give a dynamic impact on the underside of the blade, that is to say,
a negative torque representing a water turbine.

Area Ratio of Impeller Outlet Passages to Casing Throat

The author's investigation in 1938 (Ref 2 and Appendix A) commenced with the Euler
equation $H = uw/g$ and showed from the geometry of the outlet velocity triangle that the
generated head at best efficiency point was a direct function of the area ratio and that the
impeller outlet angle had a negligible effect on the pump performance.

This gave a large measure of agreement between the statistical data of pump tests then
available and theoretical considerations. This correlation was later confirmed in a more
complete mathematical investigation of impeller and casing by Worster (see below).

Flow Limitation of Throat

The casing throat can be regarded as the throat of a venturi meter wherein the velocity
cannot exceed the absolute velocity corresponding to the outlet velocity triangle (Fig 6.1)
except on turbine operation. In this manner, the casing is a major determinant of the flows
through the pump at best efficiency point and at the zero head point.

Classical Theories

It is appropriate at this stage to consider the various theoretical investigations that have
been made into pump behaviour.

Euler equated the rate of change of angular momentum to the torque on the pump shaft
and thereby obtained the equation:

$$\text{Head equals } uw/g$$

where u is the peripheral velocity of the impeller, w is the velocity of whirl and g is the
acceleration due to gravity.

Busemann (Ref 3), Stodola (Ref 4) and Pfleiderer (Ref 5), suggested that firstly, the impeller outlet angle and secondly, the number of vanes, controlled pump performance; but these investigators apparently concentrated on the impeller alone and ignored the effect of the casing in controlling flow at best efficiency point. In order to explain disparity between the above formula and actual tests, they assumed that, at best efficiency point, flow would coincide with the impeller vane angle when an infinite number of vanes were provided. The author's attempts to correlate theory and experience (Ref 2 and 6) suggested that the point of which relative flow coincided with the vane angle represented zero head, since a larger flow would necessarily represent a turbine.

Worster's Theoretical Investigation

R.C. Worster (Ref 7) equated the free vortex behaviour of the casing to the flow through the impeller and thereby succeeded for the first time in providing a theoretical explanation of pump behaviour which was in close agreement with actual fact as shown by statistics of pump tests (Figs 6.2 to 6.5).

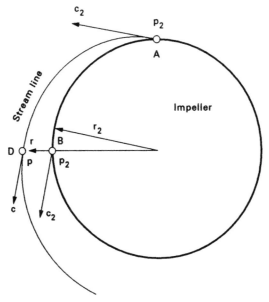

Figure 6.2
Definition sketch for force equilibrum in volute flow.

His correlation of impeller to casing produced a mathematical relationship containing area ratio as a major term so that it is hardly surprising that close agreement occurs between his calculations and the empirical data of Stepanoff (Ref 8) and Anderson (Ref 9) when plotted on an area ratio basis.

Theoretical Analysis and Actual Tests Compared

Fig 6.4 shows as test points the behaviour of several thousand pumps plotted according

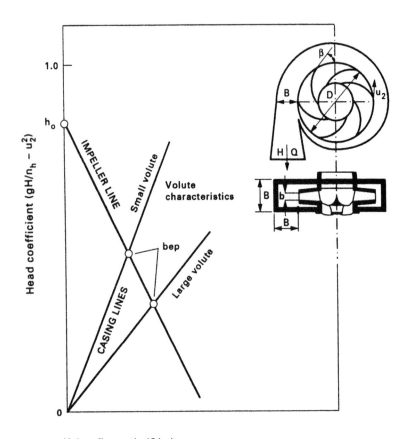

Volute:

$$\left(\frac{gH}{\eta_h \cdot u_2^2} \right) = \frac{2B/D}{\log_e \left(1 + \frac{2B}{D} \right)} \cdot \frac{1}{B^2} \cdot \left(\frac{Q}{u_2} \right)$$

Impeller:

$$\left(\frac{gH}{\eta_h \cdot u_2^2} \right) = h_o - \frac{1}{\pi\, Db\, \tan \beta} \cdot \left(\frac{Q}{u_2} \right)$$

Figure 6.3
Intersection of volute and impeller characteristics
giving best efficiency point.

to the area ratio basis, these points having been illustrated previously as a basis for design (Ref 6). The fine lines on this chart show Worster's investigations for various values of B/b where B is the square root of the throat area and b is the impeller outlet width between shrouds. Worster deduced these curves by determining that point at which the descending

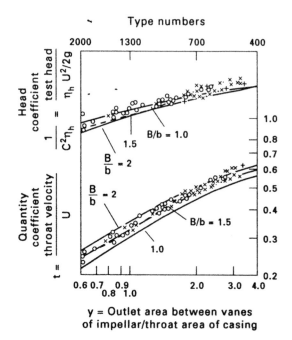

Figure 6.4
Test points of numerous pumps compared with findings of
Worster (fine lines) and Stepanoff (broken lines). β = 20°

impeller line crosses the ascending casing line on the dimensionless charge of head and
quantity coefficients (Fig 6.3).

It will be seen that there is a very close agreement between the actual test points on Fig
6.4 and Worster's theoretical lines. For further comparison the heavy dotted lines show
Stepanoff's test data (Ref 8) transposed to an area ratio basis. Here again it will be seen
that Stepanoff's data lie within the scatter of the test points and Worster's lines, although
Stepanoff (Ref 8) states that the outlet angle is the most important single design element.

Effect of Outlet Angle

Fig 6.5 shows Worster's investigation of the change of outlet angle. For example, an outlet
angle change from 16° to 24° would produce only a negligible effect on the behaviour of
the pump.

This insignificant effect of outlet angle change is in line with the experience of analysis
of tests on area ratio basis since a wide impeller of small angle can give comparable results
to a narrow impeller of large angle, the outlet angle and the impeller width thereby being
flexible to give optimum casting facilities, for example on small duties, as described later.

Effect of Number of Impeller Vanes

Change of number of vanes is limited by efficiency requirements and therefore has only
a minor effect on performance. The test data of Fig 6.4 are based upon normal designs,

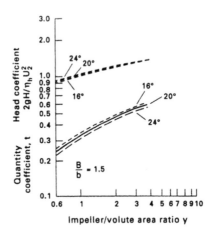

Figure 6.5
Effect of impeller blade angle on best efficiency
point.

generally with eight vanes, and include single stage volute and diffuser pumps, multistage volute and diffuser pumps, bowl pumps, single entry and double entry pumps.

Determination of Area Ratio

The test data of Fig 6.4 are based on

$$\text{Area ratio} \quad = \quad \frac{\eta \, D_b \sin \beta \times 0.95}{\text{Throat area}}$$

Worster's data are based upon

$$\text{Area ratio} \quad = \quad \frac{\eta \, D_b \tan \beta}{\text{Throat area}}$$

If the test data had been based upon $\tan \beta$ instead of $\sin \beta$ the test points would have agreed more closely with Worster's theoretical lines.

$$\text{Area ratio} \quad = \quad y \quad = \quad \frac{BC}{AC} \qquad\qquad AB \quad = \quad 2 \left[\frac{u^2}{2g} \right] \qquad\qquad \text{(see Fig 6.6)}$$

The contention of earlier investigators that the outlet angle is the major arbiter of pump performance arose through consideration of impeller only. It is certainly a fact that, for the same impeller width and casing throat, a large angle would produce a higher head, but this higher head is due to the increase of impeller outlet area or of area ratio. A similar increase of head could be obtained by widening the impeller for the same casing and the same outlet angle, thus virtually eliminating the impeller outlet angle as a critical factor in pump design.

In fairness to these earlier investigators it must be appreciated that they would have only a few pumps available on which it is easier to alter outlet angle than casing throat, whereas a designer now has access to several thousand pumps of varying designs.

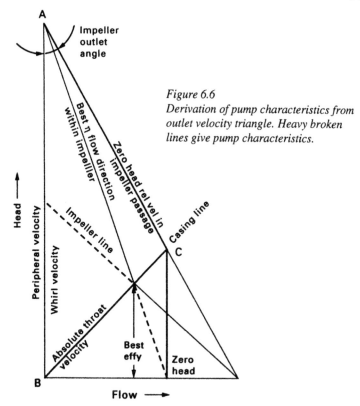

Figure 6.6
Derivation of pump characteristics from outlet velocity triangle. Heavy broken lines give pump characteristics.

Backward sloping impellers result, theoretically, in a head which falls from zero flow to zero head, but the best efficiency point on this falling impeller curve is determined by the casing throat Fig 6.3; that is, the pump behaviour is determined by the combination of impeller and casing and not by either one component. The zero head flow is determined by drawing the outlet velocity triangle for relative flow through the impeller in agreement with the outlet vane angle as in Fig 6.6 (Ref 6) which also shows the best efficiency flow.

Shape Number

Worster also proved, by correlation of impeller flow and free vortex flow in the casing throat, that the shape number is a function of the ratio of square root of throat area to impeller diameter. This is a confirmation of the simplified explanation (1938) (Ref 2, Appendix A and Chapter 7) of shape number as being proportional to ratio of throat diameter to impeller diameter.

References

1. DAUGHERTY, R.L., 'Centrifugal Pumps' (McGraw-Hill) 1915.
2. ANDERSON, H.H., 'Mine Pumps', Jl.Min.Soc, Durham University, July 1938. (See Appendix A).
3. BUSEMANN, A. 1928 Z. angew. Math.Mech., vol 8, p372, 'Das Forderhohenverhaltnis radialer Kreiselpumpen mit logarithmisch-spiraligen Schaufeln'.

4.   STODOLA, A. 1927 'Steam and Gas Turbines' (McGraw-Hill).

5.   PFLEIDERER, C. 1932. 'Die Kreiselpumpen', second edition (Springer, Berlin).

6.   ANDERSON, H.H., 'Centrifugal Pumps, An Alternative Theory'. Proc. IME 1947 Vol 157 We 27.

7.   WORSTER, R.C., 'The Flow in Volutes and its Effect on Centrifugal Pump Performance', Proc. l.Mech.E. 1963 Vol 177 No 31.

8.   STEPANOFF, A.J., 'Centrifugal and Axial Flow Pumps' (Wiley) 1957.

9.   ANDERSON, H.H., 'Hydraulic Design of Centrifugal Pumps and Water Turbines', 1961. ASME 61 WA 320.

SECTION 2

Basic Data

SHAPE NUMBER, FLOW AND EFFICIENCY

STATISTICAL ANALYSIS OF PUMP PERFORMANCE

ENTRY CONDITIONS AND CAVITATION

THE DETERMINATION OF PUMP SPEED AND DIMENSIONS
FOR A GIVEN DUTY

SHAPE NUMBER, FLOW AND EFFICIENCY

The preceding chapters considered, in a very preliminary way, the relationship between the dimensions of a pump and its performance. Since the generation of head is directly affected by the efficiency of the pump, consideration will now be given to pump efficiency.

Efficiency and Shape Number (See Appendix A)

The efficiency of a centrifugal pump depends upon size, speed and shape number. The last term is an expression to define the shape of the pump.

Shape number N_s is equal to

$$\frac{\text{rev/min}\sqrt{\text{l/s}}}{\text{Head in m}^{3/4}}$$

The shape number of a pump indicates the shape of pump and is sometimes defined as the speed at which the pump would run if reduced geometrically to give 1 l/s at 1 m head. As this is rather difficult to visualize, the following explanation is suggested. The shape number is regarded as a ratio defining the shape of the pump, colloquially described as long and thin or short and fat. For a certain design of pump, the ratio of the diameter of the casing throat to the diameter of the impeller will be obtained as follows.

The head is proportional to peripheral velocity^2, which, in turn, is proportional to impeller diameter^2 x rotational speed^2. For a given head H we can, therefore, say impeller diameter is proportional to $\sqrt{\text{head}}$ divided by the speed.

The absolute velocity leaving the impeller is regarded as the volute throat velocity or diffuser velocity, which, in turn is proportional to the impeller peripheral velocity, that is to say, is proportional to the $\sqrt{\text{head}}$.

The volute throat is the largest section of the scroll preceding the expansion cone to the delivery branch.

The quantity pumped is equal to the throat velocity multiplied by the area of the throat and, therefore, is proportional to

$\sqrt{\text{head}}$ x the area of the throat

Hence, the area of the throat is proportional to Q divided by $\sqrt{H}$ where Q is quantity, and the diameter of the throat is proportional to $\sqrt{Q}$ divided by $H^{1/4}$.

Therefore, the diameter of the throat to the diameter of the impeller is proportional to

$$\frac{\sqrt{Q}}{H^{1/4}} \quad / \quad \frac{\sqrt{H}}{\text{rev/min}}$$

This, in turn, is proportional to

$$\frac{\text{rev/min} \times \sqrt{Q}}{H^{3/4}}$$

that is to say, is proportional to the shape number.

It is, therefore, possible to indicate the change of type of pump by the varying ratio of throat diameter to the impeller diameter. From the above, it is seen that this ratio is dimensionless and is identified with the shape number.

Small quantity high head pumps will have a relatively large impeller diameter with narrow outlet to give small flow areas and small quantity. Efficiency will, therefore, be prejudiced by excessive loss due to disc friction. This is because the large impeller will have a large disc friction compared with the small liquid power consequent upon the small quantity pumped.

This machine will have a low shape number and a comparatively low efficiency. At the other end of the range, a pump for a large quantity and a low head will have a wide impeller and a relatively small diameter, so that although the disc friction is not great, there will be prejudice to efficiency by the lack of guidance in the short impeller passages. This pump will have a high shape number.

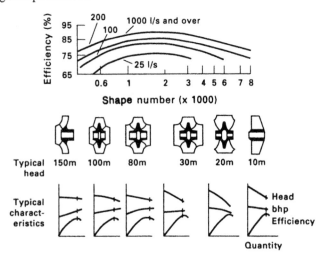

Figure 7.1
Speed and Efficiency Characteristics.

Between these limits, the efficiency will reach an optimum. It is, therefore, possible to plot the effect of shape number on efficiency, thus permitting the designer to determine the optimum shape number and, consequently, the running speed for a given duty. Compromise may be necessary here to suit the driver, for example, because alternating current operating speeds in Britain conform to 50 cycles frequency. Furthermore, for certain duties it may be necessary to pump against very high heads. Some sacrifice of efficiency may, therefore, be necessary to obtain a reasonable number of stages or to avoid the running of two pumps in series.

Fig 7.1 (Ref 9) illustrates the efficiency of centrifugal pumps plotted against shape number as abscissae with lines representing various quantities in litres per second.

The chart also shows the shape of impeller for the various shape numbers and the type of head volume, power and efficiency characteristics expected. It will be seen that the low shape number pumps at the left hand end of the charge have a head characteristic which does not rise very greatly between best efficiency point and zero flow, whilst the high shape numbers at the right hand end of the chart have a head characteristic which rises very steeply from best efficiency flow to zero flow (Ref 10).

The reason for this change of characteristic is due to the change of area ratio consequent upon change of shape number. This was introduced in Chapter 5 and will be discussed in detail later.

Variation of Efficiency with Size and Speed

In addition to illustrating efficiency variation with shape number, Fig 7.1 illustrates how an increase of size results in an increase of efficiency. The efficiency improvement consequent upon increase of size is dependent upon the Reynolds number already discussed in the first chapter in very general terms. (Ref 11).

More specifically it can be said that a pump is a dynamic device for producing head which depends on velocity[2], but it suffers from losses depending according to Reynolds number on velocity to a power in the neighbourhood of 1.7 to 1.8.

Increase of size and speed will therefore, increase efficiency since the dynamic head, depending on velocity[2], increases at a greater rate than the losses, which are varying in the ratio of velocity to the neighbourhood of 1.7 to 1.8.

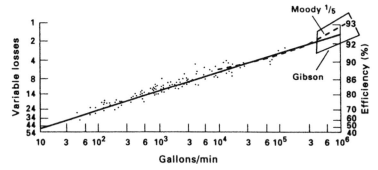

Figure 7.2
Original chart of efficiency at optimum specific speed

It has been found over the years by analyzing true Reynolds number for the various passages within the pump and the direct flow quantity, that the latter is a sufficiently accurate arbiter for the efficiency of a group of pumps.

We, therefore, plot efficiency directly against flow quantity at best efficiency point for a series of various shape numbers, thus obtaining, in effect, a group of tests of various flows and shape numbers from which Fig 7.1 can be constructed.

This, in effect, means that the efficiency of a small pump at high speed can be the same as a large pump at low speed, providing the shape number and the flow are kept the same.

This is the reason why Figs 7.1 and 7.2 can, in general, show efficiency against quantity and shape number only without reference to physical size or operating speed. Where pumps operate against very high heads, the large shaft diameter and run dry clearances considerably reduce the efficiency below this charge.

Efficiency Step-up Formulas

We have already considered the variation of efficiency with size and shape number. Consideration will now be given to efficiency step-up formulas, the effect of surface roughness on efficiency and the value of efficiency.

Efficiency Formulas for Variation of Size

The efficiency charts illustrated in Figs 7.1 and 7.2, based upon many years' test experience up to the 1950s, incorporate the aggregate performance of several thousand pumps, and were published in 1960. (Ref 10).

Flow measurements varying between 2 l/s and 3000 l/s were taken on eight different instruments, any two of which could run in series to provide a check of accuracy.

The resulting curve of efficiency against quantity for any given shape number was found to be consistent and also could be correlated to the investigation on how pipe friction varies with change of Reynolds number. The pump losses could be separated into constant losses, eg bearings, glands, disc friction, and flow passage losses, which were analogous to pipe friction losses.

The passage losses could further be split into a constant portion and a portion varying as the Reynolds number, which, from the efficiency aspect, could be regarded as varying according to the quantity pumped.

The pump maker, therefore, could produce a formula covering the whole range of his pumps from the smallest to the largest. This gradual formula containing constant losses can, therefore, never reach 100% efficiency, experience suggesting that the efficiency of infinitely large pumps would be of the order of 94%.

This suggested that the losses for an infinitely large pump would only be 6% but, unfortunately, the most accurate flow measurement would still show a permissible error according to Standard Specifications of the order of $1\frac{1}{2}$% to 2%, which is a large variation in the remaining 6% between 94% and 100% (Refs 11 and 12).

Moody Formula

The Moody formula (Ref 13) for water turbines generally uses an index of one-fifth, and therefore, becomes:

Losses of turbine efficiency, that is to say, 100% minus efficiency, are proportional to the inverse one-fifth power of the major diameter of the turbine.

This formula is generally based upon the turbine practice of a small model (laboratory tested) being compared with a full-size unit (site tested with considerably less accuracy).

The Moody formula is reasonably in agreement with the pump formula described above for the quantity range between 1000 and 20 000 l/s. Since it includes no constant loss, the Moody formula would suggest that a turbine of infinite size would have 100% efficiency.

The investigation of pump and turbine efficiencies suggests that an infinitely large turbine would have an efficiency of 95% against 94% for a pump, since the liquid power for a turbine is greater than that of a pump for the same physical dimensions, the proportion of losses thereby falling, in the case of the turbine, with consequent efficiency improvement.

The Moody formula is mentioned in this discussion of pump efficiency because very many more large turbines have been made than large pumps, so that there is a tendency to use the Moody turbine formula for large pumps, until the pump formula is established.

Note that here again, in the case of turbines, the site flow measurement has a large error, probably $1^{1}/2\%$ to 3% in comparison with the minimum loss figure of 5%.

Surface Finish

On the curve of efficiency compared with quantity, the small quantity losses are largely skin friction whilst the large quantity losses are almost entirely eddy friction.

When comparing a group of pumps from, say, 50 mm branch to 750 mm branch, it can be assumed that the whole group will have a uniform surface finish, for example, bronze, cast iron or mild steel dressed smooth. Here the size of any surface error can be considered as uniform throughout the group, as, for example, a CLA surface of 4 microns.

This roughness has quite a large frictional effect on a 50 mm pump but a negligible effect on a 750 mm pump, the consequent change of efficiency due to Reynolds number between these two sizes being show in Fig 7.1, all other aspects being equal or eliminated by appropriate corrections.

Departures from uniformity of surface such as may be caused by bad matching of casing halves, etc., or by noduling consequent upon handling corrosive waters, may reduce the efficiency of the pump below the value shown in the charge by as much as 5-10% or even more.

On the other hand by adopting the superior finishes of vitreous linings, by manufacturing the pump parts of plastics etc, or by the physical polishing of the water passages, a small quantity pump may be given a smaller absolute value of surface roughness with respect to diameter. Such a pump would have a slightly higher efficiency, perhaps 2% or more.

It is, therefore, important to ensure that the states of surface smoothness of model and full size pumps are alike before any general step-up formula such as mentioned above can be applied.

Value of Efficiency

Very roughly, it may be said that 1% increase of efficiency is comparable to the price of the whole pump. Certain contracts now include monetary penalties and bonuses to ensure that efficiency guarantees are met on site.

For example, a three MW waterworks' pump running 12 hours a day would, in twenty years, consume electricity to the value of 2 000 000. 1% of this is 20 000, which is very roughly comparable to the cost of the complete pumpset.

At the other end of the scale, the efficiency of a fractional kW pump for a domestic washing machine, running perhaps half an hour a week, may at first sight be considered unimportant. It was, however, reported by one maker that careful research had improved pump efficiency so as to reduce the size of the driving motor and save 50 pence on the production cost of each set. In manufacturing a large number of such pumps a week, this improvement would represent a substantial monetary saving over the years.

Efficiency, Quantity Charge

Fig 7.2 shows the original chart of the efficiency of pumps at optimum shape number plotted against quantity in gallons per minute. Each of the many test points was corrected to the efficiency it would give at optimum shape number according to Fig 7.1 so that shape number is eliminated as a variable from this charge.

Initially, the gal/min-efficiency curve was plotted and its mathematical formula evaluated. It was then replotted as a straight line on logarithmic paper by separating the constant and the variable losses, as in Fig 7.2.

To the right of Fig 7.2 a scale of efficiency appears but the true logarithmic scale of Fig 7.2 is the variable loss scale at the left, 4% variable loss corresponding to 90% efficiency, 14% variable loss corresponding to 80% efficiency and so on. Each point on Fig 7.2 is the average of all the pumps made of that particular size and shape number, the charge representing a record of several thousand pumps.

Efficiency, Quantity Formula

The curve of Fig 7.2 is similar to the variable portion of Reynolds number pipe friction curve.

The solution of the gal/min-efficiency curve can be mathematically expressed as follows:

$$1.0 \text{ minus efficiency} = (\text{gal/min})^{-0.32} + 0.06$$

or $\qquad$ $0.94 \text{ minus efficiency} = (\text{gal.min})^{-0.32}$

By plotting (0.94-efficiency) against gal/min, a straight line graph can be obtained on logarithmic paper as in Fig 7.2. This suggests that 0.06 or 6% of losses are constant, independently of gal/min or of Reynolds number. The above equation can be transposed to:-

$\qquad$ $1.0 - \text{efficiency} = \text{total losses} = 0.03$ fixed losses (disc friction glands, bearings and leakage)

$\qquad$ $+ 0.03$ constant friction head loss$+ (\text{gal/min})^{-0.32}$ variable friction head loss

The last two terms

$\qquad$ $0.03 + (\text{gal/min})^{-0.32}$

covering the pipe friction portion of the loss are of similar form to the expression for Reynolds pipe friction curve.

$$\frac{f}{2} = 0.0009 + 0.0765 \text{ (Reynolds number)}^{-0.35}$$

where f is the friction coefficient for smooth pipe flow.

We thus separated the various losses in the pump, the major loss being passage friction (ie pipe friction).

Comparison with Water Turbines

The chart in Fig 7.2 was extrapolated to one million gal/min and contains also, as a dotted line, the Moody Efficiency Formula for turbines using an index of 1/5.

The trapezium drawn at the right hand end of the curve shows the field of water turbine efficiencies determined by the Gibson method (Ref 14) of flow measurement. In this method, flow is determined by measuring the pressure rise for a specified rate of closure of water turbine gates.

The trapezium represents 310 tests of water turbines over the 35 years from 1920 to 1955 corrected to optimum shape number. As mentioned, above, water turbine efficiencies are expected to be about 1% above pump efficiencies but, on the other hand, many of the earlier turbines had a lower efficiency and size than those of more recent design.

Bearing these aspects in mind, it will be seen that the author's formula for efficiency when extrapolated to the larger sizes showed reasonable agreement with water turbine practice. The scatter and slope of the Gibson tests are comparable to those of the author's formula.

Zanobetti (Ref 16) shows 350 turbine efficiencies 10 to 100m²/s generally grouped about the centre line of this chart with very little efficiency variation for shape number.

Effect of Surface Finish on Step-up Formulas

The author's formula given above assumes that from the smallest to the largest pump, the finish is consistent in its absolute value of roughness, namely, cast bronze or stainless steel surfaces filed smooth in the case of the impeller or runner, and diffusers with iron or steel for the casing or scroll. This consistency of finish permits any pump to serve as a model for any other.

It is, however, more usual in the large turbine filed to provide a model with a very high surface polish and to forecast the efficiency of the full-size unit from this model with a modification to the step-up formula.

The higher surface finish of the model places it above the curve of Fig 7.2, with the result that an arbitrary correction to the step-up formula is required. For example, the Moody formula (Ref 13) or the various continental formulas are used with further empirical coefficients to allow for the change of absolute surface finish.

Scatter of Test Points

Fig 7.2 represents a sufficient number of tests to eliminate measurement error and to show only variation of pump performance.

The spread of test points at any given flow is seen to be proportional to the variable losses, which is to be expected. Comparison may be made here to BS2613 for electrical machines, which allows an efficiency tolerance of 10% of (100%-efficiency) ie 10% of losses. In contrast, ISO 3555, pump test code allows an efficiency tolerance of $2^1/_2$% of the actual efficiency specified which is intended to cover errors of test observation and not scatter of pump performance. ISO 3555 tolerance therefore is comparable to pump performance scatter on the larger sizes, but is only a small proportion of pump scatter on the smallest sizes.

Variations of geometry, single stage, multistage, single entry, double entry, single volute, double volute or diffuser give negligible efficiency change.

Present Position of Efficiency Analysis

Subsequent to the investigation of the 1950s which covered flows from 2 l/s to 3 m²/s, further experience was gained from 0.4 l/s to 16 m²/s, and overseas makers published pump and turbine efficiencies on flows up to 400 m²/s. All this more recent data fully confirmed the 1950s forecast of efficiencies of very large machines.

Efficiency Majoration Formula for Fluid Machines (Refs 10, 11 and 12)

Expressions for variation of efficiency with size are essentially empirical, with guidance as to the general form of the expression from theoretical sources.

Since statistical data are the main basis here it follows that the greater the number and range of machines in the survey, the more accurate will the formula become.

The following record, show in Fig 7.3, of some fifteen thousand pump tests and several hundred turbine tests covers flows from 0.4 l/s to 400 m²/s, and provides an expression for efficiency which is relatively simple. The distribution is gaussian with scatter proportional to variable losses, 75% of points within one sixth, 95% of points within one third of variable losses. The flow range is from one to one million.

Available Data

The formula originally published in 1960 was based on several thousand pump tests (Ref 10).

In 1961 it was suggested that an infinitely large turbine would have an efficiency of 95% (Ref 9).

The machine with the largest flow in the survey (400 m²/s) shows an efficiency of 95.4%, the deviation from formula efficiency being comparable to the tolerance of Class A measurement, which also appears on the chart.

All the machines in the survey are assumed to have the same absolute surface finish so that every machine is a model of every other machine of the same shape number.

Pump and Turbine Formulas

The average efficiency η of the 15 000 centrifugal pumps (Refs 9,10,15,16), of Pelton turbines (15) and of 310 earlier turbines, 1920 to 1953 (14), is given by the expression

$$0.94 - \eta = [\ 13.2\ l/s\ ]^{-0.32}$$

The average efficiency of 670 of the latest turbines (Refs 15 and 16) is given by

$$0.95 - \eta = [\ 13.2\ l/s\ ]^{-0.32}$$

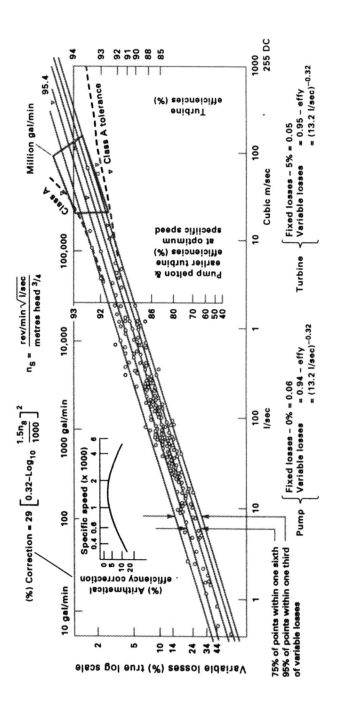

Figure 7.3 – Pump and water turbine efficiencies.

This is 1% higher than the pump formula since a turbine handles a greater liquid power for the same losses as a pump of the same geometry.

Majoration Formulas

Pumps and Pelton turbines

$$\frac{0.94 - \eta_a}{0.94 - \eta_b} \quad \left[\frac{\text{Flow}_b}{\text{Flow}_a}\right]^{0.32}$$

Water turbines

$$\frac{0.95 - \eta_a}{0.95 - \eta_b} \quad \left[\frac{\text{Flow}_b}{\text{Flow}_a}\right]^{0.32}$$

Pump Data

Each of the 200 odd circles on the chart of Fig 7.3 represents the average efficiency of all pumps of that size and duty, the number of pumps for each circle varying from a few to several hundred, according to the number made and tested of that particular size and duty. Speeds generally are normal for the duty.

Shape number (N_s)

The pump efficiencies are corrected to optimum shape number by the following formula:-

$$\text{Efficiency correction} = 0.29 \left[\log^{10} \frac{1400}{N_S}\right]^2$$

The Author is indebted to Dr D.J. Myles for the suggestion during discussion on Ref 12 that losses on the left of the optimum of the N_s–efficiency curve are due to disc friction and that losses to the right are due to the lowering value of the head coefficient. This leads to a rational dimensionless formula for efficiency and shape number with a roughness function and a through flow function.

Reversible Pump Turbines

Compared to normal machines a fall of 1% efficiency occurs here due, when pumping, to the presence of gates and, when turbing, to the relatively larger runner diameter than that of a normal turbine. A 1% efficiency increase is therefore allowed here before plotting.

Use of Model with Superior Finish

A highly polished surface on a model will result in a higher efficiency than the chart value since this model becomes a geometric replica of a larger machine which has the normal finish of the machines on the chart. For example, the efficiency of a full size machine, having a surface finish comparable to 4 microns, and a duty of 20 000 l/s is forecast by the test of a one tenth model of 200 l/s having a surface finish comparable to 2 microns. This model is therefore a geometric replica of a twice size machine which has the normal finish of 4 microns and a duty of 2^2 x 200 l/s or 800 l/s. The efficiency of the full size machine is therefore given by stepping up from 800 l/s to 20 000 l/s (instead of from 200 l/s) in order to allow for superior model finish.

The efficiency relationship now becomes

$$\frac{94 - \eta \text{ full size}}{94 - \eta \text{ model}} = \left[\frac{800}{20\,000}\right]^{0.32}$$

This correction for finish has been fully confirmed by tests of homologous pumps of 200, 300, 500, 1 000 and 3 000 mm branches with various surface finishes.

Examples of Use of Charge:

Plotting of tests

At the actual test speed, the best efficiency in litres per second, the corresponding efficiency and the shape number are determined. The efficiency correction from actual shape number to optimum shape number, as shown on the chart, is added arithmetically to the test efficiency and the resulting efficiency is plotted at the litres per second involved.

Example: 300 l/s 700 shape number, 84.2% efficiency on test.
Correction for 700 shape number is 2.6%.
Total 84.2% plus 2.6% equals 86.8% efficiency which is plotted at 300 l/s.

Determination of Efficiency from the Chart

Example: 1000 l/s is 2700 shape number.
Efficiency from chart at 1000 l.s equals 89.2%.
Correction for 2700 shape number is 2.3%.
Efficiency at 1000 l/s 2700 shape number, is 89.2% minus 2.3% equals 86.9%.

See also Ref 83, ASME prediction conference.

References

9.   ANDERSON, H.H., 'Hydraulic Design of Centrifugal Pumps and Water Turbines', 1961 ASME 61 WA 320.

10.   ANDERSON, H.H., and CRAWFORD, W.G. 'Submersible Pumping Plant', Proc IEE.Vol 107, Part A, No 32, April 1960.

11.   ANDERSON, H.H., 'Efficiency Majoration Formula for Fluid Machines' Int Assn Hyd Res 7th Symposium Vienna 1974.

12.   ANDERSON, H.H., 'Statistical Records of Pump and Water Turbine Efficiencies'. Conf 'Scaling for Performance Prediction in Rotodynamic Machines', Stirling 1977, IMechE.

13.   MOODY, L.F., 'Propeller Type Turbine'. Trans American Soc C.E. 1926,Vol 89.

14.   GIBSON, N.R., 'Experience in the use of the Gibson Method of Water Measurement for Efficiency Tests of Hydraulic Turbines'. ASME Paper 58 - A - 78. (Trapezium of Fig 7.3).

15.   MULLER, Hans Paul. 'Uberblick uber Ergebnisse von Wirkungsgradmessungen an Turbinen und Pumpenanlagen'   Pumps / Peltons Δ on Fig 7.3
Voith Forschung and Konstruktion   Francis / Kaplan ∇ on Fig 7.3
Heft 15 (May 1967)

16. ZANOBETTI, D. 'Characteristics of Water Turbines', Water Power, March 1959.
83. ANDERSON, H.H. 'Prediction of Head, Quantity and Efficiency in Pumps - The Area Ratio
     Principle'. ASME 22nd Fluids Conf New Orleans 1980. Book No 100129.

STATISTICAL ANALYSIS OF PUMP PERFORMANCE

Large scale economic production of any product demands quality control, and this is especially true of pumps wherein cast surfaces and the vagaries of water flow cause random variations in performance.

Given sufficient numbers, these random variations can, however, by analyzed statistically to provide accurate data for forward factory planning and for performance guarantees on unknown machines. To perform this analysis a gaussian curve is drawn from numerical tabulation of pump head, quantity and efficiency data showing the departure from mean anticipated performance.

Surface Quality

The internal flow passages of centrifugal pumps generally have cast surfaces which cause variation of flow pattern and of friction losses. since the head generated by the pump and its overall efficiency are functions of the dynamic head due to the peripheral velocity of the impeller and the frictional head due to turbulence and surface friction, it follows that the surface quality has an important effect on performance. Whilst changes of performance due to casting variation give rise to difficulty in forecasting the performance of any one machine, an analysis of the tests of a year's production permits a very accurate forecast of the following year's production. This forecast gives, of course, an overall and not a detailed view.

Commercial Aspects

Pumps must be produced — sometimes under financial penalties and rewards — at an economic cost to meet exacting performance requirements The total cost to the pump user over a life of say twenty years is made up of first costs, power supply cost and maintenance cost. The power cost is generally so important that pumps tend to be sold on efficiency. For example, on certain duties 1% efficiency is equivalent to the price of the pump. Competition will force the maker to quote the best efficiency that he cane safely achieve, and the natural scatter of performance will result in a few pumps being unacceptable at first test, require modifications or new parts and re-testing before they are made acceptable.

Thus, a fine balance must be achieved between too low an efficiency quoting level, which would lose orders, and too high a level which would involve excessive modifications, new parts and re-tests to the prejudice of economic production. For forward planning it is helpful to know that, next year, factory provision must be made for $x\%$ of pumps requiring modifications and re-tests, although it is not possible beforehand to say which particular pumps will be faulty.

Improvement of the Breed

Over the years pump performance slowly improves as new methods are introduced — for example, greater knowledge by model tests of flow patterns, surface improvements due to shell and investment moulding techniques, higher speeds and pressures due to improved metallurgy. These improvements will be reflected in the general efficiency level throughout the industry.

Statistics

Pump efficiencies vary from about 10% on a very low quantity high head duty, to 93% on a very large optimum shape number duty. This variation follows a rational pattern deduced from past statistical analysis and is illustrated below. It can, however, be eliminated from current year by year analyses by referring the test efficiency of any one pump to the efficiency quoted to the purchaser, since the latter takes account of the variation due to size, shape number and nature of duty. The difference between the test efficiency of a particular pump and the quoted efficiency therefore represents the departure from the norm.

Efficiency Variation with Flow Quantity and Shape Number

Figs 7.1 and 7.3 provide a reasonable representation of the efficiency level currently quoted and achieved so that it is only the departure from this level that concerns this analysis.

In order to maintain performance of all pumps and to forecast the performance of untried pumps we need very accurate tests of all pumps that the factory can test, a thorough system of statistical analysis and a step-up formula based upon constant absolute surface finish.

Pump Test Codes (see Chapter 25)

As mentioned above, Test Code tolerances are for errors of measurement; no allowance is made for variation of pump efficiency. For example, ISO 3555 allows $2^1/4\%$ tolerance on 90% efficiency where scatter of plus or minus $1^1/4\%$ occurs and allows only 1.9% tolerance on 75% efficiency where scatter of plus or minus 6% occurs. BS2613 tolerances of 10% of losses for electrical machines are more logical since scatter of tests is proportional to 100% minus efficiency. It is submitted that where several thousand tests are available the scatter is mainly variation from pump to pump since testing errors tend to be averaged out.

It should be borne in mind, however, that ISO 3555 (and its successor BS3516) do not provide for pump efficiency scatter which can be quite large on smaller sizes (see Chapter 25).

Tolerance on Head Quantity Characteristic

ISO 3555, BS3516 provide $2^{1}/_{2}\%$ efficiency tolerance eg $2^{1}/_{4}\%$ on 90%, 1.9% on 75% efficiency, but also provide a manufacturing tolerance of plus or minus 4% on flow quantity and 2% on head.

Forecast of Phenomena

Just as the Registrar General is able to forecast very accurately the average behaviours of human beings in any future year without being able to say who will fail to conform to accepted standard, so the designer having a statistical analysis of several thousand pumps is able to forecast that next year's 500 pump tests will result in X% of pumps having faulty head characteristic, Y tests per pump and Z% of pumps being low in efficiency and requiring modification or new parts, without being able to say which pumps will be faulty. The knowledge of X, Y and Z is therefore vital to the works manager's planning, and the relationship between Z and the quoting level is essential in making any guarantees of pump performance.

In practice, the economic production of large numbers of pumps is impossible without this analysis, which under good conditions generally results in 95% probability of attaining guarantees of head characteristic and efficiency at first test and slightly over one test per pump.

Current Analysis of Efficiency Tests

A typical gaussian curve of a year's tests referred to quoting level appear in Fig 8.1. It will be seen that approximately 4% of the first tests show pumps low in efficiency and requiring modifications or new part and re-tests in order to ensure the quoted performance. In order to keep the rejects to a figure that permits economic manufacture (for example, below 5%) it is necessary to tender an efficiency which is 2% below the true average of all tests. It will of course be appreciated that to quote, without tolerance, the exact average of all tests

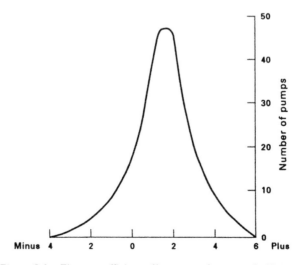

Figure 8.1 – First test efficiency % compared to quoted efficiency

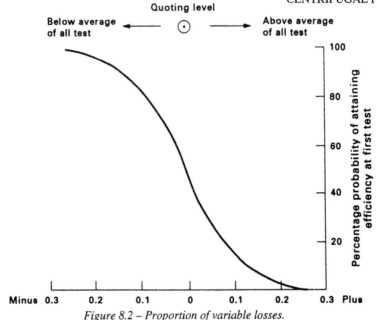

Figure 8.2 – Proportion of variable losses.

would result in 50% rejects, since 50% of tests would be above average and 50% below.

The probability of meeting any particular efficiency is show in Fig 8.2, plotted against quoting level. The distribution is shown in terms of the variable losses, $(13.2 \text{ l/s})^{-0.32}$ since this is the true basis of the scatter of Fig 7.3, and omits the tolerance of ISO 3555.

Variations in Generated Head

The head variation from specified value follows a gaussian curve comparable to that of efficiency. In general, an impeller that is high in head can be reduced in diameter, but an impeller that is low in head, beyond the 1% or so that can be regained by dressing the vane tips, will have to be replaced. Hence, it is usual to provide a margin on diameter to minimize risk of replacement, with the result that a small impeller reduction is the most frequent reason for repeated tests.

Variations in Characteristic from Specification

Here, again, any necessary minor corrections of characteristic can generally be achieved by suitable dressing of impellers and diffusers. However, a careful record of all such cases, with charts of characteristic before and after modification, is essential to build up experience of the particular product.

Plotting and Analysis of Pump Tests

The test results on each size and type of pump are plotted as follows:

> All tests are corrected to a basic speed (for example, 1 470 rev/min) depending upon the size of the pump, and a family of curves obtained for the head quantity characteristics of the pump at various diameters of impellers. These curves will be similar to the variable speed curves, but showing diameter changes instead of speed

changes. It is convenient to group the tests into the relatively small number of diameters, making a diameter correction in similar manner to speed correction and keeping different colours for different diameters. The efficiency curves are then transposed by increasing the l/s of each test reading proportional to the impeller diameter so that all efficiency curves irrespective of diameter have the same value of l/s at their peak. It is thus much easier to view the performance with the colour range than would be the case if the efficiency curves crossed (see Chapter 22).

From these charges iso efficiency lines can be drawn on the head quantity curves for the various diameters.

Cavitation tests are plotted in the same manner, although of course, they are little affected by diameter change.

The pump frames are designed as shown in Chapter 13 to cover a wide range of heads and quantities, so that it is economic to vary the impeller diameter on each frame to avoid gaps in the field to be covered.

After sufficient tests are plotted to give the true characteristic, it is only necessary to record the average efficiency and head over each successive year in order to ensure that performance is maintained (metal patterns are advisable in this respect). Each point in Fig 7.3 thus represents the average performance of up to a few hundred pumps.

Current Records

Each quarter (with summation each year) the following would be recorded for each type of pump:

Number of tests per pump: This should be maintained as close as possible to one test per pump.

Reasons for rejection at first test as follows:

Number of pumps: High in head
                   Low in head
                   Low in efficiency
                   Having faulty characteristic
                   Having mechanical faults.

Extreme spread of efficiency readings from pump to pump.

The application of this rigid quality control will minimize production costs by bringing any performance deterioration to light immediately it occurs.

The purpose of this exercise is to achieve over the years a continual improvement in pump performance and to ensure that this improvement is maintained.

ENTRY CONDITIONS AND CAVITATION

The Determination of Pump Speed and Dimensions for a Given Duty

Since the cavitation aspects determine the running speed which, in turn determines the basic dimensions of the pump, cavitation will first be discussed.

Analysis of Cavitation Characteristics

Between the inlet branch and the eye of the impeller a pressure drop occurs due to increase of flow velocity and to friction.

This pressure drop is referred to as NPSH (Net Positive Suction Head) and is the difference between the inlet head (including the velocity head in the inlet pipe) and the saturation pressure of the liquid handled.

NPSH has two values:

(a) the lower value when cavitation occurs

(b) the upper value which contains a margin of pressure to ensure operation sufficiently free from cavitation so that efficiency is not prejudiced.

Water turbine experience suggested that NPSH is proportional to total head and stated the relationship

Sigma = NPSH/Best efficiency head

Here a slightly greater flow quantity would involve a higher NPSH.

Cavitation performance is, however, more concerned with quantity and speed than with total head, and appreciable errors can occur if the sigma value is applied to a point only slightly removed from best efficiency head. For example, in a pump, in contrast to a turbine, at a slightly greater quantity a lower head would obtain and the NPSH, according to the above formula, would be lower, whereas, in fact, a larger quantity would involve a greater NPSH. In order to avoid errors which had occurred in the use of the sigma coefficient, the speed and flow quantity were investigated as suitable references for a cavitation parameter. The NPSH for complete cavitation at best efficiency point is plotted

(Ref 17) against the product of the litres per second and the square of the revolutions per minute, referred to as l/s (rev/min)^2. The flow refers to one side of a double entry impeller, or to a single entry impeller. This is shown in Fig 9.1. Test points result in a straight line on logarithmic paper, the equation to which is

$$\text{l/s (rev/min)}^2 \,/\text{NPSH}^{1\frac{1}{2}} = \text{constant}$$

This constant is seen to be the square of the shape number using, however, the NPSH instead of the head. The inlet shape number now becomes

$$N_{ss} \text{ equals } \frac{\text{rev/min }\sqrt{\text{l/s}}}{\text{NPSH}^{\frac{1}{2}}}$$

and forms a reference for cavitation studies.

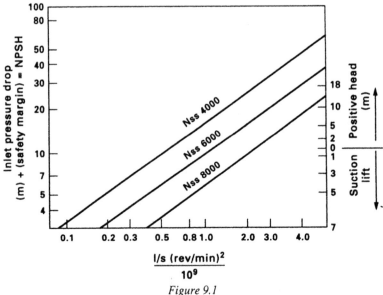

Figure 9.1

Cavitation Tests

The lines of best efficiency points in Fig 9.1 can be labelled with appropriate inlet shape numbers as follows:

N_{ss} 4 000 for operation of normal pumps

N_{ss} 6 000 for cavitation failure of normal pumps

N_{ss} 6 000 for operation line of special pumps having a large impeller eye, special inlet design and/ or large physical size

N_{ss} 8 000 for cavitation failure line of special pumps

It will be seen that the margin of safety between operation and cavitation failure is 2000 inlet shape numbers, corresponding approximately to 50% of the NPSH.

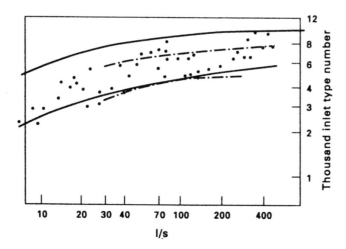

Figure 9.2 – Cavitation tests and flow quantity.

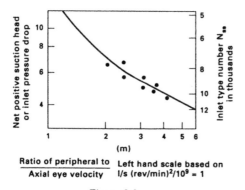

Figure 9.3.

For convenience, in referring NPSH values to the site conditions at atmospheric pressure, a scale of positive head and suction lift is shown to the right of Fig 9.1, the water barometer of 10.2 m NPSH being equivalent to zero suction lift.

Fig 9.1 is a general chart showing the relationship between certain typical inlet shape numbers, NPSH and l/s (rev/min)^2.

Fig 9.2 shows the effect of size on inlet shape number.

The effect of the larger eye is shown in Fig 9.3 where inlet speed number is plotted against the ratio of peripheral eye velocity P to the axial flow velocity A through the eye of the impeller, a larger eye giving improved inlet shape number and lower NPSH.

Fig 9.5 shows ratios of outlet diameter to eye diameter plotted against shape number for various pumps.

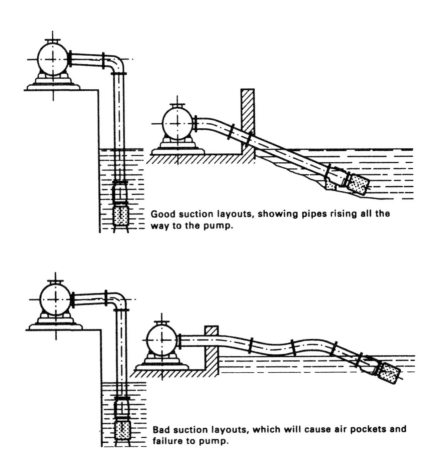

Figure 9.4 – Suction layouts to give optimum inlet conditions. Pumps drawing from overhead vessels at saturation pressure should similarly have inlet piping falling progressively from vessel to pump with no vapour pockets. A vapour bubble release pipe (a quarter the diameter of the inlet pipe) should rise progressively from the top of the inlet passage near the impeller to the saturation vessel above the liquid level.

Determination of Impeller Inlet Dimensions

For a given quantity and inlet condition, therefore, it is necessary to determine from Fig 9.1 the value of the l/s (rev/min)^2, from which the maximum permissible running speed can be obtained and a decision made as to whether a normal or special design is involved. The running speed must be suitable for the driver, which, in the majority of cases, is an a.c. electric motor.

The value of P/A can now be determined from Fig 9.3 from which the actual eye diameter can be calculated. Final checking up on Figs 9.2 and 9.5 will be necessary.

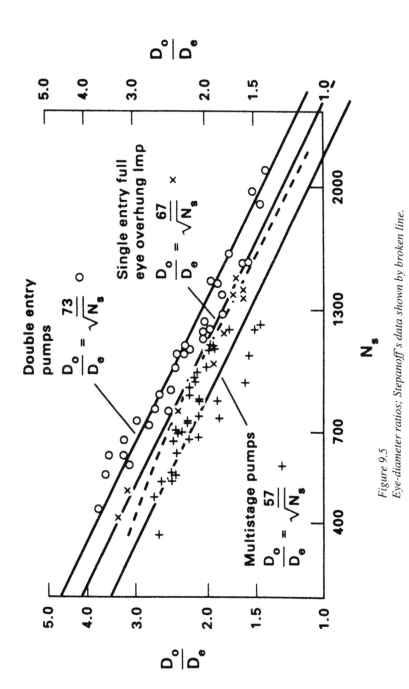

Figure 9.5
Eye-diameter ratios; Stepanoff's data shown by broken line.

General Observations on Cavitation Phenomena

The data shown in the above charts are necessarily generalized and relate to average conditions of very many pump tests.

In addition to the major arbiters of performance, described above, the following additional factors will affect cavitation performance.

The shape of the suction passage, the diameter of the hub, whether double or single entry, the blade angles at pick-up and the blade length will all have a direct bearing on cavitation performance. It is, therefore, usual to group together in cavitation analysis pumps of similar geometric shape.

The temperature, pressure, viscosity and saturation pressure of the liquid are important factors, concerned more with the application of the pump than its initial design, since the original designing and testing of a pump or range of pumps is usually carried out on cold water, and the curves shown above are drawn on this basis.

The operating speed of the pump and its size will affect the inlet shape number but, in general, the flow in l/s at best efficiency point can be referred to Fig 9.2 where inlet shape number is plotted against flow in l/s. Papers in Ref 18 deal thoroughly with the subject of cavitation.

Vaporizing Liquid Duties

If the liquid in the vessel supplying the pump is at boiling point, the free surface of the liquid must be at a level above the impeller inlet vanes corresponding to the NPSH in order to ensure that the liquid can pass into the impeller without vaporizing.

This condition can serve to control the pump flow so that at all times it is equal to the flow into the vessel. A reduction of flow into the vessel will lower its level until the pump by partial vaporization reduces its flow to correspond to the lower NPSH obtaining.

Vaporizing liquid pumps must operate at relatively low speeds of rotation and flow so as to minimize any cavitation attack to an acceptable level.

Inducers

Where high heads are involved it is essential to operate in the safe zones of 3000 to 4000 inlet shape numbers. this generally involves a booster pump at lower speed or an inducer at full speed to avoid harmful cavitation in the main pump.

An inducer fitted to the inlet of the first impeller of the main pump resembles a screw which can have a constant or variable lead and a parallel or tapered periphery or hub. Inducers were first developed for the fuel pumps of space rockets and operated at inlet shape numbers up to 40 000, but here the life need only be a few minutes and inlet shape numbers of this order are not possible on continuously operated pumps.

In general, the provision of an inducer can increase the inlet shape number of a high-speed pump by two to three times, thus giving a generous margin of safety in respect of permissible NPSH. Inducers are, however, less able to accept off peak flow conditions than conventional impellers and here vibration difficulties can occur.

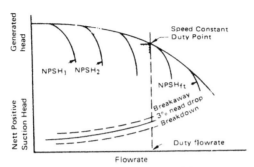

(a) Typical pump performance characteristic

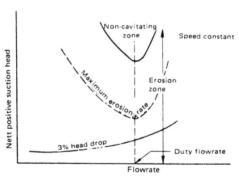

(b) Erosion zone boundary

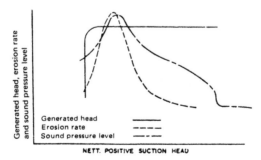

(c) Variation of head, erosion rate and
sound pressure level with NPSH

Figure 9.6
Cavitation tests.

Pump with supercavitating Inducer (Ref 19)

Here the inducer vanes are sufficiently short and run at a high enough speed to ensure that the cavitation bubbles do not recondense until they are clear of the vane and therefore can do no harm since they collapse in the liquid. The inlet shape number attained by a supercavitating inducer is comparable to that of a normal inducer but the margin of safety in respect of life is greater. Several pumps with supercavitating first stages are now in operation for chemical duties.

Cavitation Tests

The incidence of cavitation is detected by the following in order of appearance.
- (a) Stroboscopic observation of bubble appearance and size.
- (b) Sensitive acoustic tests of cavitation noise.
- (c) Reduction of efficiency and/or head of 3% which is the basis of ISO and BSI test codes in determining NPSH and inlet shape number.

Long Term Basis

The major factors causing cavitation erosion are high relative vane velocity, adverse angle of attack and, at impeller periphery, close proximity of moving and stationary vanes. NPSH should increase approximately as velocity to the power 3.2 (not the square) to avoid long term erosion.

Cavitation erosion is assessed by depth of attack or by weight of metal removed or by the cost of weld repair. Mild steel impellers and runners are often protected at vital parts by stainless steel weld deposit.

Temperature Effects

At the critical temperature of 374°C water and steam densities are the same. Hence there is no volume change or bubble collapse. In practice, cavitation erosion is negligible above about 300°C but increases progressively as temperature falls.

References

17. ANDERSON, H.H., 'Modern Developments in the Use of Large Single Entry Centrifugal Pumps'. Proc IME 1955, Vol 169 No 6.
18. Conference on 'Cavitation' Edinburgh, 1974, IMechE.
19. PEARSALL, I.S., 'Pumps for Low Suction Pressures'. Symposium on Pumping Problems. Univ Coll Swansea. IMechE, IChemE, March 1970.

THE DETERMINATION OF PUMP SPEED AND DIMENSIONS FOR A GIVEN DUTY

In the preceding chapters we mentioned that a pump speed is controlled by the cavitation aspect. We therefore investigated cavitation before considering dimensions, because dimensions depend upon speed which, in turn, depends upon cavitation performance. Having decided the speed we can now consider the basic dimensions.

Analysis of Test: Head Quantity Characteristic

Daugherty (Ref 1) referred the peripheral velocity of the impeller, the radial flow velocity of the water through the impeller, and the flow velocity of the water in the casing throat to the spouting velocity corresponding to the operating head of the pump at best efficiency point.

This has remained the conventional method of designing a pump, the velocity coefficients corresponding to these values being plotted against shape number, for example by Stepanoff (Ref 8) 1957.

This approach, however, does have the disadvantage that the calculations for the flow velocity within the impeller and the casing throat are affected by any error in measuring the operating head of the pump, and are also affected by the change in head due to change in hydraulic efficiency when a change of pump size is involved.

This is because the expression $\sqrt{2gh}$ is the basis of all speed calculations, so that errors due to variation of 'h' permeate the whole design investigation.

The use of $\sqrt{2gh}$ has been derived from the water turbine practice where, due to large physical size, the efficiency change is small, but it is very misleading and incorrect to use this expression as a basis for centrifugal pump design since their efficiencies may vary according to size and shape number from, say, 50% to 90% or more.

In order to avoid the above errors, the author referred all flow velocities to the peripheral velocity of the impeller, an independent variable, which can be determined with great accuracy since it comprises only the diameter of the impeller and its rotational speed.

Similarly, the operating head was referred to the velocity head corresponding to the impeller peripheral velocity. This eliminated errors of head measurement from flow and area calculations. As a further refinement, the generated head was divided by the hydraulic efficiency, the latter being determined by the assessment of losses described below.

The test head divided by the hydraulic efficiency was referred to as the Newton head, and represented the head generated by the dynamics of the pump at 100% hydraulic efficiency, thereby eliminating all losses due to the friction in the passages and bringing a 50 mm pump with an efficiency of, say, 65% to the same dynamic standard as a 500 mm pump with an efficiency of 90%.

These two steps were sufficient to reduce the scatter of test results to a figure comparable to the tolerances of the pump test code, ISO 3555.

The plotted results are shown in Figs 6.4 and 10.1. Each point represents the average

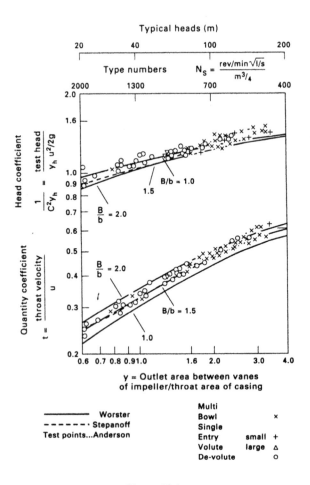

Figure 10.1.

of all pumps of that particular area ratio, the charge representing average behaviour of several thousand pumps (Refs 6 and 17).

Effect of Area Ratio on Pump Characteristics (Appendix A)

It is found that the head, quantity and power characteristics of a pump were primarily decided by the ratio of the relative flow area between the outlet vanes of the impeller to the casing throat area. These areas are shown in Fig 5.1. The reason for this is again emphasized as follows:

A small impeller outlet area and a large casing throat area result in a steep falling head-quantity curve with a power curve that reaches a maximum at the best efficiency point, since the relatively higher backward velocity between the outlet blades gives a lower whirl velocity and hence a lower head at best efficiency.

As a typical example, the slope of the head quantity curve may be 45° on logarithmic paper at best efficiency point. Mathematically, this point is then the peak of the power curve since: (a) the product of head and quantity, ie liquid power; and (b) efficiency, are both constant for an infinitesimal change of quantity.

Conversely, a large impeller outlet area with a small throat area will give higher head and relatively flat head curve, having very little variation of head for a changing quantity, since backward flow velocity is low and, therefore, the whirl velocity remains high. The power curve will, therefore, rise beyond the maximum efficiency point since the slope of the head quantity curve at best efficiency point will be less than 45°.

Head and Quantity Coefficients

Since the quantity pumped is related to the peripheral velocity and the head to the square of the peripheral velocity, tests were analyzed by plotting:-

$$\frac{\text{Flow velocity in casing throat}}{\text{Peripheral velocity}}$$

and

$$\frac{\text{Head x 2g}}{(\text{Peripheral velocity})^2 \text{ x Hydraulic efficiency}}$$

against the ratio of the impeller and casing flow areas, thus taking into account the two major components of the pump (see Figs 5.1 and 5.2).

The flow velocity and the head are chosen at the best efficiency point of the pump.

General Survey of a Whole Range of Pumps

The head and quantity coefficients of Fig 10.1 are the result of an analysis of a very large number of pumps of various shape numbers and sizes. Each test point in Fig 10.1 represents the average of tests on all pumps of that area ratio. The relatively small scatter of points of Fig 10.1 illustrates the fact that the area ratio is a reliable means of forecasting pump performance, and of showing how variation of characteristics can be obtained at a given shape number.

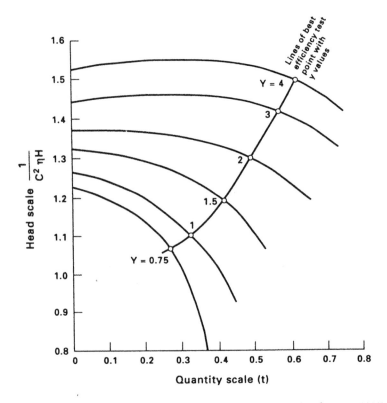

Figure 10.2 – Centrifugal pump characteristics on dimensionless basis at 100%
hydraulic efficiency for best efficiency point. This chart may also be regarded as
showing change of performance for various impeller areas in the same casing.
Closed valve zero flow head is NOT corrected for hydraulic efficiency.

The area ratio analysis has, therefore, the advantage over the conventional analysis of being able to provide, within limits of course, any characteristic at any given shape number.

Effect of Various Dimensional Changes in the Pump

Fig 10.1 can be transposed as shown in Fig 10.2, where the ordinates represent the head and the abscissae represent the quantity (Refs 6 and 17). On a proportional basis, the rest of the head quantity test characteristics can then be drawn in, so that a family of curves is obtained showing the whole range of pumps between the limits of an impeller of small area in a casing of large throat and an impeller of large area in a casing of small throat. Best efficiency points are shown as circles labelled with the appropriate value of the area ratio.

Fig 10.2 may also be regarded as showing the change of performance for change of

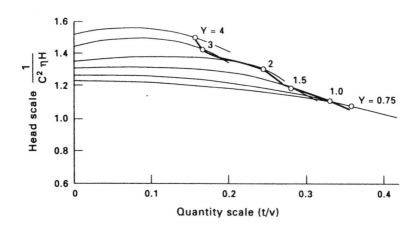

Figure 10.3 – Centrifugal pump characteristics. This figure can also be regarded as showing the best efficiency performance for various casing throats with the same impeller, e.g. a gated pump.

impeller width or angle where impeller diameter and casing throat area remain unaltered. The impeller width would then be proportional to the value of the area ratio y shown on the chart. Quantity is represented by velocity through the unchanging casing throat, that is, by the coefficient t. The line of best efficiency points corresponds to the casing line in Worster's investigations (Fig 6.3).

The effect of change of casing throat area is shown in Fig 10.3 which is a transposition of Fig 10.2, quantity being represented by velocity through the unchanging impeller outlet area, that is by t/y (Ref 17).

Fig 10.3 will then represent a pump whose throat area can be adjusted by gates, as in a water turbine, and where the gate area is inversely proportional to the area ratio y shown on the charge. A larger gate results in a larger quantity but a lower head at best efficiency point. The line of best efficiency points represents Worster's impeller line (Fig 6.3).

Correlation of Area Ratio and Shape Number

The true references of Fig 10.1 are the head and quantity coefficients and the area ratio. At the top of Fig 10.1 is inserted a scale showing the heads and shape numbers at which the respective area ratios are more generally used. For example, a shape number of 2000 would usually adopt an area ratio of 0.6.

It is, however, possible to use an area ratio of 0.6 at the extreme left-hand side of the chart for shape numbers as low as 600, but a ratio of 0.6 when used on such shape numbers would give a steep falling head-curve, a larger diameter impeller than normal, with, in consequence, a more expensive pump, and by reason of higher disc friction, a somewhat lower efficiency. Such a pump would, however, have the advantage of a power curve which reaches a peak at the maximum efficiency point of the pump (that is to say, a non-overloading characteristic). See Appendix A.

Technical and commercial considerations of cost of pump, cost of motor and versatility of characteristic for duties demanded by the market, help the designer to decide upon a suitable area ratio for any pump, or group of pumps.

Determination of Impeller Outlet and Casing Dimensions – General Summation

For a given head, the typical shape number and area ratio can be taken from Fig 7.1. Knowledge of the shape number will permit calculations of the running speed (duly checked by the cavitation limitations already described, and arranged to suit the driver).

The overall and the hydraulic efficiencies can be determined from the equivalent Reynolds number in the form of litres/sec and the shape number, N_s 1 400 giving the optimum efficiency. (See Figs 7.2 and 7.3).

From the area ratio curve Fig 10.1 or Figs 10.2 and 10.3 the values of the head and quantity coefficients can then be read off.

The steepness of the head-quantity curve is determined from Fig 10.2. Site conditions may, however, demand a steeper curve than is normal for the head and shape number involved - a lower value of area ratio must therefore be taken from Fig 10.1, which will result in a larger impeller for the same duty with larger disc friction and slightly lower efficiency.

As already mentioned, the reason for the lower head coefficient with lower area ratio lies in the fact that backward swept impeller blades are used. The smaller impeller area results in a lower velocity of whirl (since the relative backward flow is greater) and the lower whirl gives a lower head for a given impeller diameter.

References

1. DAUGHERTY, R.L., 'Centrifugal Pumps' (McGraw-Hill), 1915.
6. ANDERSON, H:H., 'Centrifugal Pumps, an Alternative Theory' Proc IME, 1947, Vol 157, WE 27.
8. STEPANOFF, A.J., 'Centrifugal and Axial Flow Pumps' (Wiley), 1957.
17. ANDERSON, H.H., 'Modern Developments in the Use of Large Single Entry Centrifugal Pumps'. Proc IME 1955, Vol 169, No 6.

SECTION 3

Practical

PRECISION MANUFACTURERS OF HYDRAULIC
PASSAGE SURFACES FOR FLUID MACHINES

PUMP TYPE VARIATIONS

PLANNING A RANGE OF PUMPS

LOSSES IN CENTRIFUGAL PUMPS

PRECISION MANUFACTURE OF HYDRAULIC PASSAGE SURFACES FOR FLUID MACHINES

The efficiency of a fluid machine is limited by the frictional and turbulent losses due to imperfections of flow passage surfaces. It is therefore essential to produce surfaces as near as possible to the geometry specified by the designer and with the highest possible surface finish. The novel conception of the impeller as a gear wheel is of great value in this task since it permits the passage surface of the metal or plastic corebox to be produced by generation as a gearwheel in a standard machine tool to within 0.025-0.050 mm.

The pump maker has thereby given the best possible start to the foundry which can contribute by various processes of precision castings, shell moulding, investment casting, etc, towards a perfect component.

Low shape number impellers for small quantity high head duties have relatively large diameters and small axial widths. Where it is essential to keep first cost to a minimum — even at the expense of efficiency — these impellers often have plain radial flow and their vanes have single curvature which can readily be withdrawn from the mould. For medium and higher shape numbers, however, the impellers are generally of double curvature, variously described as mixed flow or Francis impellers. the liquid, normally water, enters the impellers almost axially and is turned in the impeller to radial flow, hence the term mixed flow. Impeller passages have generally be produced by the error triangle, conformal transformation or aerofoil methods (Refs 8 and 20).

These methods involve development of, say, three streamlines from periphery to inlet or eye, cutting off the ragged edge at the eye, transferring the shapes to planes normal to the axis, pricking through on to wooden pattern boards and then carving the boards to the prick-marks. Very large errors arise in all these operations, and the wooden corebox will certainly distort in use. There is no knowledge whatsoever of the true acceleration in three dimensions imparted to the water in its passage through the impeller. The impeller form at the eye is quite indeterminate, since the eye angles chosen will be altered by removing the ragged edge. These conventional methods aim merely at a smooth change from inlet to outlet, but this does not necessarily give the optimum acceleration rate at all radii.

Generation Methods

In contrast, the generation procedure rotates the impeller as if it were in the pump and simultaneously defines by a cutting tool the flow of water through the impeller with appropriate accelerations in three dimensions (Ref 21).

The variation of inlet angle from hub to eye periphery to allow for the varying radii of streamlines is thereby automatically ensured and the large drawing and pattern- making errors of the conventional methods entirely eliminated.

The requirement of maximum hydraulic mean depth of passage at all radii is readily arranged by this method, thus improving performance. The provision of a small helix angle at the blade periphery to minimize noise and vibration is also facilitated.

A major advantage of generation is the fact that its geometry is entirely defined by numerical data given to the machine tool operator. A larger or smaller pump merely requires that all these numerical values should be multiplied by a size coefficient to produce any new frame. In contrast, on conventional methods the whole development must be repeated, and since this depends on the skill of the designer draughtsman and the pattern maker there is no certainty that an alleged geometric replica pump is, in truth, an exact replica. This aspect is vital in model to full size reproduction.

Transmission of designs to associated firms abroad is exact and there is no need, as in the past, to send cloth development drawings with elaborate precautions against shrinkage in the post.

Flow Conditions

The path of the water through the impeller takes the form of a quadrant, usually of varying

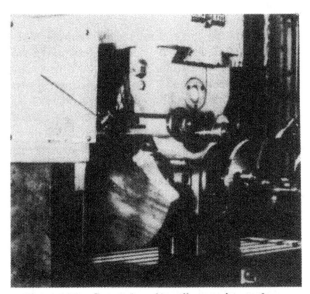

Figure 11.1 – Generation of impeller core box surfaces.

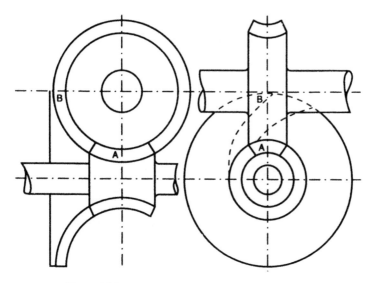

Figure 11.2
Worm wheel and worm extended to represent impeller.

width. The combination of impeller rotation and water rotation about this quadrant from inlet to outlet can therefore be said to resemble a gear drive.

With pure axial inlet the combination of impeller rotation and water flow will result in purely helical inlet vanes. At outlet the flow will be purely radial which, combined with impeller rotation, will result in the outlet vanes being purely spiral.

Gearing Analogy

Fig 11.2 shows a worm and a worm wheel. In order to increase surface contact the worm envelops the worm wheel and the worm wheel envelops the worm. The amount by which the worm extends in radius to a line through the worm wheel axis parallel to the worm axis.

With respect to the worm axis, the teeth at the smallest diameter of the worm (A) represent a pure helix and the teeth at the largest diameter of the worm (B) represent a pure spiral. The tooth profile of the worm between A and B represents an impeller vane and the tooth profile of the worm wheel represents the water passing between the impeller vanes in the quadrant A-B.

Purely Generated (Hobbed) Impellers

Some small impellers for rocket pumps were generated in a hobbing machine as described above, but the process was not economic and, moreover, involved a loose shroud, which had to be hydrogen platinum brazed or welded to the vanes. The cutter shape presented difficulties of attaining adequate strength and the process limited the design to a constant meridional velocity.

Numerical Control of Passage Shape

Although the above work proved to be uneconomic, it gave rise to a successful method of

Figure 11.3 – Generated impeller passage surface.

producing corebox surfaces by machine tools. The desired geometry was defined by a mathematical expression which was tabulated and fed into a conventional machine tool so that the metal surface of the corebox was shaped to the exact dimensions required by the designer which accuracy comparable to the turning in a lathe of a shaft or disc.

The arc of a milling cutter represented the advancing wave of water through the impeller.

Stationary Components

The generation procedure is applicable to other hydraulic passage components (for example, diffusers, radial guides etc) and in particular to the more complex three dimensional double curvature surfaces.

Over several years production from the smallest to the largest pumps for power stations and other duties, the generation process has shown a materially higher efficiency than the conventional processes. Moreover, the resulting smoother flow has minimized NPSH and erosion attack. Design and development work is simpler and quicker than with the older methods, and the manufacture of the metal or plastic corebox in the factory has proved to be quite straightforward and very economic.

References

8.   STEPANOFF, A.J., 'Centrifugal and Axial Flow Pumps'. (Wiley) 1957.

20.   KOVATS, A., 'Centrifugal and Axial flow Pumps and Compressors', (Pergamon Press) 1964.

21.   ANDERSON, H.H., 'Improvement of Reliability in Thermal Station Pumps', Reliability Conference, Loughborough, 1973, IMechE.

PUMP TYPE VARIATIONS

Choice of Speed and Number of Stages

For large units above, say, 200 kW, the physical dimensions of the pump are such that the impellers, diffusers, casings, etc, can be moulded, cast and dressed with reasonable ease and accuracy, and furthermore, there is a sufficient choice of 50 cycle motor speeds to permit operation at or near the optimum shape number as far as efficiency is concerned. In consequence, the heads per stage are as shown in Figs 7.1 and 13.2 for units above about 100 l/s, namely the normal limit of 100 m per stage.

Smaller units are limited by the fact that working to the optimum shape number may demand an actual operating speed which is too high, either by reason of exceeding the 2 pole 50 cycles speed of 3000 rev/min (synchronous speeds are given for simplicity – actually a small slip will generally occur) or because 3000 rev/min operation may give rise to relatively high frequency noise, undesirable in pumping stations near residential districts.

For small multistage pumps, therefore, the stage head may be as low as 10 m where the total head may be only a hundred meters, for example, at 4 l/s 1500 rev/min.

Conversely, a large waterworks pump may take advantage of the flooded inlet to generate 100 m or more in one stage, delivering say $2^{1}/_{2}$ m^{3}/s and driven at 750 rev/min by a motor of 3000 kW, giving a shape number which is near optimum and a complete pump and motor unit which, by virtue of high speed, is relatively small and, therefore economic.

A geometric replica of the above pump designed for a duty of say 25 l/s at 100 m head would need to run at 7500 rev/min, which is impracticable electrically and which may give rise to undesirable high frequency noise.

High Pressure High Temperature Pumps

At the other end of the scale for dealing with high pressures, a boiler feed pump of 10 l/s 25 bar would operate with ten stages at 3 000 rev/min. A larger boiler feed pump for a 60 megawatt turbo alternator would have seven stages and also run at 3000 rev/min for the duty of 100 l/s 90 bar with a head per stage of about 120 m and a driving motor of 1000 kW. Such a pump would, of course, require a relatively high inlet pressure to permit operation without risk of cavitation.

The pressures and flow quantities of boiler feed pumps have hitherto increased reasonably in step, for example, 25 bar for a 5 megawatt turbo alternator and 130 bar for a 120 megawatt turbo alternator, so that suitable pumps can be designed for all these duties at 3 000 rev/min, the latter operating at 200 m per stage.

On the other hand, a high pressure, 250 bar with small quantity, 100 l/s involves a geared pumping set in order to give a reasonable number of stages and good efficiency.

To sum up, the question of choice of number of stages and of head per stage rests largely on the duty, on the inlet conditions, on the nature of drive and on the size of the unit. For small units, advantage is taken of multistaging to operate at reasonably low speeds using cast iron or mild steel casings with bronze impellers and diffuser. For large units and high heads, the designer is forced to operate at relatively high heads per stage and high peripheral speeds for the sheer physical limitations of size, cost and available speed. Here, the use of stainless steel permits such high heads with economic life.

Series Operation, High-pressure High-temperature Glands

It is sometimes necessary to run two pumps in series, for example, where the head is higher than one pump can efficiently operate against or where hot liquids are handled requiring high inlet pressure to avoid vaporization. The chief problem associated with series sets is the provision of a reliable high-pressure gland. The incidence of high temperature with high pressure such as occurs on boiler feed or hot oil duties demands particular care in gland design. it is possible to provide a fine clearance bush with an injection of cooler, higher pressure liquid from the upstream pump, and to discharge the injected liquid at a lower pressure. In addition, there may be an intense but uniform cooling of leakage liquid so that ultimately the packing is subjected to minimum pressure and temperature conditions. A trickle of water may be used to quench any wisp of steam from gland leakage on high-temperature feeders, whilst a jet of steam is usually provided to prevent fire or toxic hazards on hot oil duties (See also Boiler Feed Pumps, Chapter 33.

Recent developments tend to replace series sets by geared sets at higher speeds.

The Choice of Single or Double Entry Impellers

In general, double entry pumps with horizontal shafts and horizontally split casings are used for single stage duties on medium heads, say up to 100 m. The inlet and outlet branches are arranged in the lower half of the casing to permit the upper half to be removed for examination of the pump and removal of rotating element without disturbance of pipe joints.

For heads up to 200 m two stage pumps embodying two single or double entry split casing stages are adopted.

Single entry pumps are, in general, used with horizontal shafts and vertically split casings for small single stage duties and for pumps where low cost is a first consideration.

Multistage pumps, in general, adopt single entry impellers for reasons of convenience in construction, the end thrust being taken on a hydraulic balancing disc or drum which both carries thrust and acts as a pressure reducing device preceding the gland. The reason for single entry impellers on the very small stage pumps is primarily a question of cheapness, and for single entry impellers on multistage pumps a question of convenience of arrangement of flow passages from stage to stage.

All charts (except cavitation) treat a double entry impeller as a single entity.

Large Single Entry Pumps

There has lately been a tendency to use single entry pumps for medium and large duties up to 200 m head. This is largely dictated by reasons of mechanical strength and compact arrangement of pump and motor, and is discussed in greater detail under the mechanical design section. (See Chapter 16).

From a purely hydraulic aspect, a single entry impeller is preferable for very small duties at low shape numbers since a better hydraulic mean depth and an easier casting is thereby obtained. For the higher shape numbers the single entry is less attractive, particularly where there is a shaft obstructing the impeller eye since the larger impeller eye required gives a shorter blade length. Such single entry designs are, therefore, confined to duties with a shape number below 2 000 units and to the higher heads and the more severe duties.

A single entry impeller, having a simpler inlet flow and being suitable, when on vertical axis, for mounting nearer to water level, has a suction performance little inferior to that of a double entry impeller with complex inlet passages. The motor, stuffing-box and impeller are lifted from the casing to permit examination without disturbing pipe joints. (See Chapter 34).

The introduction of single entry pumps for large heads and powers follows the precedent of Francis water turbines, which now have single exit runners. (See also Chapter 35.)

Diffusers on Multistage Pumps

The water discharging from the impeller of a pump is collected in a volute casing in a single stage pump, or a diffuser (which resembles a number of small volutes grouped around the circumference of the impeller) in a multistage pump. The main purpose of diffusers is convenience in slowing down the water velocity and conducting it to the next stage with a minimum pump diameter.

Hydraulically, there is no difference between a volute and a series of diffusers, except that the latter has a smaller hydraulic mean depth. (See Chapter 16).

In general, multistage pumps having a lower shape number have a relatively high flow velocity when leaving the impeller, so that renewable diffusers of superior metal, for example bronze or stainless steel, are generally used.

PLANNING A RANGE OF PUMPS

General

The basic principles of geometric designing were described briefly in Chapter 4. These principles are now repeated in this chapter for ready reference.

Change of speed for the same size of pump — For a given operating point on the characteristic of the pump, flow varies as rev/min, head varies as (rev/min)^2. Since power is the product of flow and head divided by efficiency, it follows that power is proportional to (rev/min)^2. A small change of impeller diameter gives similar variations.

Geometric change of size for the same rev/min — Flow varies as diameter^2, head varies as diameter^2, power varies as diameter^2 subject to efficiency changes.

Geometric change of size with inverse change of speed — Flow varies as diameter^2, power varies as diameter^2, rev/min varies inversely as the diameter. The above is subject to minor changes consequent upon change of efficiency with size and speed.

The following are constant: generated head; margin of safety above cavitation conditions; stresses in rotating and stationary parts, and margin of safety above critical speed of shaft.

Dimensionless Designing

It is obvious that in the designing of a range of pumps, many calculations would be repeated, merely using different values in a similar equation to cover a different size of pump. It is therefore helpful to carry out the design calculations on a general or dimensionless basis and thereby establish generated head, working pressure, stresses and relationship to critical speeds for a range of constant shape number pumps as a whole. For this range, head will be constant, flow in l/s will vary as the square of the size whilst speed in rev/min will vary inversely as the size.

Variations of Shape Numbers

If we consider, first of all, the driving arrangements having complete freedom of choice of speed, eg belt drive, gear drives, etc. it is found that a geometric range of pumps can be designed with constant shape number, the operating speed being chosen by gear or belt to suit the size of the pump.

This is the least complex range to design, since all pumps can be strictly geometric, apart from minor adjustments of dimension to suit standard threads, ball/roller bearings, flanges, etc.

Small, cheap single entry pumps, as described in Chapter 19, can be geometrically designed in this manner with almost constant shape number.

An approach to this constancy of shape number also occurs in the case of extremely large pumps or turbines where operating speeds are below 1 000 rev/min and where there is, in general, a large choice of alternating current motor speeds.

In this range, geometric designing is generally practicable on the basis of constant shape number.

For the majority of small and medium size pumps, however, the normal drive is by a.c. motor with 2-pole or 4-pole speeds corresponding to 3 000 or 1 500 rev/min.

It will be appreciated that 3 000 rev/min can give rise to objectionable electrical and/ or hydraulic noise, so that 1 500 rev/min is often used for quite small duties, thereby introducing very low shape numbers.

For example, a range of pumps to generate 30 m head at 1 500 rev/min will require very

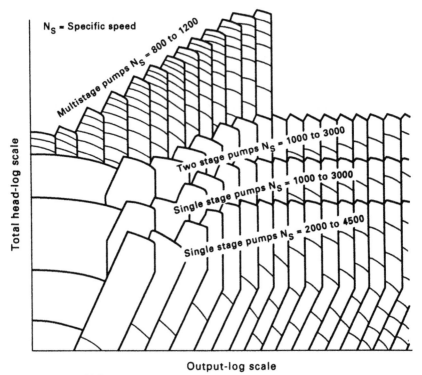

Figure 13.1
Chart showing method of planning a range of single, two and multistage pumps to cover the field of duties economically and efficiently.

narrow impellers for the 50 mm size. These would be difficult to cast and therefore the 40 mm, 50 mm, and 80 mm pumps may well be designed to generate heads of the order of 15 m, 20 m, or 25 m, followed by larger pumps up to say 300 mm generating 30 m head.

In this case, shape numbers will vary from perhaps 300 to 2 000. Such arrangements are generally made for single stage and two stage pumps, the single stage pump having one, two or three ranges of head for a given branch diameter and quantity and the two stage pumps having only one range of head.

On multistage pumps, where flexibility occurs by varying the number of stages, it is generally possible to adhere closely to a constant shape number, but of course, on constant speed, the head per stage will increase as the size increases.

A typical chart showing the planning of a range of single stage, two stage and multistage pumps is shown in Fig 13.1. (Author 1932)

This illustrates how a given region of quantity and head can be efficiently covered by suitable ranges of pumps. The upper line of each small section or 'tombstone' in this chart represents the head curve at full diameter at the particular speed involved. Lower duties in the area of each pump performance are obtained by reduction of impeller diameter where fixed speed is involved.

The fine lines on the chart in respect of the single and two stage pumps represent lower speeds, for example 1 000 rev/min instead of 1 500 rev/min etc, and on the multistage pumps, the fine lines represent the separation between different number of stages. Fig 13.1 can be compared with Fig 13.2, chart of general pump duties.

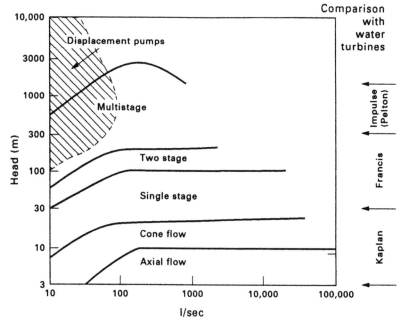

Figure 13.2
Chart of pump duties. (Displacement duties occur in the shaded area.)

Choice of Area Ratio for a Given Duty

General

The preceding paragraphs showed how a range of pumps could have a very low shape number at the small end, gradually increasing to a normal shape number, which could be maintained for the larger sizes throughout the range. The reason for this is the fact that true scale speeds for the smaller sizes would be too high to be practicable. We now consider the implications of this fact in the design of pumps and, in particular, in the choice of area ratio of a given duty.

Typical Example of a Range of Pumps for 30–50 m Head

In general, the determination of head for a range of pumps is based upon a market survey of enquiries and potential business, for example, in the waterworks, chemical and general industrial fields. A head of 50 m represent a round number covering a large proportion of general purpose pumps and therefore it is not surprising to find that practically every manufacturer has a range of pumps for this head.

Design Details

The smallest pump of a typical range could have, for example, 50 mm branches and a duty of 10 l/s 30 m head at 1 450 rev/min resulting in the relatively low shape number of 350. At first sight, this would involve an impeller of relatively large diameter and small width, such that a satisfactory casting could not be made. A built up impeller is generally ruled out on account of expense and therefore, for these duties, it is usual to distort the design by using an extremely high area ratio purely to facilitate casting. Alternatively, it may be found economic to sacrifice head, reducing the 30 m of the range to 22 m or so for the smaller sizes.

As the size increases, for example to a 250 mm pump for a duty of 300 l/s 30 m 1 470 rev/min, where the shape number is 1 600, the impeller will increase in width so that it can be easily cast and dressed and, at this shape number, will give a good efficiency. From the 250 mm pump upwards there is generally a sufficient choice of 50 cycle operating speeds to permit a range of pumps at 30 m nominal head to have a good shape number and, in consequence, good efficiency.

The range could comprise larger units geometrically similar to the 250 mm pump wherein dimensions will be proportional to a.c. motor pole numbers. For example, a 250 mm pump will be driven by a 4-pole motor at 1 470 rev/min, a 500 mm pump by a 8-pole motor and a 750 mm pump by a 12-pole motor and so on. The area ratio for these larger pumps is in no way limited by casting problems and its choice is discussed below.

Area Ratio and Shape Number

Fig 13.3 shows a general experience chart of area ratios that have been used by the author over the years, plotted against shape number. For example, with a shape number of 800, the area ratio could range between 0.6 and 4, a very considerable amount of variation. Pumps have been made at each extreme and have given economic efficiencies. The pump having an area ratio of 0.6 will have a larger impeller diameter, a large casing throat, a 50%

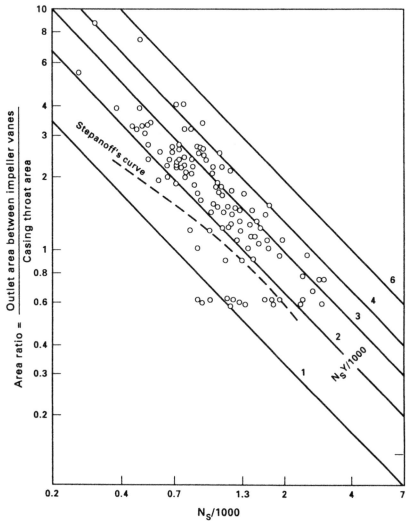

Figure 13.3
A chart of area ratios plotted against type number that have been used by the author over the years.

increase of pressure from best efficiency point to zero flow, with a stable head curve and a power curve that reaches a peak at best efficiency point — that is to say, a non-overloading power characteristic.

On the other hand, the pump having an area ratio of 4 will have a very much smaller impeller diameter and a smaller throat diameter, thus providing a very much cheaper pump. The pump efficiency will be higher since the smaller impeller will have a lower disc friction. The pressure rise from best efficiency point to zero flow point will be small, probably 5% or 10% and the curve may be unstable. The power characteristic will still be rising at best efficiency point, so that, if this pump is required for a head which is likely to vary downwards, a larger powered motor will be used to prevent overloading. (See Appendix A.)

Between these two limits, for example on area ratios of 1½ to 2, an optimum efficiency would be attained, the design being intermediate between the two examples mentioned above, but the optimum efficiency will be very little greater than the efficiency of the two extremes in question.

Fig 13.3 also shows the curve transposed from pump design data given by Stepanoff (Ref 8) which represents an intermediate value between the two limits described above. The advantage of the area ratio approach is the fact that at any given shape number a complete range of characteristics is attainable, the experience range of which is shown in Figs 10.2 and 13.3, instead of the single shape of characteristic at each particular shape number as shown by Stepanoff.

It will be noted that, as mentioned above, the very low shape numbers associated with small sizes can use particularly high area ratios. Rectangular hyperbolic lines show the product of shape numbers and area ratio and it will be seen that the majority of points in the experience curve seemed to be grouped along such lines.

In his paper 'Dissimilarity Laws in Centrifugal Pumps and Blowers' (Ref 22), Stepanoff refers to area ratios other than those on his curve (Fig 13.3) as representing 'mismatched impellers and casings' which are difficult to predict. The area ratio approach gives a simple explanation of these 'mismatched' pumps and thereby extends the understanding of design to the wide field of useful pumps shown in Fig 13.3.

Briefly, it may be said that the choice of area ratio on the very small low shape number sizes may be dictated by casting problems. On the medium and larger shape numbers and sizes, choice of area ratio is dictated by the characteristics required and by economics of manufacture.

It is important to note that the higher values of area ratio can give rise to off peak vibration on stage heads above 300 metres.

References

8.   STEPANOFF, A.J., 'Centrifugal and Axial Flow Pumps', (Wiley) 1957.
22. STEPANOFF, A.J., and STAHL, H. A., 'Dissimilarity Laws in Centrifugal Pumps and Blowers'. ASME Paper 60-WA-145.

LOSSES IN CENTRIFUGAL PUMPS

The efficiency of a centrifugal pump naturally falls below 100% and it is the difference between the 100% figure and the actual efficiency that is referred to as the total loss. The total loss is made up of the following three major losses: power losses, head losses and circulation losses.

Power Losses

These comprise the friction due to rotating the shrouds of the impeller in a chamber of water, the gland and the bearing friction. In the case of the multistage pump, there are, of course, several impellers and the balance disc to incur disc friction.

Disc friction, although a hydraulic phenomenon, is made manifest by a retarding torque on the shaft and therefore is correctly grouped under power losses.

There is reason to believe, however, that the water in the space between the impeller shroud and the chamber, where in consequence of disc friction is rotated at a speed slower than the impeller, eventually passes into the main stream of water in the volute and gives a slight addition to pumping efficiency, particularly on volute pumps (Fig 16.4). This aspect is, however, neglected in the normal split calculations of losses.

Investigations of the power absorbed by rotating a disc in water were made by Gibson & Ryan (Ref 23). Their data has been transposed into a convenient form for designers in Fig 14.1 where disc diameter and speed is plotted against power absorbed.

The disc friction is generally of the order of 1 to 2% of the total power on a single stage pump and slightly larger due to presence of the balance disc, namely about 2% on a multistage pump, depending upon shape number. As mentioned above, the reduction of efficiency as shape number is reduced below the optimum of 1 400 in l.s m and rev/min units, is primarily due to the fact that on the lower shape number the liquid power for a given impeller diameter is reduced and therefore the disc friction becomes a greater proportion of the total power, thereby reducing efficiency.

Gland and bearing friction on medium and large pumps each absorbs approximately $1/2$ to 1% of total power. On very small pumps, particularly with high inlet pressures, the gland friction can considerably exceed these figures.

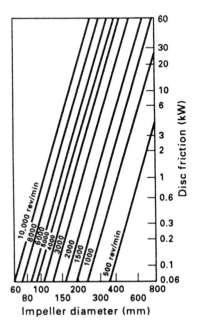

*Figure 14.1
Disc friction
losses.*

Head Losses

On the inlet side, the head loss between the inlet branch and the inlet passage of the impeller may be deduced by knowledge of the cavitation behaviour. For example, referring to Fig 9.2, the NPSH is given as $K_1 A^2/2g + K_2 P^2/2g$ where K_1 and K_2 are constants determined from test. A is the axial flow velocity into the eye, P is the peripheral velocity of the impeller eye and g is the acceleration due to gravity.

$A^2/2g$ represents the velocity energy of the water entering the impeller, in the impeller eye, which is not a loss.

The term $K_1 P^2/2g$ represents the shock loss of the water entering the inlet portion of the impeller. Knowledge of the NPSH and of the changes in velocity between pump branch and impeller inlet permits, therefore, determination of head losses. The inlet loss now becomes

$$(K_1 - 1) \frac{A^2}{2g} + K_2 \frac{P^2}{2g}$$

Head losses due to friction in the impeller passages are somewhat difficult to assess but an estimate can be made by treating them as pipes of rectangular or square shape and tapering dimension as shown by Gibson (Fig 14.2). Here a difficulty arises since Gibson's investigation dealt with straight pipes and the impeller vanes are curved, with the result that, for an expanding passage, the curved pipe will have a greater loss than Gibson's straight pipe. Such losses are therefore generally estimated after all other losses have been assessed, the difference between 100% and the sum of the efficiency and the calculable losses being spread over those losses which are difficult to calculate.

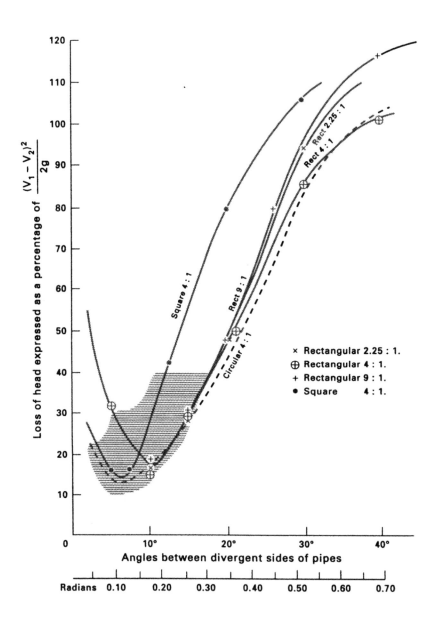

Figure 14.2
Percentage loss of head in straight taper pipes of
square and rectangular section.

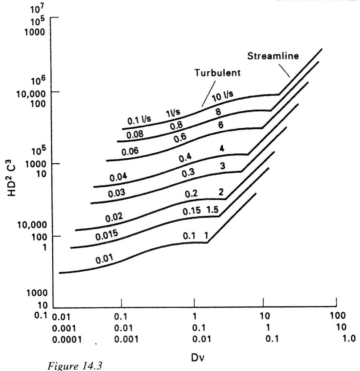

Figure 14.3
Frictional resistance through fine clearances.

Shock losses between impeller discharge and volute are also difficult to calculate since the exact velocities of the water leaving the impeller and water within the volute are not easily determined.

The losses in the taper between the throat of the casing and the discharge branch are determined by the Gibson data, Fig 14.2 (Ref 24).

Leakage Losses

The direct leakage losses of neckring, balance disc (in the case of a multistage pump) etc are dependent upon the actual clearance in the cylindrical space between the impeller neck and the neckring; upon the length of this clearance axially; upon its diameter, upon the pressure difference and upon the viscosity of the liquid. Investigations by R.J. Cornish (Ref 25) at Manchester University, 1924, were transposed by the author to a convenient form for use by designers, as shown in Fig 14.3. Over the past 50 years, this chart has been used in the turbulent region for low speed pumps and in the streamline region for oil servo pistons for water turbine governors and large valves. (See Chapter 33, for effect of high speed).

Neckring leakage in pumps is generally of the order of 1-2% of the total flow, depending on shape number. Balance leakage on a multistage pump is similarly of the order of 1-2% of total flow.

As with disc friction, the leakage losses increase with the reduction of shape number.

V = *kinematic viscosity in cm*/s, stokes (water 0.01 at 15°C).*
H = *loss per 10 mm clearance length in metres of liquid passing.*
D = *diameter of rotor in mm.*
C = *clearance on diameter in millimetres.*

Note three scales for 1 to 10 litres per second
0.1 to 1 litres per second
0.01 to 0.1 litres per second

referring respectively to the three ordinate and abscissal scales and in descending order. Scales can be extended either way by suitable multiplication by 10 or 100.

Dr Myles in Ref 26 describes his system of evaluation of pump performance parameters by summation of losses. The determination and the minimizing of each loss in order to attain optimum efficiency is carried out in a computer programme.

The author's area ratio method, on the other hand, relates to overall performance and the efficiencies recorded here were attained by a logical and statistical approach before computers came into general use.

Dr Myles' methods and the author's methods are therefore complementary systems of designing and both are essential in the future improvement of pump performance.

References

23. GIBSON & RYAN, 'Disc Friction Investigation': Proceedings Inst CE 1910, Volume CLXXIX, Page 313.
24. GIBSON, A.H., 'Hydraulics and Its Applications'. Constable, 1920.
25. CORNISH, R.J., 'Investigation on leakages through fine annular clearances with inner and outer boundary wall stationary and rotating. Manchester University, 1924.
26. MYLES, D.J., 'Analysis of Impeller and Volute Losses in Centrifugal Pumps'. Proc IMechE 1969/70 Vol 184 Part 1.

SECTION 4

Materials

CONSTRUCTION MATERIALS

MECHANICAL DESIGN OF CENTRIFUGAL PUMPS

CRITICAL SPEEDS, VIBRATION AND NOISE

BEARINGS

PRESSURE VESSEL ASPECTS

MECHANICAL SEALS AND OTHER
SHAFT SEALING DEVICES

CONSTRUCTION MATERIALS

Normal Construction Materials

General service pumps for low and medium pressures (eg 10 bar on single stage split casing pumps and 36 bar on cellular type multistage pumps) have cast iron casings, high tensile steel shafts with bronze rotating and wearing parts.

The material specifications are as follows:-

Cast iron BS1452.

High tensile steel shaft BS970

Bronze wearing parts BS1400 LB2, a soft lead bearing bronze.

Bronze impellers, diffusers, sleeves, balance discs, etc. BS1400 LG2 gun metal, or with a greater portion of tin to lead, LG3 or PB3 zincless bronze, according to severity of duty.

For higher pressures, the casings are of cast, forged or welded steel and, for the highest heads per stage, stainless steel impellers and diffusers are used. The casing steel forging specification is 14A in general cases, but variations of this occur according to the tensile strength and other properties required. See also BS970 for EN steel.

The stainless steel rotating and stationary parts are made to specifications BS1630, 13-chrome steel castings, BS970 for 16-chrome $2^{1}/_{2}$% nickel forgings and BS1631 for austenitic stainless steel 18-chrome 8 nickel castings or forgings.

High temperature high pressure pumps, (eg boiler feed pumps and hot oil pumps) have forged steel pressure vessels, forged stainless steel shafts with cast stainless steel impellers, diffusers, etc. (See Chapter 33).

For temperatures above 120°C, bronze and austenitic stainless steel are unsuitable owing to their low strength and their coefficient of expansion (50% greater than that of mild steel or other stainless steels) which causes difficulty in maintaining correct fits and running clearances at higher temperatures. Bronze is, however, used for the very low stress wearing parts, (eg chamber bushes of multistage pumps).

Corrosion - Erosion

Two distinct problems arise in choice of material for pumping duties. The first concerns the provision of material whose strength is sufficient to withstand any anticipated stress loading without fracture or undue distortion. The second concerns the avoidance of surface pitting due to a combination of chemical attack with the erosion of high speed flow, which may on certain applications contain solid particles.

There is reason to suppose that, in general, all normal metals used in engineering construction are attacked by all liquids normally pumped. This attack gives rise to a film of oxide or the salt corresponding to the acid impurities in the liquid pumped, and this film acts in many cases as a protection to the parent metal against further attack under static conditions or low flow velocity conditions, (up to, say, 7 m/s).

When, however, the flow velocity due to generation of high head per stage increases to the order of 60 m/s, the protective film is eroded away, exposing fresh metal to attack. Under these higher velocity conditions consequent upon high head pumping and particularly where corrosive elements are in the liquid and where solid particles may occasionally be handled, it is essential to increase the quality of the material and thereby increase its resistance to attack.

It is felt that corrosion and erosion proceed together so that a metal that has high resistance to corrosion is as important as a metal that has a high tensile strength. Local cavitation due to imperfections of surface and of flow can further aggravate these corrosion-erosion problems.

The stainless steels possess a combination of high tensile strength and high resistance to corrosion and therefore they are of particular value in dealing with high temperature, high pressure and high head per stage duties.

For pump duties, liquid velocities must therefore be considered in any corrosion-erosion-cavitation problem.

Special Construction Materials

Special duties can be described as those concerned with severe chemical attack, with very high speeds and high stage heads.

Severely Corrosive Duties

Pumps for handling very corrosive liquids, (eg various acids, etc, used in chemical and process work) are often manufactured in ceramics, plastics or the more noble metals. Such duties normally involve relatively low heads up to 40 m or so, where the problem is entirely corrosion, since flow velocities are not sufficiently high to cause the combination of erosion and corrosion mentioned above to any marked degree. Some simplicity of pump design is often adopted in order to permit moulding or fabrication in these somewhat special materials. Many pumps are made with metal impellers and casings which are afterwards coated with rubber, plastic or ceramic, but in these applications it is essential that the covering should be entirely free from any flaws or cracks, as otherwise its purpose will be defeated and the liquid will have access to the more vulnerable metal frame. (See Appendix B for materials on chemical duties).

Titanium is a most valuable pump material for corrosive duties. At the present time titanium is regarded as a forged or rolled material capable of being welded; small titanium pumps are therefore cut from solid metal and medium sized ones are fabricated. Cast iron can be poured into a mould lined with titanium to produce a pump which has titanium at all parts in contact with the corrosive liquid and a cast iron outer shell to withstand the forces involved in tightening pump branches on a pipe flanges which are slightly misaligned. A good bond is obtained between the cast iron and the titanium.

The combination of high strength, lightness and weldability renders titanium ideal for space rocket pumps. It is clear that the casting of titanium is becoming a commercial proposition which will be of great value to the pump and chemical industries.

Materials for High Stage Heads (See Ref 27)

High heads involve high speeds both of impeller and of liquid flow, so that high metal stresses and severe erosion- corrosion necessarily occur. An attractive material for these duties is martensitic precipitation hardening stainless steel, for example FV520 containing 14% chromium and 5% nickel. This steel has the corrosion resistance and weldability of austenitic stainless steel 304 and the strength and acceptable expansion coefficient of the non- weldable high strength stainless steels 420 and 431.

A similar martensitic precipitation hardening stainless steel contains 17% chromium and 4% nickel.

Ferralium, 25% chromium is of particular value on offshore duties (see Table II of Appendix B).

These steels have assisted the attainment of stage heads up to 2 000 m and show promise of greater heads.

For the future, glass and carbon fibres show promise of even greater strength and resistance of erosion.

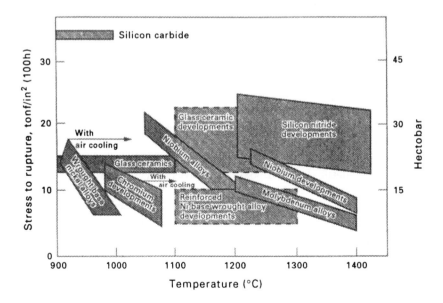

Figure 15.1

Brittle materials, for example cemented carbides and ceramics, have applications in pump duties since their hardness is valuable for bearings lubricated by abrasive fluids (Fig 15.1, Ref 28).

Plastics

Plastics offer considerable attractions for chemical services since many of the man-made materials based on synthetic resins are notable for their chemical inertness and thus high resistance to chemical attack. This is particularly true of many of the more recently developed thermoplastics, and thermoset plastics in general. Certain limitations apply to all plastics, notably:

(i)   Low mechanical strength compared with metals.

(ii) Lower moduli, particularly with thermoplastics.

(iii) Low creep resistance, particularly with thermoplastics.

(iv) Limited top service temperature.

Some families of plastics have further limitations which are less apparent at first sight but show up in service such as: lack of dimensional stability, a susceptibility to stress, corrosion cracking in contact with chemically active fluids, and high permeability. To some extent these can be reduced, or even overcome, by modifying the plastic, but the best practical solution is to use a newer plastic which does not exhibit the same limitations. New materials with better properties continue to be evolved, and in recent years there has been a considerable increase in the number of commercial plastics with 'engineering' properties; that is to say, they can be considered as direct alternatives to metals for some applications.

The first plastics used on any scale for chemical and process pumps were PVC and other vinyl resin compounds, and polythene, chiefly for linings and coatings. The rigid form of PVC was, and still is to a certain extent, employed for moulded components and casings for small pumps where high strength and rigidity were not required. Neither material fulfilled its expectations and the use of polythene linings, in particular, gave trouble because the material is permeable and allows liquid to penetrate to the underlying metal, causing corrosion and disruption of the bond.

Since then the engineering plastics have progressed and the all-plastic chemical pump has become a practical proposition, at least in smaller sizes. The materials favoured now are polycarbonate and epoxy. The only metal component necessary is the shaft, which can be stainless steel or plastic sleeved. Epoxy resins have better engineering properties since they are thermoset rather than thermoplastic materials. The mechanical properties can easily be raised by filling with, say, glass fibre. Thus, epoxy is coming into greater use for impellers in the larger chemical pumps and even for casings.

Owing to the multiplicity of plastic materials with potential applications — some with proven limitations and others still little explored — it is impossible to deal with plastic constructions comprehensively. Table III Appendix B summarizes the chemical resistances of the plastics which bear consideration for chemical duties. This can be read as a general rather than specific guide in most cases since some acids or solvents may attack a material shown as suitable. Table IV summarizes the leading physical properties of the same materials and shows their current applications.

The above description of plastics was taken from an article by R.H. Warring (Ref 29).

References

27. MORLEY, J.I. 'An improved Martensitic Stainless Steel', Proc BISRA — 151 Conference, Scarborough, 2-4 June 1964.

28. DUNCAN, P. and MORTIMER, J. 'Ceramic Turbines — Why Britain is Leading the Race'. The Engineer.26.2.70 and 26.3.70.

29. WARRING, R.H., 'Materials for Chemical and Process Pumps'. Pumps Pompes-Pumpen, Jan 1969. See Appendix B.

MECHANICAL DESIGN OF CENTRIFUGAL PUMPS

The design of a centrifugal pump involves a determination of a pressure vessel having very complex shape, which may be subject to sudden pressure variations and, in the case of a high temperature pump, to sudden temperature variations which bring in their train stresses and distortions resulting from differential thermal expansion.

The rotating element must be designed to transmit the necessary power from the driver to the impellers, must withstand the various axial loadings involved, and must also be capable, in respect of direct stress and reversing stresses, of accepting the radial loading due to the variation of pressure around the impeller periphery at duties other than best efficiency flow. (Fig 16.1). See Appendix D.

A single stage pump impeller generally delivers the water into a scroll casing having a single volute; however, for higher heads, a single stage pump may be fitted with a double volute.

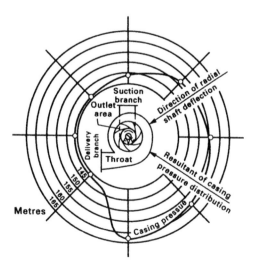

Figure 16.1
Casing pressure distribution and
radial shaft deflection for
model pump zero flow.

A multistage pump impeller delivers into a series of diffusers which resembles a number of volute collecting portions arranged around the circle of the impeller. In the case of the single stage pump, the single volute is the most convenient way of collecting the flow and taking it to a single branch at the pump delivery, whilst in the multistage pump, the number of diffusers around the periphery of the impeller is a convenient means of guiding the water into the next stage with a minimum of carcase diameter. (See Figs 16.2 and 16.3).

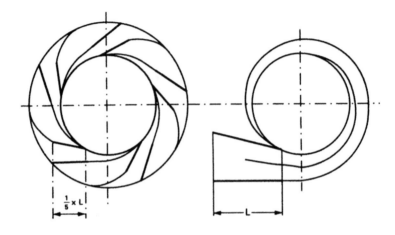

Figure 16.2 Diffuser   *Figure 16.3 Double volute*

When a pump is operating at its best efficiency point, the flow into the volute is fairly uniform and the pressures around the circle of the impeller are reasonably balanced. At duties other than the best efficiency point, however, these pressures show considerable variation since the flow in the volute is very much disturbed.

The flow within the volute at duties other than best efficiency point is turbulent and indeterminate, giving rise to dynamic loadings on the impeller. The combination of the pressure variation at the impeller periphery and the dynamic loadings due to disturbed flow, gives rise to shaft deflections which are illustrated in a typical case in Fig 16.1. These shaft deflections are electrically measure to provide data for design of shaft.

It will be seen from Fig 16.1 that the resultant of casing pressure distribution is at a different angle from the measured shaft deflection, the latter pointing approximately towards the delivery branch.

Radial Thrust Formula

The magnitude of this radial thrust varies considerably from pump to pump, but the formula given below is a reasonable one for design purposes. The direction of the loading in Fig 16.1 refers to operation at closed valve; at duties beyond best efficiency point, the direction of this thrust is reversed. The magnitude of the thrust is approximately proportional to the departure from best efficiency point.

For heads up to 300 metres per stage the approximate value of the thrust in hecto newtons is given by the following formula:-

$$0.3 \times \frac{(impeller)}{(dia\ m)} \times \frac{(impeller\ discharge)}{(passage\ width\ mm)} \times (pressure\ in\ bar)$$

The constant 0.3 indicates that the area of the impeller subject to the radial loading, namely diameter by width, is unbalanced to the extent of 0.3 of 30% of the head.

Double volutes (Fig 16.2) or diffusers (Fig 16.3) are effective in reducing these thrusts, but do not entirely eliminate them. Measurements of pressure variations around the casing and of radial deflections of shaft are given in Ref 30.

Axial Thrust

Axial thrust due to unbalance of pressure areas on single entry impellers are more readily assessed. It is usual, however, on all impellers, to make an empirical allowance for axial unbalance due to uncertainty of cast surfaces.

Further axial thrust loading is that caused by the change of flow direction when water entering a single entry impeller in the axial direction is turned 90° to discharge in a radial plane. There is also the thrust involved in accelerating the water as it enters the inlet blade.

These thrusts can be calculated from assumed terminal conditions by Newton's Laws, but the confirmation of the accuracy of assumed terminal conditions can be made by actual tests with a pressure cell on the pump shaft (see Chapters 33 and 36).

For stage heads materially exceeding 300 metres imperfections of geometry and the wayward nature of water can, on off peak duties, give rise to radial and axial forces many times the above values. Accurate tests of instantaneous loads are advisable.

Example of Mechanical Design of Combined Motor Pump
(Ref 19, Chapter 34)

The axial and radial loading phenomena described above form an interesting study in the case of a single entry overhung pump combined with its driver, since we have the additional problem of the effect of deflections of the shaft under hydraulic and magnetic loads on the critical speed of the combined unit.

Large Pumps and Water Turbines

Where the hydraulic unit, pump or water turbine, is very large, it is usually tailor-made for its particular duty and environment. There is, in these cases, considerable advantage in designing the whole unit, electrical and mechanical, as one combined entity using, for example, a motor or alternator having only two hearings with the pump impeller or turbine runner overhung, the flow being single entry or single exit respectively.

The intimate construction of two units with a common shaft requires very careful design to ensure that the double duty with one set of bearings and shaft is performed adequately without undue deflection or risk of excessive war. The single entry impeller offers considerable advantage here since it allows unlimited increase of shaft diameter. The water enters the pump at one end and the shaft enters at the other end, neither prejudicing the flow area nor the shaft diameter. For economy in space, power and expenditure of metal, it is logical to place the rotor and stator of the electrical unit between the two bearings required by the shaft carrying the overhung impeller.

CENTRIFUGAL PUMPS

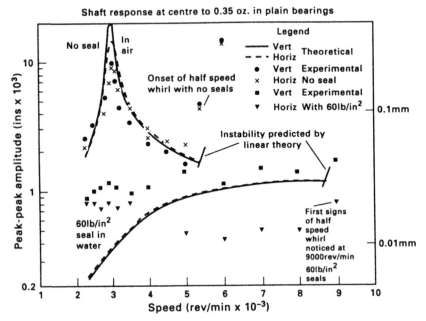

Figure 16.4
Critical speed of rotor in air and in water.

Shaft and Bearing Loadings

Here we must consider the forces due to gravity which are present, both stationary and in rotation, and the loads produced by rotation. The critical speed of the combined unit which must be avoided during normal or emergency operation is derived from the relationship between the mass of the rotating parts, the elasticity of the shaft and the electro-magnetic forces. (See Fig 16.4, Critical speeds in air and water and Fig 16.5, Vibration limits).

Axial Loadings

Electrically, the axial loadings of the rotor can be regarded as negligible since the rotor is generally set in its magnetic centre with respect to axial float.

Axial thrust will however arise in the pump as follows:

1.   If the inlet pressure differs from atmospheric pressure, for example, a suction lift, there will be a thrust, in tens of newtons, away from the motor equal to the area in square centimetres of the circle corresponding to the sleeve diameter multiplied by the suction lift expressed in bars.

2.   If the inlet pressure exceeds atmospheric pressure, a similar thrust occurs, in this case towards the motor.

3.   The pump is fitted with two neckrings in order to separate the high pressure from the low pressure regions. Neckring leakage of the inlet or low pressure side of the impeller flows into the inlet passage of the pump. Neckring leakage at the driving side of the impeller is returned to the low pressure side via an external pipe so as

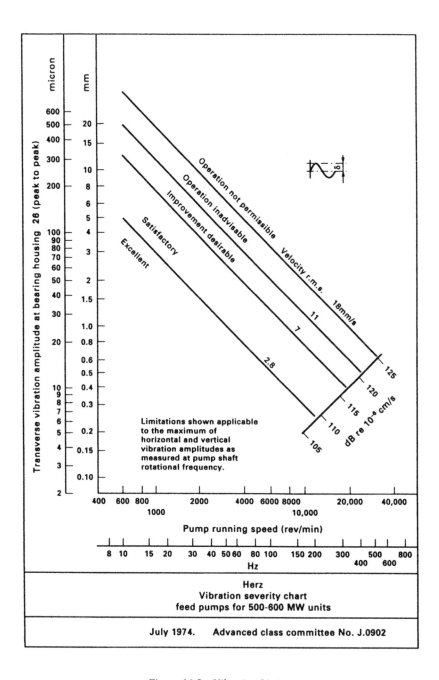

Figure 16.5 – Vibration Limits

to effect balance. On mass produced small pumps, the leakage is returned via holes in the impeller shroud, but on large tailor-made units, such holes are replaced by an external pipe in the interest of efficiency. It is therefore possible in a study of overall efficiency to measure the back neckring leakage and thereby deduce the total leakage. Occasions may arise when the leakage from the neckring at the driving side of the impeller is led to a discharge point at a different pressure from that of the pump inlet. This will give rise to a thrust equal to the area of the annulus between the diameter of the neck of the impeller and the diameter of the sleeve, multiplied by the pressure difference between the inlet pressure and the point of disposal of the neckring leakage. This thrust may be positive or negative according to the relative values of the pressures concerned.

4. Theoretically, the pressure of the sides of the impeller shrouds above the neckring diameter should be equal, since they are both in communication with the volute of the casing, thus giving theoretically true axial balance. However, in practice, it is necessary to make an empirical allowance in view of casting or fabrication roughness, inaccuracies of impeller and casing, and the vagaries of flow.

5. The water entering the impeller is turned through 90° before discharging, giving rise to a dynamic thrust in the direction of the motor, varying as the square of the quantity.

6. In the case of vertical machines the weight of the impeller and the weight of the rotor give rise to an axial thrust on the shaft and on the bearing. The thrust bearing, (ball or Michell type) on the motor, must carry the algebraic sum of these axial thrusts. In a horizontal machine, however, only the impeller thrust loadings described above must be carried on an axial thrust bearing, since the journal bearings will deal with the static weights of these parts.

Centrifugal Stresses

Impellers rotating freely in thin air, carrying mainly centrifugal stresses and negligible dynamic stresses, have burst at speeds corresponding to stage heads of 25 000 metres. Since present stage head achievement on liquid pumps is about 2500 metres, it will be appreciated that the primary stresses on impellers are due to dynamic forces, especially at off peak flow and under cavitating conditions.

References

17. ANDERSON, H.H. 'Modern Developments in the use of Large Single Entry Centrifugal Pumps'. Proc IME, 1955, Vol 169, No.6.

30. WORSTER, R.C., 'Flow in the Volute of a Centrifugal Pump and Radial Forces on the Impeller'. BHRA,RR543, 1956.

31. BLACK, H.F., BROWN, R.D., FRANCE, D., and JENSSEN, D.N., 'Theoretical and Experimental Investigations relating to Centrifugal Pump Rotor Vibrations', Conf: Vibrations in Rotating Systems 1972 (February), IMechE London.

CRITICAL SPEEDS, VIBRATION AND NOISE

The preceding chapter considered the radial loadings on a rotating impeller and the axial loadings of a single entry pump. We now consider the effect of the radial loadings, hydraulic and electrical, on the critical speed of a combined motor pump. This unit has one shaft held in two bearings between which is placed the electrical unit, the impeller being overhung at one end of the shaft.

Electrical Radial Loadings

The armature of the electrical motor is rotating in a magnetic field. The effect of tolerances in measurement of bearing settings and of static deflection of the rotor, involves operation some distance away from the theoretical magnetic centre. This gives rise to an unbalanced magnetic pull (referred to an UMP) at the rotor, tending to deflect the shaft in a radial direction. Electrical motor designers conventionally determine the UMP for a standard deflection of 10% of the air gap, and for the purpose of this analysis, the magnitude of the UMP may be taken as proportional to the deflection from the centre of the air gap.

Shaft Deflection

The pump has an impeller rotating in a water gap (the neckrings) and the motor rotor rotates inside the air gap. Under all conditions of operation it is vital that the impeller neck does not make contact with the stationary neckrings across the water gap and the rotor does not make contact with the stator across the air gap.

In determining the deflection of the rotor, the forces on the rotor and the forces on the impeller transmitted through the shaft, causing further rotor deflections, must be considered as a whole, with similar consideration in respect of the impeller.

Critical Speed of the Rotor of the Combined Unit

The deflection of the shaft at the impeller under radial loading will have no effect on the critical speed of the pump portion of the shaft, which is determined in the usual manner. In the motor armature, however, considerably different conditions prevail, in that the UMP is proportional to the deflection from the magnetic centre. Since centrifugal force is also proportional to the deflection from the centre, both these loadings must be added together in the determination of critical speed. The loadings, however, must be added under equal deflection conditions. They can be conveniently added, for example, at unit deflection or, more conveniently in this case, at the static deflection of the motor under gravity, which is used in the determination of critical speed.

The deflection of the rotor to the static deflection point is caused by the sum of the UMP at the static deflection and the centrifugal force. The centrifugal force required to produce static deflection is lower where UMP is present.

The critical speed of the motor portion of the rotating element is therefore lowered by the presence of UMP to a value which is determined by dividing the original critical speed by the square root of the ratio of the rotor weight to the rotor weight minus the UMP at the static deflection. The UMP at the static deflection is proportioned down from the UMP at 10% of the air gap – the usual specified value of the UMP.

The method of integration of weights and deflections described in dynamics text books then gives the critical speed of the complete rotor, using an artificial deflection of the armature portion to simulate the aforementioned analysis.

Rigidity of Anchorage between Pump and Motor

In addition to the previous consideration of rigidity of the shaft, we must investigate the rigidity of the stationary members which carry the reactions of the shaft loading. The moment of inertia with respect to shear and with respect to bending of the anchorage of pump and motor are therefore determined and care taken to ensure that these values have an adequate margin above the moment of inertia of the shaft so as to ensure that the deflections of the stationary parts are negligible.

Account must be taken of bearings elasticity in critical speed determination.

Size of Shaft

Since the shaft diameter does not interfere with the passage of water entering the impeller, there is no need to restrict it and as a result the shaft diameter of an overhung single entry impeller on such a unit is usually equivalent to one-quarter of the impeller diameter. A similar diameter ratio is obtained in respect of large water turbines. The torque stress on the shaft is therefore very low, since the shaft is designed more for minimum deflection under radial loadings.

In comparison, the shafts of medium duty multistage pumps and double entry single stage pumps tend to be appreciably smaller than a quarter of the diameter so as to avoid undue prejudice to inlet flow. For severe duties, say above 300 m (984 ft) per stage, larger shafts are involved.

Previous paragraphs have considered the general problem of radial and axial loadings on the rotating shaft of a pump. We now consider the rigidity of the shaft in order to resist such forces with a minimum of deflections.

Rotating Dynamic Machinery

This group comprises machines where the shaft is free to rotate and where torque only occurs in the shaft as a result of dynamic forces of fluids, *ie* liquids or gases against some member attached to the shaft; for example, the impeller or turbine blades. Such machines handling gases would suffer damage if accidental contact occurred between rotating and stationary parts. In these cases, therefore, the shaft rigidity must be such that internal contact of moving and stationary parts cannot occur. This calls for an extremely heavy shaft and for very careful deliberations so as to determine accurately the critical speed of the shaft which must be avoided in the machine may occur, but this is enerally negligible in effect.

On machines handling liquids, *eg* centrifugal pumps, water turbines, *etc*, the pumped liquid has, to a greater or lesser degree, lubricating properties, which means that accidental contact of rotating and stationary parts, *eg* sleeves, chamber bushes, *etc*, can take place without severe damage. In the case of multistage pumps and split casing double entry pumps, this acceptance of possible internal contact permits the use of the shaft which is designed primarily to withstand torque and end thrust loadings (see *Reference 31*, Figure 17.1 and Shaft Centering Forces).

Multistage Pumps on Medium Duties

These duties, would, in general, involve pumps having heads per stage up to 200 m (656 ft), the number of stages on a single shaft having a limit of about 10 or 12. For such pumps, the natural deflection of the rotor, when assembled fully with impellers and under the influence of gravity, would exceed the radial clearance of the internal chamber bushes. That is to say, the pump rotor is supported by two ring-oil bearings and by a bronze water lubricated bushing at each stage. Since the surface load on these internal bushings is relatively low, and since there is a pressure difference to ensure a passage of lubricating water, such an arrangement is entirely satisfactory and results in a reasonable life.

Within the head per stage and stage limitations mentioned earlier, it will be found that there are very many pumps giving satisfactory operation at speeds which would correspond to the critical speed of the shaft if supported in air between the ring-oil bearings. This is due to the fact that the radial deflection of the shaft at its nominal critical speed in air is limited when in the pump by the viscous drag of the water, by the pressure of water in the clearance of the chamber bush and by the actual position of the chamber bush itself. Operation at such nominal critical speed for these modest duties is therefore entirely satisfactory. There is indeed a history of multistage pumps on these medium duties up to 15 or even 16 stages on a single shaft, adequate life being obtained from the water lubricated internal bushings.

Multistage Pumps on More Severe Duty

Where stage heads materially in excess of 200 m (656 ft) are invovled, it is essential to avoid operation at the critical speed. It is usual to design the shaft so that the operating speed is well clear of any critical speed.

Such a pump may still have a deflectio permitting accidental contact on the stage bushings and here again a reasonable life is obtained owing to low contact pressure and adequate lubrication.

When such a pump, *eg* generating 400 m (1312 ft) per stage at full speed, is operated at half speed, it is still safe for it to run upon its apparent critical speed because at this half speed it will only be generating 100 m (328 ft) per stage. The rigidity of shaft, of bearings, and of general frame for a pump to give 400 m (1212 ft) per stage is such that operation on the nominal critical speed corresponding to 100 m (328 ft) per stage is entirely safe as described previously for the more modest duty.

Shaft Centering Forces

Stage heads of 1000–2000 m (3280–6560 ft) can be developed by submersible and glandless pumps and it is here that consideration must be given to the very large radial hydraulic centering forces that can arise from liquids in the clearance of stage bushings, neckrings and oil lubricated bearings.

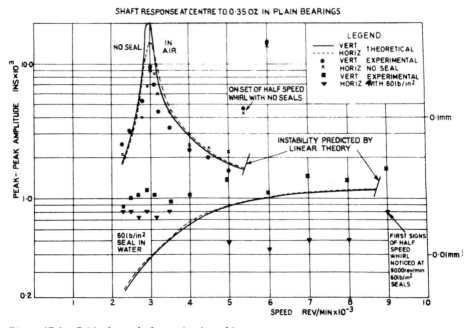

Figure 17.1 – Critical speed of rotor in air and in water.

Figure 17.1 shows a 17 fold reduction in amplitude and a trebling of critical speed in the change from air to water within the pump. A similar, but limited improvement occurs on submersible motors.

A run-dry pump which is in the mechanical sphere of a compressor will be more expensive and suffer a slight sacrifice of efficiency compared with a slender-shaft pump, but is advisable where loss of inlet flow may occur on vital plant. It should be emphasized however that reliable continuous operation has been obtained from slender shaft pumps over the decades.

Rotor

The pump shaft is made of stainless steel and is subject to a maximum stress at the first

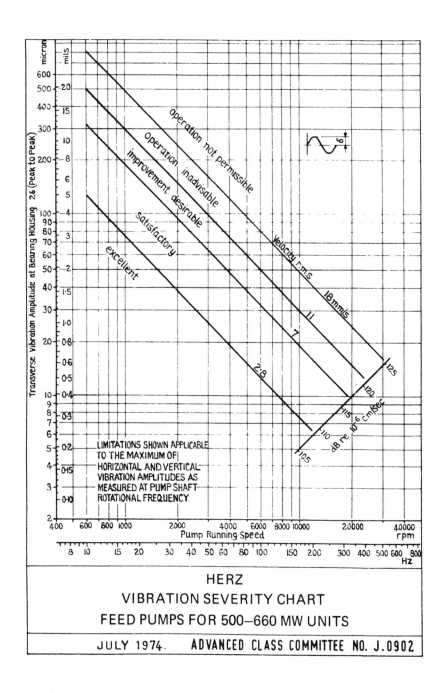

Figure 17.2
Vibration limits

impeller, the stress here being the combination of the total torque and the end thrust of the first impeller.

Since each impeller has, for simplicity of fitting, its own diameter on the shaft, the end thrust reaches a maximum at the balance disc where the torque stress is nil. This gives the advantage of economic use of the shaft material.

The operating speed of the pump is between the first and second critical speed, since it would be uneconomic to run below the first critical speed except on run dry pumps.

Analysis of general data from earlier designs gives the shaft size and critical speed which is checked by the following analysis:

(i)   the usual graphical integration in the drawing office to determine deflection of shaft aided by computer.

(ii) a checking of this deflection by mounting the built-up rotor on the surface able and measuring its deflection.

(iii) an oscilloscope test of the natural period of vibration of the shaft in aid by hitting the centre with a light hammer and holding an oscilloscope pick-up at this point.

(iv) a similar check of the second critical speed at the one-quarter and three-quarter points along the length of the shaft.

(v)   an oscilloscope test during the running up of the pump from rest to full speed, and during the running down after power is switched off.

All these test give reasonably comparable results, the last test on the finished pump being, of course, the most important.

Minimizing of High-Frequency Vibration and Noise

As each impeller blade approaches a difuser, a small hydraulic shock occurs which is of relatively high frequency, for example, seven or eight times the running frequency, but of relatively small amplitude. In order to reduce this shock and the consequent alternating stress on the casing to a minimum, the number of impeller blades is prime to the number of diffusers, and the impeller blades individually and as a group are of helical form, giving the whole rotor a skewed effect so as to reduce this high-frequency hydraulic shock and noise to a minimum.

Vibration limits appear in Figure 17.2 and 17.3.

Relationship of Shaft Mass to Pipe Rigidity on Suspended Sets

Several vibration problems have arisen from the fact that the shaft has been too large in diameter for the rigidity of the support pipe so that the tail wags the dog.

In a borehole this is unseen, but in a well the amplitude with nodes at every few bearings may appear alarming. If the resulting stress exceeds safe limits it can be cured by the provision of smaller diameter shafts.

Recirculation – Fraser's Fundamental Solution of the Vibration Problem

Economic forces in the limit and on severe duties demand the highest liquid power from the smallest expenditure of metal. This tends towards the highest speed, the highest area

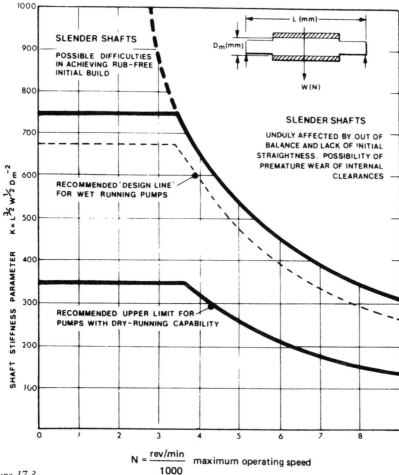

Figure 17.3
Pump rotor rigidity from Reference 32

ratio to give the smallest impeller outlet diameter and casing throat and the lowest NPSH involving the largest impeller inlet diameter, all leading to the shortest vane.

The author in 1944 (see *Reference 6*) suggested that a pump head comprised a centrifugal field, *ie* acceleration to the centre and a dynamic head, *ie* tangential acceleration.

Fraser, independently, initiated the concept of two terms to solve the problem of rapid destructive vibration by predicting the exact poit of inception of vibration as flow isreduced. He showed that when the centrifugal pressure field exceeds athe dynamic head the pressure gradient is from inlet to outlet and operation is stable. When the dynamic head exceeds the centrifugal pressure field the gradient reverses from outlet to inlet and destructive recirculation can occur. In *Reference 34* Fraser shows examples of damage by recirculation and demonstrates how to predict its point of incidence from geometric

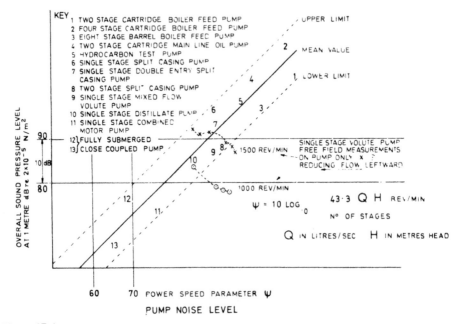

Figure 17.4
Typical noise levels of pumps from Reference 33

impeller data and from suction shape number. He shows clearly how such vibratio can be avoided in any given situation.

Pump Noise

An analysis of noise levels of a very large number of pumps appears in Figure 17.4 the powes ranging from 50 kW to 20MW. Several of the pumps are identified by type and by the power-speed parameter. Acoustic protection for operators local to the plant is advisable on duties above about 500 kW. In general it can be said that up to about 500 kW the pump noise is likely to predominate, above 500 kW the motor noise will be greater, depending on the methods of motor cooling. Submersible noise down a well is generally remote from the operators.

Noise philosophy and methods of noise measurement are given in *References 33 and 35*.

References

6.   ANDERSON, H. H., *Centrifugal pumps, an alternative theory*, Proc IME 1947, Vol 157, We 27.

31. BLACK, H. F., BROWN, R. D., FRANCE, D., and JENSSEN, D. N., *Theoretical and experimental investigation relating to centrifugal pump rotor vibrations*, Conf Vibrations in rotating systems, 1972, London IME.

32. DUNCAN, A. B., and HOOD, J. F., *The application of recent pump developments to the needs*

of the offshore oil industry, Conf Pumps and compressors for offshore oil and gas, Aberdeen 1976 IME.

34. FRASER, W. H., *Recirculation in centrifugal pumps*, ASME winter meeting 1981 Washington, D.C. Reprinted in World Pumps May 1982.
33. FRANCE, D., *Noise, a philosophy for pump makers*, Fourth Tech Conf BPMA, Durham 1975.
35. SAXENA, S. V., *Measurement of airborne noise emitted by liquid pumps*, European Pump Maker's Conference, Gleneagles 1984.

BEARINGS

Centrifugal pumps, being high speed machines, are entirely dependent on adequate bearings. The type of bearing adopted is dictated by a combination of load and speed and by economic factors, rolling contact bearings being the most convenient for the large majority of pumps.

Deep groove ball journals have the advantage of carrying radial and axial thrusts and represent the most popular bearing for pump shafts. The bearing at one end of the shaft is provided with an axial location, whilst the bearing at the other end is free to slide axially in its housing, to allow for differential expansion. Alternatively the second bearing may be of the roller type, particularly if heavy radial thrusts occur at one end; for example, thrusts arising from a belt drive or from the radial unbalance of a single volute.

Rolling contact bearings can be grease lubricated at the normal speeds of the majority of pumps, thereby offering the simplest and most easily maintained arrangement. For speeds and loads that are more severe, oil lubricated rolling contact bearings can be used, thus providing a short rigid construction with less complication and space than a sleeve bearing.

Ball/roller bearings with grease lubrication are of particular value on vertical spindle applications since oil retention problems are thereby avoided. For special applications, ball and roller bearings can be supplied with split housings, and stainless steel bearings can be used with special precautions on water lubricated duties.

Sleeve Bearings

These are oil lubricated for the majority of duties but it is often convenient to lubricate by the pumped liquid, eg in the case of liquid filled motor pumps, combined steam turbine pumps and borehole pumps. The lubricating liquid is introduced by oil rings or by pressure feed, and the rotation forms a wedge of oil or other liquid which separates the shaft from its bearing. The bearing surface can either be fixed or formed of tilting pads to assist the wedge action. In certain applications the shaft can be held entirely free of the bearing surface by pressure zones of fluids which are controlled by three balanced leakages from the pressure source, any radial shaft deflection throttling the liquid supply in the particular sector concerned and thereby raising the local pressure to provide a centralizing force.

This, of course, is a hydrostatic bearing, whilst the oil wedge represents a hydrodynamic bearing. A further development is the arrangement of a number of lobes in the bearing surface so as to provide more than one oil wedge around the circumference of the journal. This has the effect of minimizing the actual free movement of the shaft within the metal confines and tends to reduce oil whirl.

The dynamics of these multilobe bearings are discussed by Cameron in Ref 33.

Tilting Pad Thrust Bearings

The axial load on a rotating shaft is often carried by an external oil lubricated thrust bearing of the Michel or Glacier type. Such bearings are used on boiler feed pumps (see Chapter 33), on vertical pumps (Fig 16.4) and on any multistage pump where violent pressure fluctuations occur. Provision must be made to carry away any excess heat by a coil of water cooling pipes in the bearing housing or in a separate chamber. Extreme care is necessary to avoid any ingress of water into the oil.

In the case of vertical shafts, a tube must be placed within the bore of the elongated thrust element of the rotor, to ensure that the thrust pads and the journal portion are always immersed in oil, whether standing or running.

Bearings Lubricated by Pumped Liquid

Totally enclosed combined motor pumps and steam turbine pumps must necessarily use the pumped liquid as a bearing lubricant.

This arrangement is invaluable when dealing with precious, toxic, scalding, radio active or flammable fluids since glands and leakages are eliminated. It is also essential on borehole and well pumps for water works, as oil contamination would be unacceptable. Here the main problem is to establish an adequate liquid wedge to separate the journal of the shaft from the fixed bearing surface. Arkless (Ref 34) gives design data and the following formula for the relationship between speed, viscosity of fluid, loads and bearing dimensions: Absolute viscosity times rev/min/specific load must be comparable to that of an oil lubricated bearing.

A pressure difference between the two ends of the bearing promotes a fluid flow which assists cooling and lubrication.

Thrust Bearings Lubricated by the Pumped Liquid

The multistage pump balance disc (to be described later) is a typical example of a liquid lubricated thrust bearing. Here the principle is hydrostatic (in contrast to the hydrodynamic wedge described previously) since the pressure of the last stage of the pump achieves positive separation of the thrust bearing faces on the basis of controlled leakage. Theoretically, there should be no contact but, in practice, transient starting conditions and abrasive liquids cause inevitable wear. A balance disc on very clean water may run for twenty years before replacement, but abrasive materials in the pumped liquid can result in a very short balance disc life. In the latter case, an external oil lubricated tilting pad bearing will be used to carry the thrust and to keep the balance disc clear of the fixed face. The disc now acts only as a pressure reducing device to avoid high gland pressure, and the oil lubricated thrust bearing must carry the full thrust load from the impellers.

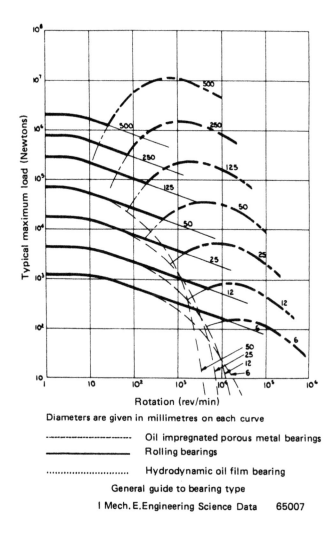

Diameters are given in millimetres on each curve

---------------------------------- Oil impregnated porous metal bearings
─────────────────── Rolling bearings
.... Hydrodynamic oil film bearing

General guide to bearing type

I Mech. E. Engineering Science Data   65007

Figure 18.1.

Non-Metallic Bearings

Internal bearings lubricated by the pumped liquid often use non-metallic materials or metal-plastic combinations. for example, carbon, lignum vitae, PTFE and other plastics, asbestos, etc are very suitable for water lubricated duties.

Plastic materials have very low friction coefficients but are poor heat conductors. Several successful bearing designs use a thin plastic layer to give low friction, with a bronze backing to carry away the heat. The plastic layer may be attached to the metal or may be a loose strip. Combinations of porous metal with plastic filler are described in Ref 34.

Secondary Bearings

The foregoing describes the main bearings holding the pump rotor. In multistage pump there is, in addition, a shaft busing at every stage. When the shaft deflection due to its weight exceeds the radial clearance of the stage busing (for example on a long pump), it follows that the shaft will be partly supported by the stage bushing which becomes secondary bearings. this additional support is a material factor in the satisfactory operation of long pumps, whether horizontal or vertical shaft. The stage bushings of metal or plastic are necessarily lubricated by the pumped liquid. (See Fig 16.3 and Appendix B).

References

33. CAMERON, A., 'Principles of Lubrication' (Longmans) 1966.
34. ARKLESS, G.F., 'The Development of the Water Lubricated Feed Pump', Proc I Mech E, 1963 Vol 177, No 691.

PRESSURE VESSEL ASPECTS

A pump is a pressure vessel which must accept all operating conditions without undue deflection.

Single stage pumps have a very awkward shape in respect of pressure stresses and so are generally cast with fairly thick walls, ribbed where necessary for higher pressure ranges.

Pumps split normal to the shaft axis require a cover to close the casing after the impeller is admitted. This cover, and the corresponding casing wall, act as diaphragms and must also hold the volute sections from opening out. The discharge cone, generally of the same wall thickness as the rest of the pump for casting reasons, is relatively lightly loaded since its basic diameter is small.

Pumps split by a flange containing the axis represent a more complex vessel, since the irregularly shaped flanges must be thick enough to prevent distortion and leakage. On higher pressure ranges — above 40 bar — steel casings with deep bosses to permit an adequate number of bolt facings are necessary, particularly on multistage pumps.

Cellular multistage pumps present a more attractive example of use of metal, since the basic cylindrical shape involves more uniformly stressed components. In general, the ring section or throughbolt pump involves several stages clamped between end covers by heavy bolts. This permits a geometric design basis for bolt and cover dimensions.

For example Type 100 for 100 bar pressure, cold, could adopt a ratio of 8 between end view pressure area and bolt area, involving a bolt stress of

$$8 \text{ (ratio)} \times 1.5 \text{ (gasket allowance)} \times 100 \text{ (pressure)} = 12 \text{ hectobar}$$

Using bolts which permit 15 hectobar maximum stress gives a margin of 3 hectobar to cover thermal stresses of shock heating or cooling from 100°C. For higher temperatures, for example 200°C, a larger thermal stress allowance of 6 hectobar would limit the safe working pressure to 75 bar.

Barrel casings are used for temperature conditions beyond the capacity of throughbolts. Here the inner pump is under external pressure and can be cellular or two piece horizontally split, since the weaknesses of the irregular casing shape do not arise with external pressure. Barrel stresses are more readily ascertained than single stage casing stresses, since the barrel is predominantly cylindrical. Branches in the barrel involve adequate strengthening.

Temperature pressure charts for various pressure shell and bolting conditions appear in Chapter 33.

Economical Use of Metal on Single Stage Machines

Medium sized pumps may have fabricated casings with external ribbing to prevent the trapezium section of the volute from opening out under the pressure loading. This permits the use of thinner metal, eg where stainless steel or stainless cladding is used (see Chapter 34). Very large pumps may be fabricated in similar manner to water turbines, the scroll section being formed from a series of C plates of increasing circumference, each welded to a vaned annular ring. The metal thickness then corresponds to the diameter of the C plate, the vanes in the annular ring preventing the C plates from opening out under the pressure loading (see Chapters 34 and 35).

Space rocket pumps are fabricated from very thin titanium sheet, as described previously, since the weight must be reduced to an absolute minimum and cost of fabrication is a secondary consideration.

Joints involve the provision of adequate pressure to prevent leakage under all operating conditions. The minimum pressure between the joint faces must be equal to the pump casing pressure times a gasket factor depending on the type of join, varying from one for rubber sealed joints to five for certain metal-to-metal joints.

A selfseal joint is illustrated in Fig 19.1. Here a seal ring is ovalled within its elastic limit to enter a barrel mouth smaller than the outer diameter of the ring. When the ring is inserted and relaxed to revert to circular shape, it can hold captive a previously inserted plate and withstand internal pressure (Ref 35).

In order to provide the self-tightening feature, the endless metal seal ring of heavy section is fitted into a groove within the bore of the pressure vessel. At first sight this seal ring cannot be fitted into its groove since the outer diameter of the ring exceeds the bore of the vessel. The ring, however, is so designed that, well within its elastic limit, it can be deflected into an oval, rotated out of the plane normal to the vessel axis and inserted through the vessel bore into the groove which is made wide enough for this purpose. The section of the ring must, therefore, be so designed that it is heavy enough to resist the crushing and shear forces resulting from the pressure in the vessel and thin enough to deflect safely into the necessary oval shape for insertion within the groove and finally to return to circular shape. This introduction of the ring through a hole of smaller diameter ensures that the ring, once it is within its groove, cannot possibly be forced out by pressure, and effectively provides the simplest and safest of all joints, namely a self-tightening one.

The end plate closing the vessel presses against the inner half of the endless ring of which the outer half fits within the groove in the vessel.

Bolt loading is merely required to ensure initial sealing and is, therefore, only a small proportion of the bolting for a conventional joint.

The new joint does not require a gasket but the term gasket factor or joint factor is still applicable to describe the ratio of joint interfacial pressure to hydraulic pressure being sealed.

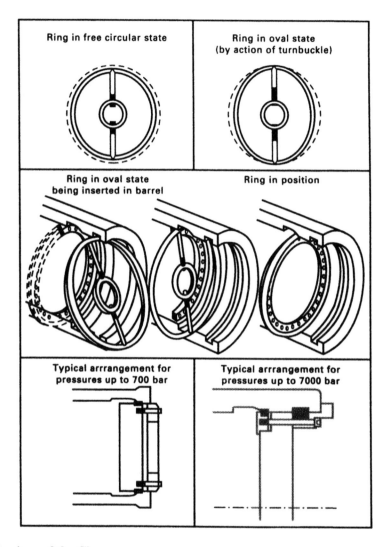

Two Versions of the Closure

The joint can embody a single oval fitting ring only, or can perform the duty of sealing, while a segmental shear ring takes the greater part of the hydraulic end load. (See Chapter 33).

The single ring is suitable for pressures up to 700 bar in cases where it is possible to move the end plate further into the vessel for a distance of one sixth of the diameter, to provide clearance for insertion of the ring (see Appendix D).

References

35. ANDERSON, H.H., 'Self Seal Join for Very High Pressures'. ASME 67WAPVP5.

MECHANICAL SEALS AND OTHER SHAFT SEALING DEVICES

Where a rotating shaft enters a pressure vessel, clearance between shaft and pressure wall is essential for safe running and threfore some form of seal is necessary to prevent excess leakage of fluid.

For the majority of liquids, a simple seal is satisfactory, but where the liquid is precious, toxic or flammable, a more elaborate seal is provided to minimize leakage or, in the case of radio-active liquids, to prevent entirely any leakage by enclosing completely the pump and its driver, *eg* a canned motor pumpset.

The shaft sealing of a submersible or glandless pump involves the minimizing of leakage from one pressure vessel to another since both pump and driver contain liquid. In general the pressure differences are modest, the chief concern being to keep the clean driver liquid in the closed box, free from the contamination of the pumped liquid.

A mechanical seal is the ideal solution since with the good alignment conditions of submersible and glandless pumps a long life with minimum outage is attained. Where a barrier liquid or a thermal insulatio chamber is required double mechanical seals are involved, arranged back-to-back.

The seal must allow for the thermal expansion of the motor liquid as operating conditions change, an elastomer storage vessel accepting the surplus liquid. Some small flow across the seal faces is essential to provide lubrication and a circulatory flow in the region of the seal is needed to provide cooling.

Gas, oil, water or steam turbine driven pumps permit limited leakage of clean bearing fluid through fine annular clearances thus avoiding the cost and length of more elaborate seals.

Hydrodynamic Seals

On certain duties it is possible to keep the shaft free of liquid by means of rotating vanes or by use of the impeller disc so that the liquid forms a rotating cylinder with air in the centre. Some provision however must be made to prevent leakage when the pump is standing and, in general, such a hydrodynamic seal is only suitable for low pressures.

Flexible conical seals can prevent standing leakage and can be held clear of contact at speed by centrifugal force.

Mechanical Seals (Figures 20.2, 20.3 and 20.4)

A mechanical seal involves two members, either of which can be rotating and one of which is harder than the other, the harder material having a mirror surface. Carbon, metal, plastic or ceramic surfaces are pressed together by a spring with provision for axial float and for wear of surface and, in some cases, embodying bellows. On nearly all duties, it is necessary to provide some form of cooling to remove the frictional heat.

Figure 20.1
Crane Packing type 1A mechanical seal
for minimum radial cavity depth

Mechanical seals are used for leak-tight sealing of virtually all fluids from air to concentrated acids and flammable liquids at temperatures from $-210°C$ to $+600°C$ and pressures up to 180 bar, including fluctuating pressures and temperatures.

The axial direction radial face-type mechanical shaft seal consists of five basic components. In a typical installation there is a stationary seat sealed to the pump end plate by a gasket, or O-ring; against the stationary seat runs the rotating seal face, the opposing sealing surfaces being lapped to a high degree of flatness. The rotating seal face is mounted on a flexible member which also seals on the shaft. Contact betwen the radial sealing surfaces is sustained by means of spring pressure, either from a single sprint (as shown), or from a series of small springs mounted in a retainer, combined witht he hydraulic pressure of the fluid being sealed. The rotating seal unit is secured to the shaft.

For general services, the flexible member may be a rubber bellows (as shown) or O-ring; for more corrosive conditions a PTFE wedge ring or PTFE bellows would be used, and for higher temperatures an asbestos based wedge ring or metal bellows, (copper, bronze, stainless steel, *etc*). One of the sealing rings is always metal with a hard mirror finished face, while the mating ring is of softer metal, carbon or plastic.

Balanced Seals

Balanced seals are used at higher opeating pressures, the contact diameters of the seal face being stepped down so that the face load is less than the hydraulic pressure of the liquid being sealed.

Coolant Circulation

Coolant circulation is necessary to provide a stable liquid film between the rubbing surfaces. All seals operate better with coolant circulation and for most conditions this is essential.

Double Seal Arrangements

For submersible pumps and for certain conditions such as heavy clay slurries and unstable liquids, it is necessary to employ a double seal arrangement, in which two seals are arranged back-to-back and clean, inert liquid is introduced from an external source at a positive pressure to form a stable interface liquid film. Double seals are also required on the rotating cup mercury sealing chambers of certain electric submersible pumps.

Examples of Mechanical Seals

Figure 20.2 shows a Crane seal with single bellows and rotating spring for general abrasive and corrosive duties. A rubber bellows seal and a simple sprint gives maximum resistane to abrasion and clogging and has ability to cover a wide range of pH values. Silicon or

Type 2 seal fitted with 'N' seat.

Type 1S combines two type 1A seals in a back-to-back double installation when the seal chamber length is restricted. This arrangement is used for sealing gases or very difficult abrasive conditions.

Figure 20.2

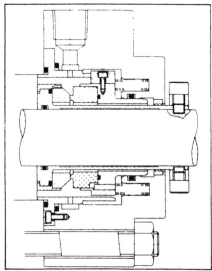

Figure 20.3
Improved high pressure single seal

tungsten carbide seal faces further extend the wear resistance. In a submersible or glandless application a rubber lip seal would be used as a secondary barrier *(Reference 37)*.

A Flexibox seal is shown in Figure 20.3 which illustrates the stationary sprint system. This design offers advantages particularly on high speed duties where pressures vary over a wide range, and minimizes power and distortion. These improvements mean that seals can be operated close to the saturation point without vaporization at the seal faces *(Reference 38)*.

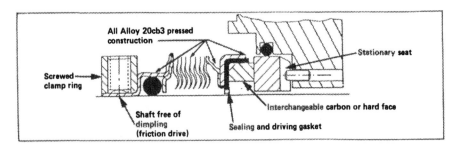

Figure 20.4
680 Concept. Mid-performance metal bellows seal with
improved metallurgy and advanced manufacturing techniques.

Figure 20.4 shows a Sealol Metal Bellows Seal for general services particularly at high speeds. Bellows have the following advantages:

(i) No moving parts or sliding elastomer in contact with the shaft and therefore eliminating wear on these parts;

(ii) This eliminates seal drag which is otherwise a cause of failure;

(iii) A bellows seal in inherently balanced thus avoiding shaft stepping.

(iv) Materials of construction can accept the widest temperature range, organic elastomers being then replaced by graphitic static packings *(Reference 39)*.

Materials of Construction

Carbon is a very attractive material for a vast range of duties as it is capable of withstanding severe conditions including occasional dry running but has a low modulus of elasticity and cannot accept tension without risk of fracturing on severe duties. Carbides of tungsten and silicon and various stainless steels are therefore used to give the longest life on submersible and glandless pumps.

Table 1 shows typical materials for sewage and general duty pumps.

TABLE 1 – MATERIALS APPLICATIONS GUIDE

Services	Metal Parts	Bellows	Springs	Face-Seat Combination	
Raw sewage	Stainless steel	Rubber	Stainless steel	Tungsten carbide v Tungsten carbide	
Raw sludge	Stainless steel	Rubber	Stainless steel	Tungsten carbide v Tungsten carbide	
Activated sludge	Stainless steel	Rubber	Stainless steel	Carbon v T/carbide	T/carbide v T/carbide
Humus sludge	Stainless steel	Rubber	Stainless steel	Carbon v T/carbide	T/carbide v T/carbide
Effluent (basically water)	Stainless steel	Rubber	Stainless steel	Carbon v Nickel iron	
Clean water	Stainless steel	Rubber	Stainless steel	Carbon v Nickel iron	
Oil	Stainless steel	Rubber	Stainless steel	Carbon v Nickel iron	

Note: Silicon Carbide may be substituted for tungsten carbide if preferred.

Radial Seals or Lip Seals

A radial seal capable of withstanding pressures in one direction is show in Figure 20.5. Double seals are used to withstand pressure in either direction.

The sealing member is an L-shaped ring (1) of elastic material (leather, rubber, *etc*), accommodated in a sheet metal housing (2). The contact pressure is usually produced by a garter spring (3) or a conical blade spring (not shown). The seal is fitted into the maching casing (4) so that the pressure diferene supports the action of the spring.

This type of seal is used for low pressures (*eg* 2 bar), generally for sealing bearings against leakage of splashed oil and against ingress of foreign matter. Somewhat higher pressures can be held by providing a metal backing to the seal ring (*eg* 6 bar), but some shortening of life may occur.

Sealing by Limited Clearance Flow

Where a suitable disposal sink is available, leakage from relatively low pressure regions can be allowed to pass through a number of annular clearances to reduce pressure to an

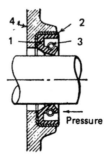

Figure 20.5 – Angus lip seal.

acceptable value where quench flows can prevent leakage of product to atmosphere. Dry
running, when input is accidentally starved, can thereby be permitted.

References

37. *Sealing trends in sewage, sludge and slurry pumping*, Crane Packing Ltd, World Pumps, June
    1983, page 220.
38. PHILLIPS, J., *Design of mechanical seals for high and ariable pressures*, World Pumps, April
    1984, page 129.
39. PLUMRIDGE, J. *Low cost metal bellows sealing*, World Pumps, April 1984, page 109.

SECTION 5

Characteristics

APPROACH AND DISCHARGE CONDITIONS

CHARACTERISTICS OF PUMPS AND PIPE SYSTEMS

APPROACH AND DISCHARGE CONDITIONS

The performance of a centrifugal pump is very largely dependent upon satisfactory flow conditions and an essential part of the design lies in ensuring that the associated equipment before and after the pump is suitably integrated into the design of the pumping station as a whole. For this reason the pump maker should be consulted in the layout and arrangement of the complete pumping station; many cases of faulty pump performance have, in the past, been traced to inadequate flow conditions preceding the pump.

Methods of Conveying Water

A pump may receive water from a sump or river below the pump centre line; from a storage tank or reservoir which may be above or below the pump; from a borehole or well; or from a pipe. The pump will normally deliver water into a pipe but may, in some cases, be immersed in a reservoir in a similar manner to certain low head small turbines.

Pump Drawing Water from a sump or Culvert

Here it is essential to ensure that the total suction lift, static plus friction, plus velocity head, does not exceed a safe operating limit with respect to the pump's cavitation performance. It is equally important to ensure that the flow of water into the culvert and into the bellmouth of the suction pipe is smooth, uniform and generous so that one pump does not take excess water to the prejudice of other pumps. It is very surprising how sensitive pumps can be to any faulty distribution of flow in the culvert, particularly where several pumps may operate singly or in parallel. Ideally, each pump should have its own culvert and be provided with some means of preventing prerotation of water at the bellmouth. For example, such prerotation is often avoided if the bellmouth is fairly close to a wall. It is most undesirable to have several pumps in a row in one culvert, so that water to the last pump will pass the bellmouths of the other pumps.

Valuable data assisting the design of suction tanks and culverts appear in Ref 37.

Inlet Conditions

The pressure difference between the 10 m absolute of the atmosphere and the partial vacuum induced in the pump is available to lift water to a certain height, overcome friction losses in the suction pipe and provide velocity energy in the inlet passage. The permissible suction lift is limited by vapour pressure by the presence of gases in the water, and by the speed of the pump. Normal limits are 5 m centrifugal, 7 m reciprocating, with higher values on special designs.

Certain pumps require an inlet pressure greater than atmospheric in order to permit operation at the designed speed. Suction lift is reduced approximately 1 m for each 1000 m of altitude owing to reduced atmospheric pressure. An increase of temperature involves the reduction of suction lift by increase of vapour pressure, requiring an inlet pressure above atmospheric in case of water temperatures of about 70°C and above. When dealing with liquids other than water, the same general rules apply. Inlet problems are more readily solved by referring all pressures and levels of metres absolute of the liquid handled, data being available on vapour pressure, specific gravity and viscosity at the pumping temperature. In general, apart from very viscous liquids, the pressure drop necessary to induce the design quantity into the pump (NPSH) may be assumed to have the same value of metres head of liquid as on the same pump operating on water.

Suction Piping

Low velocities and smooth flow layouts are essential for satisfactory operation. It is most important that the inlet pipe should be free from air leaks and from any air pockets, that is to say, the pipe must rise progressively from the sum or suction tank to the pump centre line. It is desirable to avoid even horizontal pipes because a pipe which is horizontal by design may be erected in such a manner that an air pocket of 1 cm or so could accidentally appear. Even such a small air pocket can give rise to a faulty operation, hence the suggestion that a definite slope should be ordered for all piping. Taper pipes in particular should have the to surface of the pipe flat, since the taper pipe with horizontal centre line necessarily involves an air pocket. Shar bends preceding the suction branch are undesirable. (See Fig 9.4).

Reciprocating and rotary pumps are capable of handling vapour with, of course, corresponding reduction of generated pressure. Centrifugal pumps generate a head of whatever fluid is handled. Since air has only 1/800th of the specific gravity of water, it follows that the centrifugal pump when full of air instead of water will only generate 1/800 of the pressure which, in the normal case, is insufficient to overcome suction lift or delivery head. As a result flow cannot take place.

Vaporizing Liquid Pumps

Where pumps draw from a condenser or from any vessel containing liquid at saturation pressure and temperature, it is essential to provide a difference in static level between the free liquid surface of the condenser and the pump centre line. Thus the liquid enters the pump with an excess of pressure over saturation which is able to overcome the inlet losses (NPSH) and ensure that vaporizing does not occur within the pump. On condenser pumps, this level may be used to control the flow of the pump to the flow of steam into the condenser, reduction of stream flow causing reduction of until the pump, by partial cavitation, has its own output reduced to balance the steam flow, each quantity of flow resulting in a particular condenser level. It is usual to provide an equalizing pipe from the top of the pump suction passage to the vapour space of the condenser or process vessel, so that any bubbles of vapour which have been released immediately preceding the impeller are able to pass into the vapour space, instead of through the pump.

Foot Valves and Strainers

In order to keep the water in the pump whilst standing, and to facilitate priming, a foot valve is necessary where the inlet water level may be below the pump centreline. The foot valve should be ample in size so as to reduce friction losses to a minimum. The strainer should have a total area of holes at least equal to four times the suction pipe area and the diameter of holes should be less than the smallest passage through the pump so as to avoid choking the pump.

The foot valve may be dispensed with, thus saving cost and head losses, where ejector priming is practicable. On inlet pressures exceeding atmospheric, an inlet sluice valve is provided, so as to isolate the pump for maintenance, but if operating conditions are such that a vacuum can occur at this valve, it is necessary to water seal the gland on the valve stem.

For larger pumps as for water turbines trash racks with suitable raking equipment are provided: alternatively, rotary strainers capable of continuous operation on trash laden waters can be used. Rotary micro strainer filtering down to a limit of, for example, 0.01 mm may be provided on certain duties to avoid risks of damage to fine clearances within the pump, but the capital cost involved must be weighed against the risk of damage to pump by foreign matter. In practice, the provision of such elaborate strainers for the first year of operation on a circulating system (eg, a boiler feed installation) is a reasonable compromise since, by that time, the foreign matter should be entirely removed from the system, thus permitting the straining equipment to be used for commissioning a later plant.

Delivery Valves and Non-Return Valves

A sluice valve is generally fitted to the pump delivery to permit priming, starting and isolation for maintenance. Regulation of the flow through centrifugal pumps can, in certain cases, be carried out by the delivery sluice valve, but this involves loss of power and, on a long term basis, erosion of the valve. By-passes may be provided to permit priming of the centrifugal pumps from the rising main and to facilitate the opening of the valve under pressure. A diversion of flow to atmospheric or to inlet pressures may be provided to avoid overheating the centrifugal pump.

Non-return valves are provided on the delivery to avoid flow back when the pump is standing and to protect the pump from inertia forces that may arise in the pipeline. Ball or plug valves, with seating faces protected from water erosion when in use, may be provided for high pressures. These valves open and close easily under full pressure and replace sluice and non-return valves, the non-return feature being added automatically by oil reservoirs and servo motors.

Velocity and Pressure

Velocity energy is convertible into pressure energy and vice versa by Bernoulli's theorem;

ie head due to velocity equals $V^2/(2g)$. In practice, converging taper pipes have negligible loss, straight diverging taper pipes have losses given in Ref 24, minimum losses for circular pipes occurring when the angle between the sides of the pipe is $5^{1}/_{2}°$; for rectangular pipes diverging in two dimensions $5^{1}/_{2}°$; for rectangular pipes diverging in one dimension $11°$. Heads of centrifugal pumps are evaluated in terms of $U^2/(2g)$ where U equals the peripheral velocity of the impeller. This is more accurate than referring the velocity of the impeller to $\sqrt{(2gh)}$, since errors due to size and efficiency change are avoided. Water flow velocities in a pump are referred to U not $\sqrt{(2gh)}$ for the same reason. All pipeline losses are evaluated in terms of $V^2(2g)$ where V is the pipe velocity.

Delivery Piping

The aforementioned considerations of velocity and pressure emphasize the need for straight delivery tapers between the pump branch and the pumping main. The delivery branch of a pump is generally less than the diameter of the pumping main. This is particularly the case with high head installations where high velocities are involved within the pump casing, and is obviously the case where several pumps deliver to one pumping main. It is therefore important that the velocity energy at the pump delivery branch should be converted to pressure energy with a minimum of loss, for example, by using the tapers suggested by Gibson (Ref 24).

There is also considerable advantage in using full flow valves, for example, plug or ball valves, on the pump delivery branch, followed by a long taper, thereby economizing in diameter of valve and avoiding the disturbance of flow preceding the taper, which would result from a sluice valve. It will be appreciated, of course, that the conversion of velocity energy to an increase of pressure at the final discharge is only possible in a taper pipe where the flow is free from disturbance.

Overall Picture of Pumping Station

It will be appreciated that pressure measurements taken close to an impeller are liable to considerable error due to disturbance of flow. There is, therefore, mutual advantage to pump maker and to customer in specifying that the performance guarantees for the pumping plant will be based upon pressures at the inlet and outlet to the pumping station itself, appropriate pipe and valve friction losses being allowed. This gives very much more accurate head measurement and also gives a fairer assessment of pump performance. This is due to the fact that the flow from a pump will often have a high velocity at the centre of the pipe so that the total energy in that pipe is very much greater than would be determined from a reading of pressure gauge and the allowance of velocity head, assuming uniform flow across the pipe diameter. The measurement remote from the pump will give a more accurate value of the total energy, since by mixing of the stream an improvement of flow pattern results. This is illustrated in the model tests for the 50 WW Grand Coulee pumps (Ref 38).

References

24. GIBSON, A.H. 'Hydraulics and Its Applications'. (Constable). 1920.
37. British Hydromechanics Research Association: BONNINGTON, S.T., and DENNY, D.F.,

'Some Measurements of Swirl in Pump Suction Pipes' RR 526. DENNY, D.F., 'Vortex Formation in Pump Sumps' S 436 DENNY, D.F., and YOUNG, G.A.J., 'The Prevention of Vortices and Swirl at Intakes'. S 583.

38. BLOM, C., 'Development of the Hydraulic Design for the Grand Coulee Pumps', ASME 49-SA 8.

CHARACTERISTICS OF PUMPS AND PIPE SYSTEMS

Pump Characteristics

The characteristics of head, power, efficiency and cavitation plotted against flow for a typical centrifugal pump at two speeds so as to illustrate the effect of speed on the performance of the pump are shown in Fig 22.1. The maximum efficiency line for variation of speed is clearly shown.

As will be seen, there are two lines for cavitation, namely, the operation line, which contains a margin of safety, and the cavitation line which is determined and plotted so as to ensure that the margin of safety is adequate.

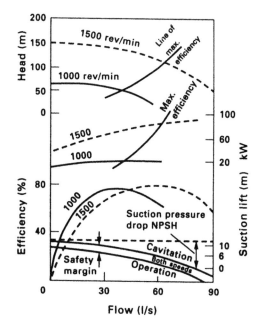

Figure 22.1
Effect of speed on
pump characteristics

For such curves showing moderate changes of speed, the cavitation performance is approximately constant for a given quantity, although, as will be seen from the inlet shape number curves, a slight change of cavitation performance does occur with change of speed. (See Chapter 9).

Consideration of Pump and Pipeline

As opposed to a reciprocating pump, a centrifugal pump generates a head which forces a certain quantity of liquid through a system. For any centrifugal pump, flow of water will only commence if the generated head at no discharge exceeds the pressure difference between the suction and delivery branches of the pump. When flow takes place the head of the system will be increased by the pipe, valve and other friction and by the velocity head. The head generated by a pump may increase as the discharge valve is opened and then fall (Fig 22.2) or may fall progressively with increase in quantity (Fig 22.1). The latter characteristic, a constantly falling one, is essential for certain pressure systems where the flow is controlled at a considerable distance from the pump, eg in boiler feeding.

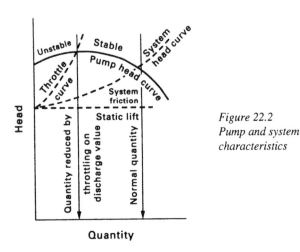

Figure 22.2
Pump and system
characteristics

Falling Characteristic

A falling characteristic is not usually necessary where valves are full open or where throttling, if any, takes place at the pump discharge. Regulation at the end of a long pipe, as in boiler feeding, demands a falling characteristic, since the change in pipe volume due to elastic deformation under pressure variation permits a momentary storage of water with consequent water hammer at the rising part of the curve (Fig 22.2). It is, however, essential on all systems to ensure that the closed valve head of the pump exceeds the static lift.

Static Head and Flow Conditions

When a pump is operating on a given system, the flow will stabilize at a point where the generated head is equal to the static head plus friction and velocity head, the operating point being determined by the intersection of the pump characteristic and the total friction characteristic of the system (Fig 22.2). It is assumed that the suction lift and suction piping to the pump are such as to permit the pump to operate well within its cavitation limit.

Physical Explanation of Pressure Oscillations

When a pump having an unstable characteristic (Fig 22.2) is operated on a long pipeline with a control valve remote from the pump, water hammer may result. This is due to the fact that the elasticity of the pipe permits a small amount of water to be stored in the pipe. The storage volume corresponds to the expansion of the pipe between the closed valve head and the peak of the characteristic.

When operating at the unstable part of the curve, the quantity increase to peak head point momentarily, the excess quantity being stored in the pipe. When peak is reached, the pipe can accept no more water and the pump flow reduces. In consequence of the characteristic, the head also reduces. The pipe then has a higher pressure than the pump and a back flow is induced, closing the non-return valve on the pump. The pump flow and head again increase to peak and the cycle repeats indefinitely. The only way to avoid this water hammer is to provide a characteristic which falls progressively with increase of quantity so that the pump is always higher in pressure than the pipeline. Such a characteristic also permits parallel operation with equally shared loads.

A falling characteristic is essential for boiler feed duties where control is by boiler valves remote from the pumps and where parallel operation over the whole characteristic is required. Where pumps operate on an accumulator, a falling characteristic is only essential if the pumps run at the top to the stroke on closed valve (the accumulators are generally close to the pump) or if parallel operation is required. On tappet control a falling curve is less important. Closed valve head, however, must exceed maximum accumulator pressure.

Where pumps discharge to nozzles, ie all friction head, and where the control valves are not likely to operate below approximately half quantity, (eg on descaling or debarking duties), a falling characteristic is not quite so important, but it is essential that the closed valve head should materially exceed the duty head.

Waterworks pumps on a static plus friction duty do not require a falling characteristic, but where more than one pump is installed, the closed valve head must exceed every possible operating head, otherwise it would be impossible to bring into load a second or third pump. This can only be determined by drawing the pump and system characteristic (as Fig 22.2) for every condition of load.

SECTION 6

Operation

INSTALLATION OF PUMPS

STARTING, TRANSIENT AND EMERGENCY CONDITIONS

TESTING AND TEST CODES

MONITORING AND MAINTENANCE OF PUMPS

INSTALLATION OF PUMPS

Erection of Centrifugal Pumps

A bedplate of steel or cast iron is provided for a small or medium pump and its driver. For larger pumps it is usual to provide a separate bedplate or soleplate for the pump. Pump and driver are generally connected by a flexible coupling of the pin type with rubber bushes for small/medium powers and of the gear or membrane type for the largest power. The coupling allows free axial movement but should not be regarded as permitting any out of alignment. Axial freedom is essential since drivers and pump have individual axial locations which may vary due to wear of balance surfaces or due to temperature effect (eg in the case of a multistage pump).

In general, pumps rest on prepared reinforced concrete bases with provision for holding down bolts. The pump and driver are supplied bolted but not dowelled to a bedplate or to soleplates. The bedplate is then lowered to the concrete foundation and very accurately levelled by steel wedges prior to the pouring of cement grout. When the cement is set, the holding down bolts are tightened and the pump set correctly aligned (using shims where necessary) and finally dowelled.

Provision must be made in alignment for any expected difference of temperature between pump and driver and between cold erection conditions and full temperature operating conditions. Where both units may operate at high temperature separate setting and dowelling of the respective bedplates may be involved.

Coupling alignment to within 0.02 mm must be attained, using a dial indicator on each shaft in turn

Pipe Loadings

On small cold pumps it is preferable that no pipe load should be imposed upon the pump branches but in practice this ideal is not always attained, especially on in-line pumps where the pipes must spring apart to permit entry of joint gaskets. Pipe and pump alignment should, however, aim at minimum stress on the casing and flanges.

For high temperature pumps considerable pipe loading is inevitable since the pipes expand when rising in temperature to normal operating condition. A cold pull up is therefore calculated and specified by the customer.

Pipes and pump branches are subject to hoop stress, thermal stress and longitudinal stress. The longitudinal stress is half the hoop stress which allows a margin for cold pull up. It is essential that the root where the pump branch is connected to the casing has a section modulus at least equal to that of the pipe, and that the bolting of pump to bedplate and bedplate to foundation is equally strong.

For vertical machines, particularly borehole and well pumps, it is essential that the borehole be plumbed so that there is knowledge of how the pump and pipes will hang. Where vertical sets have several floors of support, (eg for motor, for intermediate bearings and for pump), the exact alignment of the operating centre line can be attained by the use of piano wire suspending a heavy weight in a bath of thick oil to damp any oscillation. A micrometer and earphones to detect contact with the wire to the nearest hundredth of a millimetre provide a for the determination of an accurate datum reference.

American Petroleum Institute document API 610 recommends reasonable pipe loadings. The ISO pump committee has prepared a Pump Technical Specification which will include pipe loading recommendations and will have as reference documents the API 610 and the recommendations of European Pump makers, Europump.

STARTING TRANSIENT AND EMERGENCY CONDITIONS

Priming and Suction Arrangements

Where a pump is situated above the inlet water level, a foot valve is generally provided and priming can be effected by funnel for the initial start, after which priming back from the rising main via a bypass on the delivery valve is generally practicable. The foot valve may leak, thus requiring priming at each start, the leakage being made up from the rising main. On automatic sets, a low pressure relay may be set to prevent starting with less than a nominal pressure at the rising main.

Flooded inlet pumps are readily primed by opening the suction valve, but this valve should never be used as a means of regulating the flow through the pump. Air is removed from the pump by opening air cocks at the top of each stage and at the top of the inlet passage. For multistage pumps, it is necessary to prime a proportion of the stages representing the proportion that the static head bears to the total head with a suitable margin of, say, two further stages. It will be appreciated that where the head is wholly frictional, priming of two stages will suffice to free the later stages of air.

As mentioned earlier, it is absolutely essential that the suction pipes and suction passage should avoid any air pockets, the suction pipe to this end rising progressively to the pump. (See Fig 9.4).

When a suction foot valve is not fitted, priming may be carried out by a steam, air or water ejector at the top of the casing.

Starting of Centrifugal Pumps

In general, all centrifugal pumps must be filled with water before starting, since the internal running clearances depend upon water for lubrication. There are, however, exeptions to this rule, for example pumps having very heavy shafts with a total deflection at all times less than the radial clearance, or very large pumps under flooded conditions which are evacuated of water by compressed air in order to reduce the starting load on the motor. In this latter case, a water supply is usually provided to the neckrings to prevent risk of seizure.

The static starting torque of a centrifugal pump is relatively low, thus facilitating the use of normal squirrel cage motors. Gland friction, particularly on small pumps, renders forecast of starting torque somewhat difficult, but a reasonably cautious figure for very small pumps is 20% of full load torque. After starting, torque rapidly drops to meet the curve of torque proportional to speed squared as shown in Fig 24.1 (Ref 39).

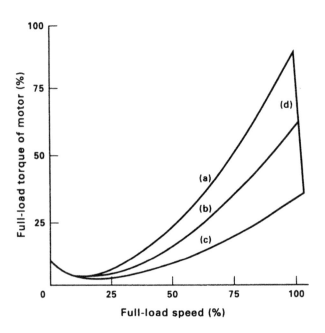

Figure 24.1 – Starting torque of centrifugal pump.
(a) valve open, low static lift.
(b) and (c) valve closed or high static lift.
(d) valve opening

(b) medium type number
(c) low type number

Starting Characteristics of Submersible Pumps – Described as an example of zero head start

Fig 24.1 shows the starting torque of submersible pumps. At zero speed the static friction torque must be overcome, for which a safe allowance of 10% of full load torque is made, since there is no gland friction. The torque taken by the pump then falls to meet the line of torque proportional to speed squared. For a zero head start, the upper curve (a) is used corresponding to immediate flow. For example, when no foot valve is provided and the water level is the same in the rising main and in the well.

When started against full head or closed valve, the pump approaches full speed before flow can start and the startin torque follows curve (b), medium shape number, or curve (c), low shape number.

Omission of Foot Valve

The foot valve is generally omitted on submersible pumps since its liablity to sticking open and difficulty of maintenance at the bottom of the well render its use impracticable.

The omission of a foot valve involves a start against zero head, which explains the need for consideration of zero head power consumption and thrust loading. Care must be taken in cases where the water stands very high in the well or shaft under certain rainfall conditions, thus demanding fairly continuous operation at low pumping heads.

Transient Conditions

The operating cycle of a submersible pump set is generally as follows:

> With the well water standing at normal level and with the pump immersed in water, the motor is started. The head at the start is zero if no foot valve is fitted. As the set accelerates, the water rises in the delivery column, imposing a head on the pump. During the first minute or so, the pump operates against a head which is less than normal. On a low shape number pump, this may impose more than normal load on the motor, although this is usually acceptable on short-term rating.
>
> After the delivery column is filled, the pump operates at normal duty. When the set is shut down by switching off the motor or by power failure, the water column and the rotor decelerate, stop and then reverse.
>
> The energy in the vertical water column is many times greater than the kinetic energy in the rotor so that overspeed in reverse is reached. The period of reversal is, however, relatively short (provided that check valves are installed on surface mains and reservoirs) before the vertical water column and the rotor finally come to rest at zero head.

Abnormal Conditions

The complete characteristics of a centrifugal pump are show in Fig 24.2 to illustrate the pump behaviour on power failure. Here the pump performance is plotted when running as a normal pump and as a reverse-flow turbine, at zero speed, and when running in reverse as a normal turbine and as a reverse flow pump. Between pairs of these phases, zones of energy dissipation occur. Submersible units have relatively small rotational inertias. For example, a typical set may have a kinetic energy in its rotor equivalent to the energy necessary to drive it at full speed and full load for only one-third second (ie the mechanical time-constant is one-third second).

The low mechanical time-constant results in the column being master of the rotor, and therefore the sequence on power failure is as follows:

Normal pump	Normal turbine
Energy dissipator	Turbine at runaway speed
Reverse flow turbine	Slow down to stop at zero head
Zero rotation	

The important stage here is the runaway speed in reverse, generally about 120% of the

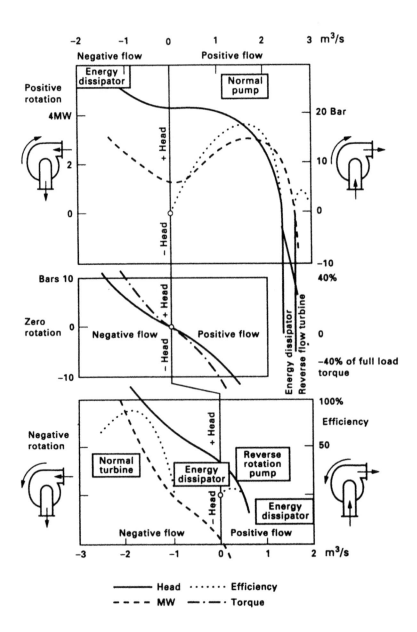

Figure 24.2 – (Reference 10) Complete characteristics of centrifugal pump.

normal speed. It is essential that the rotor shall safely withstand this speed and that the thrust bearing shall permit reverse running.

Inertias of Rotor and Water Column

The mechanical time constant is used to describe the relationship between the kinetic energy at normal speed of the complete rotor and the power taken to drive the pump at normal duty. Diving the kinetic energy by the drive energy, the mechanical time constant, in seconds, becomes

$$* \quad T = \frac{5.5 \, wk^2 \, N^2}{kW \times 10^6}$$

Inertia of Pipe Column

The stopping time of the flow in a simple pipe will be a period of seconds after power failure, expressed as:

$$\text{Stopping time} = \frac{lv}{gh_r}$$

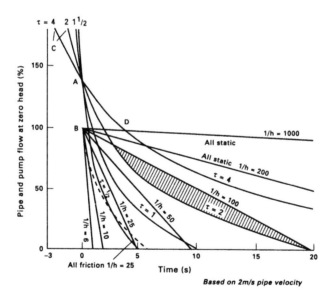

Figure 24.3 – Pump and pipeline stopping times. Shaded area represents cavity.
A. *Mean point assumed to be approximate starting point of column retardation.*
B. *Typical equivalent point of pipeflow deceleration taking into account pump energy above 100% flow only.*
C. *Points of power failure. Full pump head maintaining full column flow at these points.*
D. *Full pump flow but at zero head. Pump cannot maintain full flow of column.*
Run down time approximately 11T

* For list of symbols, see Example 2.

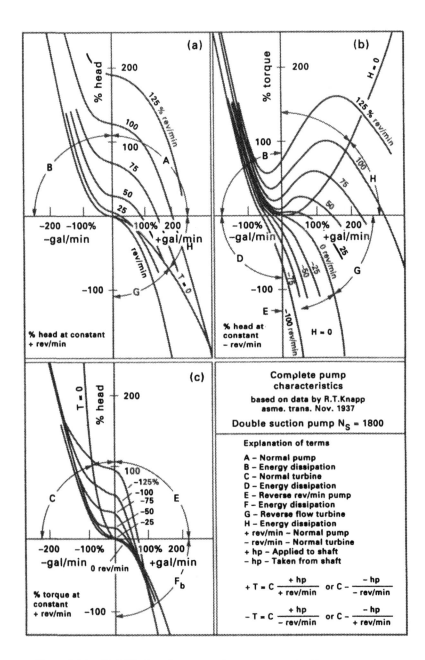

Figure 24.4 – (Ref 39) Complete pump characteristics, double-suction pump:
$N_S = 1800$ *(Us gal/min) and ft)* $N_S = 1100$ *(l/s and m).*

Fig 24.3 illustrates, on a time basis, the performance of pump and pipeline after power failure has occurred and gives a preliminary approximate idea of pump and pipeline behaviour. These problems are generally referred to a computer but the following quick appreciation still has its value and comforting assurance of reality.

The heavy lines labelled with various values of the time constant on Fig 24.3 represent the quantities that the pump is capable of delivering against zero head plotted against the time after power failure.

In general, the danger of column separation occurs when the pressure falls below zero head and approaches a vacuum of 10 m, at which vaporization of cold water occurs.

The adoption of zero head (atmospheric pressure) as a reference thus contains a margin of safety above actual water column breakage and also avoids risk of pipe collapse under vacuum where very thin pipes are involved.

The T curves were drawn by plotting the speed/time curve of the pump when it is

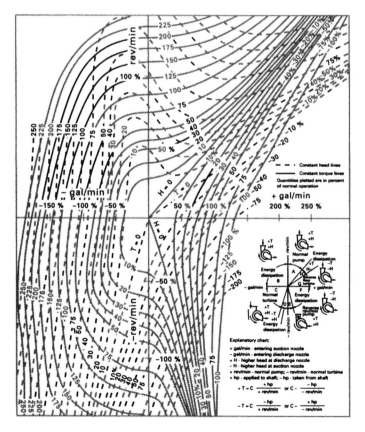

Figure 24.5 – (Ref 39) Complete pump characteristics, double-suction pump: $N_s = 1800$ (US gal/min) and ft) $N_s = 1100$ (l/s and m)

allowed to reduce from full speed to zero speed after a power failure. This is based on an integration wherein kinetic energy is converted to pumping energy. The speed/time curve of Q, starting from 100% pipe flow and falling to zero, is then multiplied by 1.8 to give the equivalent flow curve of the pump at zero head on the average pump will give a flow of approximately 180% of normal flow. The power at 1.8 times normal flow is assumed in he general case to equal normal flow power. We then have a curve of the flow capabilities of the pump on zero head.

The fine lines in Fig 24.3 represent the time taken for various columns to stop on power failure, the characteristic being, in the general case, a series of straight lines having varying values according to the ratio of pipe length to operating head. These values of l/h_r are indicated on the lines.

In Fig 24.3 the pipeline characteristics all start at a point which is to the right of the starting points of the pump characteristics. This is to allow for the energy given by the pump to the pipeline flow after power failure.

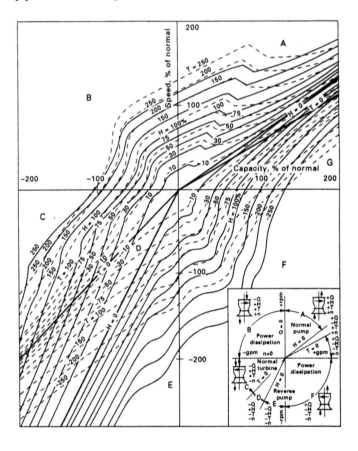

Figure 24.6 – (Ref 40) Complete pump characteristics.
Mixed flow pump: N_s 4 500 (l/s m).

Example 1

A pump set has T = 2 sec and a pipeline l/h_r = 100. Approximately 6 sec after power failure, a cavity will occur in the pipe, for which an air and water pressure tank having a water capacity equal to the shaded area in Fig 24.3 is recommended. the pressure in e tank after the water has filled the cavity should be high enough to avoid vacuum at any peak in the pipeline. This pressure is determined by drawing a straight line from the final reservoir level to touch the highest pipeline peak and to meet a vertical line from the pump house on the elevation drawing of the pipeline. The point of meeting determines the minimum safe pressure at the pipeline entry.

Example 2

A pump set has T = 0.5 sec and pipeline l/h_r = 10. Here no danger of a cavity occurring exists as the T line is well to the right of the l/h_r line in Fig 24.3.

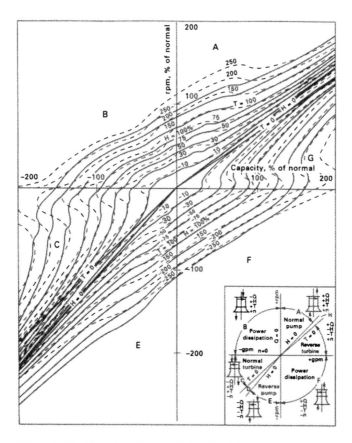

Figure 24.7 – Complete characteristics of axial flow pump N_s 8000

List of Symbols

Q = Pump flow, m³/s
T = Mechanical time constant of the kinetic energy of the rotor, sec
w = Total weight of pump and motor rotors, kg
k = Mean radius of gyration of pump and motor rotors, metres
N = Angular velocity of the rotor, rev/min
P_m = Power of motor in kW
l = Length of pipeline, metres
v = Velocity of flow in pipeline, m/s
g = Acceleration due to gravity, ms²
h = Total operating head on pump
h_r = Mean retarding head on pipeline column

Example of Filling of Rising Main

At the instant of startup and in the absence of a foot valve, the pump operates against zero
head and in consequence the motor is overloaded for the first minute or so.

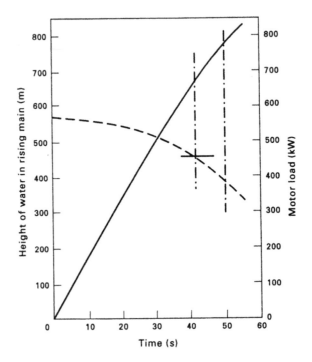

*Figure 24.8 – Variation in motor load and height of water
column with time during filling of rising main.*

——————— *height of water in rising main.*
— — — — *motor load*

Fig 24.8 shows a typical example of 450 kW motor wherein motor load and height of water in the rising main are plotted against time in seconds.

The motor is overloaded until the pump head has risen to aout 670 m in this example, so that it is necessary to calculate the time required to fill the rising main. To do this accurately, a step by step integration is necessary. Fig 24.8 shows the results of such a calculation indicating the height of the water in the rising main, slowly building up to approximately 800 m over a period of about 50 seconds, of which 42 seconds represent motor overload. The duration of this overload is short compared with the heating time consant of the motor, so the temperature rise due to this overload is of no practical significance even when the machine is warm when started up.

Complete characteristics of pump, turbine and energy dissipation phases are shown in graphs in this chapter (Figures 24.4 – 24.7).

Incorrect Assembly of Impeller

A double entry impeller can sometimes be assembled in error in the reversed position.

Subsequently, when driven at normal speed in the correct direction of rotation with respect to the casing, normal head may be generated up to about best efficiency flow, when severe vibration and surging of head from normal head to zero will commence. When rotated correctly with respect to the impeller but incorrectly to the casing, approximately half normal head will be generated at two-thirds normal flow.

References

10.   ANDERSON, H. H., and CRAWFORD, W. G. 'Submersible Pumping Plant', Proceedings Institution of Electrical Engineers, Vol 107, Part A, No 32, April 1960.

39.   KNAPP, R. T., 'Complete Pump Characteristics', Trans ASME November 1937.

40.   SWANSON, W. M. 'Complete Characteristics Circle Diagram for Turbo Machinery'. Trans ASME Vol 75, 1953.

TESTING AND TEST CODES

Testing of Pumps

When testing pumps at manufacturers' works, an artificial head is created by throttling on the discharge valve. Suction and delivery gauge readings are taken. The total head (see Fig 25.1) is made up of: suction gauge readings; level difference between point of attachment of suction gauge to pump and centre of delivery gauge; delivery gauge reading; velocity head difference. The last term is the difference between the velocity heads of suction and delivery branches, the difference being added when the delivery branch is smaller than the suction branch, since the liquid has more velocity energy at outlet than at inlet of pump.

The suction gauge readings indicate the sum of the static suction lift, the friction in the suction pipeline and the velocity head at the point of attachment of the suction gauge. For cavitation test an artificial suction lift can be created by fitting a valve in the suction pipe, thus permitting a test of the maximum quantity at various values of suction lift. Alternatively, the pump can be fed from an inlet tank of varying level of pressure. A series of tests at the same quantity are taken with varying inlet level until head and efficiency start to fall, thus giving the point of complete cavitation failure. A typical suction lift test curve is given in Fig 22.1 varying from approximately 9 m suction lift at zero quantity to 6 m suction lift at normal duty.

The temperature rise through a high head pump may be used to determine efficiency.

Test Codes

Pump tests are carried out to the terms laid down by ISO and BSI which provide three classes A, B and C with decreasing standards of measurement accuracy. There is therefore an appropriate standard of accuracy for each and every test varying from the class A for laboratory tests and vary elaborate full size tests to class C for the general run of production tests. The existence of the three classes encourages the provision of higher accuracy test installations since higher efficiencies can thereby be developed. A technical committee of the international Standards Organisation is currently preparing an international Code for Class A pump tests defined at present as materially better than Class B. Thermodynamic testing based upon the temperature rise through the pump is being covered in the Class A

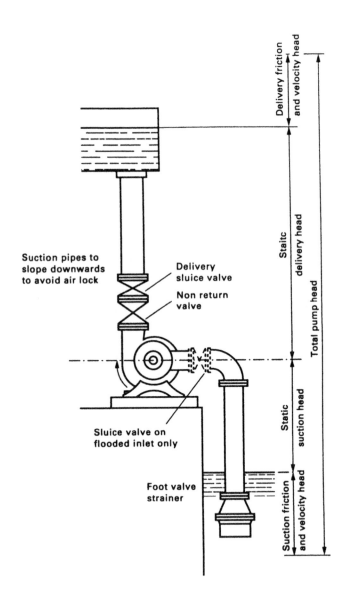

Figure 25.1
Diagrammatic representation of the work done by a pump.

document. The Class A code is intended for laboratory tests and for the more important and generally large power installations leaving classes B and C for the general run of pumps.

The standards already laid down are as follows:-

Class C - ISO BS5316 Part I

The efficiency tolerance is 5% pump only, $4^1/2$% pump and motor.

Class B - ISO 3555, BS5316 Part II

The efficiency tolerance is 2.8% pump only, 2.5% pump and motor.

Storage pumps are covered by IEC/TC4 Document 497 with a suggested efficiency measurement accuracy of 1-2%.

Dimensional Standards. It should be mentioned in passing that the 16 bar range of end suction back pull out pumps are defined by ISO 2858, BS5257.

MONITORING AND MAINTENANCE OF PUMPS

Maintenance of Pumps

Wear and tear of pump may be summed up in two words - water quality. many cases may be cited of pumps running for thirty years without wear or loss of efficiency. These are generally cases in which the water is reasonably clean and pure. At the other end of the scale are the severe waters and slurries where the life of the parts of relatively short.

Maintenance of pumps can be direct, ie servicing of the pump itself, or indirect, ie taking all possible steps to render the water harmless. Under the latter heading we may consider the settling out by a large sump of abrasive particles. such treatment is invaluable in reducing the harmful effect of corrosive waters, as the abrasion, besides causing direct mechanical damage to the protective skin is formed on the metal which prevents further corrosion. A small amount of abrasives in the water, however, would remove this skin. It is sometimes possible to dose water with inhibitors to form a skin of copper or oxide on the metal. Organic acids are liable to cause graphitization of cast iron, rendering it sufficiently soft to be cut with a knife.

Certain waters cause a heavy deposit in the internal parts of a pump, thus reducing the areas of the passages. This is harmful mainly in the case of small high speed sets.

Direct maintenance of the pump consists in replacing periodically the wearing parts - sleeves, neckrings, chamber bushes, and renewable balance plate. Impellers and diffusers generally have a long life. Provided that the set is correctly chosen for the duty, cavitation damage to the first impeller is unlikely.

Damage may be caused by a combination of all factors - corrosion, erosion, electrolytic action, graphitization, and corrosion fatigue of the shaft. Where any deflection of the shaft occurs it is very difficult to hold the sleeve and impeller hub joints, since the compression of the join varies at every revolution. The strength of the metal in fatigue is greatly affected by corrosion. Where there is any doubt of the water, a stainless steel shaft is recommended to avoid the corrosion factor.

Routine Recording

A record of regular checks on pressures and power consumed will immediately demonstrate any small change in performance so that the reason can be determined and any necessary action taken. Variations of performance can normally be expected with start up and changes of levels, pressures etc, and must be allowed for in assessing the daily records. Wear of internal parts which would cause parasitic leakage and reduction of flow would be determined fairly early by continuous monitoring of these records.

Periodic Service

Regular monitoring and periodic overhaul depending on the severity of duty are normal practice in most industries, and for this it is desirable to have in readiness spare wearing parts, bearings, seals, etc. In the case of vital plant with no standby, a complete pump or pump stage cartridges for barrel pumps should be available for substitution, leaving the servicing of the outgoing parts to be effected at leisure. This is economical, because maintenance staffs generally have the resources and time to carry out such work between the rush periods of periodic overhaul. Thermodynamic monitoring is useful here since it involves no efficiency loss.

Repair Procedures

Pump maintenance must cope with:

Infantile mortality, ie the teething troubles of installation and setting to work.

Random failure during the normal operating years.

Wear out when impellers, diffusers, running contacts or close clearance surfaces are worn beyond efficient operation.

For process and refinery duties technical obsolescence may give rise to shut down and disposal or scrapping of the plant after a few years.

On waterworks and pumped storage duties for example, pumps on clean water may run for thirty years or more without any need for repair or maintenance and it is here that daily readings are of value in confirming that efficiency is being maintained.

Rebuilding of Worm Surfaces

In addition to the actual replacement of neckrings, bushes, sleeves, etc, there are several valuable processes of rebuilding, for example, metal spraying, spray welding, electrolytic deposition and weld deposition which can help not only by providing a replacement surface but also by affording a harder or more corrosion resisting surface. It is important however to note that these processes of rebuilding can only be used where a good metal surface is available. The attack of high speed fee water on mild steel renders a depth of a millimetre or so fo the metal spongy, so that metal spraying or welding will not hold. The attacked mild steel will not show sparks when ground, or will it produce curlings when machined, but will instead machine like cast iron. It is necessary to machine or grind until the normal nature of the mild steel is demonstrated by sparks or curlings.

A further method of deposition is to use a plastic or ceramic spray, which has on occasion proved to be superior to metal on certain abrasive and corrosive duties.

Monitoring of Pump Performance

Rudyard Kipling describing engineers in 'The Sons of Martha' writes "They do not preach that their God will rouse them a little before the nuts work loose" but future monitoring may perform just this arousal.

The recent developments of electronics, computers, chemical analysis, lasers, vibration pick-ups, sonic measurements, ultrasonics and radiographics have permitted the analysis and forecast of the life of any pump component so that opening up for maintenance can be carried out at a reasonable period before expected failure rather than at routine times. There is really no point in the stripping of a machine if it is certain that it is operating satisfactorily.

The following methods will n the future minimize maintenance costs: chemical analysis of debris from bearings, etc in the lubricating oil or the balance leakage; metal wall thickness and integrity measurements by laser, ultrasonics, radiographics, magnetic or other means; vibration and noise analysis by Kurtosis, eg wave form investigation of bearing movements, monitoring of head, quantity and efficiency performance.

It is probable that the cost of monitoring will become less than the costs of routine maintenance and outages.

SECTION 7

Types

SINGLE STAGE PUMPS

MULTISTAGE SPLIT CASING PUMPS

MULTISTAGE CELLULAR PUMPS

BOREHOLE PUMPS

CONE FLOW AND AXIAL FLOW PUMPS

SEVERE DUTY PUMPS AND SMALL HIGH SPEED PUMPS

SINGLE STAGE PUMPS

The preceding chapters have considered the design of centrifugal pumps generally without reference to particular type or function.

We now consider the various types and applications of pumps in the light of the particular design features involved.

Single Stage Pumps of the Simplest Type (Fig 27.1)

A very large number of simple pumps are used throughout the world for general service tasks of a modest nature, say up to 50 or 100 kW and up to 100 m head. Such duties would include irrigation, minor drainage and the circulation of water or other liquids, the pumps being driven directly by motor or indirectly by engine and belt.

Cost Aspect

The majority of these pumps, particularly when supplied in large batches for irrigation and other general duties in countries which are undergoing development, are sold under keen world competition where a low price is paramount. Such production is in sufficient numbers to permit advanced tooling practice and the pump is designed with these aspects in view.

Materials of Construction

For minimum costs, cast iron is the most popular material and is used for impeller, casing and bearing brackets. Cat iron makes very good pressure-tight castings and has a hard surface skin which gives reasonable wear even on somewhat abrasive duties. There are many occasions where it is economic to provide a cheap pump of cast iron and replace it frequently, rather than a very expensive pump, particularly in markets where low first cost is important.

For very low quantity, high head duties, the resulting narrow large diameter impeller present difficulty in casting but it is often possible to alleviate this problem by reducing the outlet angle to permit an increase of impeller width and consequently a stronger core to the impeller.

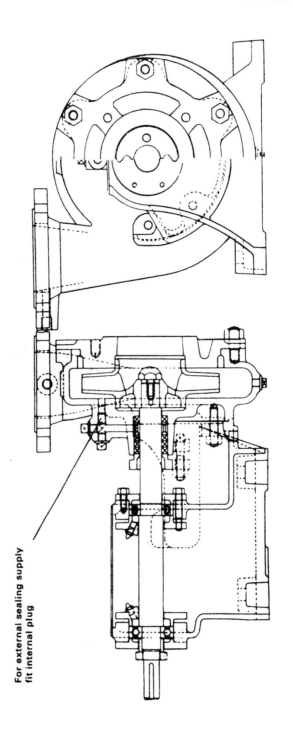

For external sealing supply
fit internal plug

Figure 27.1 – Section of small single entry pump.

Figure 27.2 – Small single entry pumps, engine driven.

The shaft material is usually 60 hectobar steel with occasional use of stainless steel on more corrosive duties. Ball and roller bearings are now universally supplied for these small simple pumps with grease lubrication and a simple thrower, for example a rubber ring, to protect the bearing from ingress of water.

General Construction Details

It is usual to omit neckrings but to provide sufficient metal thickness so that these can be added at a later date if necessary. The bearing bracket generally carries the feet, with drilled holes or cored slots for bolting to the bedplate and an extension ring to the bearing bracket on which is mounted the volute casing of the pump. On certain applications the volute casing is capable of being fitted in any of eight positions so that the delivery branch can point in a direction suitable for the piping layout.

Splitting of Casing

For these small simple pumps the casing is generally split in a plane normal to the axis of the shaft. This plane sometimes passes through the centre of the impeller and of the volute, so that a threeway joint is involved at the delivery branch. Alternatively the impeller can be inserted from the suction side with a suction plate to which is attached the inlet pipe. Further variation is the splitting of the casing at the driving side so that the bearing bracket, shaft and impeller can be removed leaving the volute casing and the inlet and outlet pipes undisturbed, but in this arrangement it is essential that the casing should carry feet to maintain alignment.

Radial and Axial Thrust

In these pumps the radial thrust at small flows is carried by the ball and roller bearings via the shaft; this infers that the shaft should be relatively large in diameter, eg one-seventh or one-eight of the impeller diameter. Axial thrust is minimized by drilling holes through the driving shroud of the impeller and fitting a neckring at the back of the impeller.

Glanding Arrangements

It is usual to omit the sleeve in the interests of cheapness, so that the shaft runs in the packing. This is entirely practicable in the case of separate pumps since the shaft can be regarded as cheap and expendable, but small combined motor pumps should have a sleeve to avoid the damage to the motor shaft, or have a stainless steel stub shaft friction-welded to the motor shaft.

For many of these small pumps mechanical seals are provided, which have the advantage, particularly on combined motor pumps, of reducing considerably the length and deflection of the shaft. (See Chapter 20).

It is usual to incorporate a universal gland sealing arrangement which can provide, without stripping of the pump, for internal seal from the volute, for external clean water seal or for a grease seal.

Applications of Small Single Stage Pumps

Municipal Water Authorities — Fresh water supply, transfer, boosting and relay duties, water treatment and filtration works.

Factories and Offices — Heating and cold water circulation.

Farms — Drainage, irrigation, spraying, farm and domestic water supplies.

Food — Refrigeration, washing, cooling and cleansing systems.

Laundries — circulation, water storage and processing.

Gas — Washing, transfer, boosting and coke quenching.

Mines and Quarries — Fresh water supplies, spraying, jetting and de-watering.

Public Authorities — Fire fighting services, parks, drainage, flood control and harbour services.

Chemical — Process and Refinery duties.

Marine duties.

Single Stage Double Entry Centrifugal Pumps (Fig 27.3)

The double entry split casing pump is used for general duties up to about 300 m and about 4 MW at the present stage of the art. These are very general limits and though they have been exceeded on occasional applications it is more usual on duties in excess of these heads and powers to provide the heavy shaft single entry pump resembling a water turbine.

The double entry feature gives the advantage of symmetrical hydraulic flow and pressure balance and by splitting the casing in a plane containing the shaft axis it is possible to examine the pump and to remove the rotating element without disturbing the inlet and outlet pipes. This design is used very largely for the highest quality pumps, eg on waterworks duties where the pump is the most important machine in the pumping station,

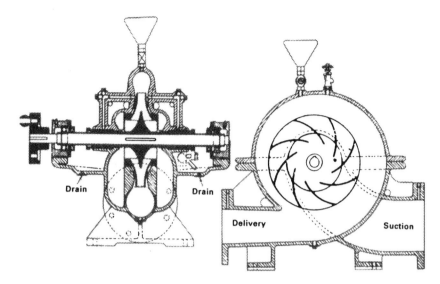

Figure 27.3
Sectional view of single stage split-casing pump.

in contrast to many other applications where the pump is only a small auxiliary. The water works industry quite naturally regard cleanliness as their primary saleable commodity and therefore waterworks pumps are finished to a very high standard with polished flanges, stainless and chromium plated fittings, etc. It is therefore in this range of split casing double entry pumps that we find the most attractive examples of the art of the pump maker.

General Construction

The pump casing up to about one metre branches will be made in two parts, namely the top and the bottom half casings. Above this size the lower half casings may be split into two parts.

The casings are generally of cast iron, normal or high strength according to pressure of the pump. Bronze or cast steel casings are used for sea water or other corrosive liquids.

The bottom half casing contains the inlet and outlet branches, the inlet passage dividing immediately in the casing to two passages, one on either side of the impeller. The casting is therefore somewhat complex but cast iron has been developed to a point where very good pressure vessels (even of complicated shape) are practicable. The double entry impeller is cast in bronze and mounted on a high tensile steel shaft and for sizes up to about one metre branch ball and roller bearings are used. above this size sleeve bearings are often used especially on the higher head and power ranges.

Provision for Wear

This type of pump carries renewable bronze neckrings fixed to the casing; occasionally the impeller necks may be fitted with renewable rings or alternatively have sufficient thickness to provide such rings when required. The shaft is sleeved so as to provide a renewable part at the glands and also to protect the shaft from attack by the liquid pumped. Renewable bronze bushes are provided at the inner ends of the stuffingboxes.

Manufacture

The double entry split casing is rather complex to machine since the internal bores for stuffingboxes and neckrings are within the casing and out of sight when being machined. Great care is necessary to ensure that all the bores for the bearings, stuffingboxes and neckrings are truly in line and that adequate rigidity is provided in the bearing brackets to prevent deflection under radial loading. The impeller is machined and dynamically balanced by removing a small amount of metal from the shrouds at appropriate positions, afterwards check balancing in the case of higher duties.

Details and Auxiliaries

The gland sealing device is similar to that on the simple single stage pumps described earlier. Air release and priming connections are provided together with drain pipes and drain collection box, attractively finished so as to improve the efficient working and appearance of the pumping station as a whole.

Figure 27.4 – Large vertical cooling water pump

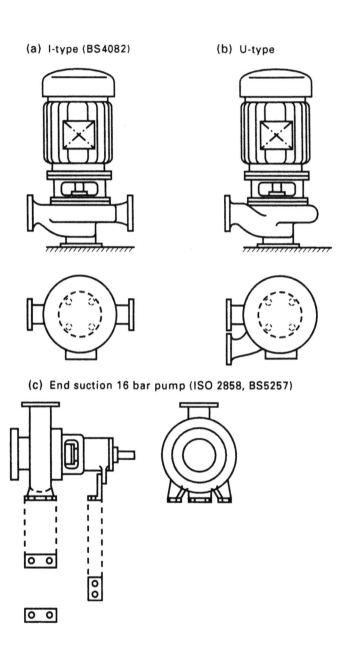

Figure 27.5
Standard pumps.

Design Problems of Single Stage Pumps

The first problem concerns the correct choice of duties for the ranges of pumps in order to meet the market demand adequately and efficiently. For the larger units, geometric designing already described is practicable, but for the smaller sizes the limitations of a fixed frequency of electrical supply tends towards lower shape numbers which are, of course, more difficult in the casting of the impeller and in the achieving of satisfactory efficiencies.

Accurate recordings of tests and statistical analysis (so that this data can be put to the best use) is essential in an industry where efficiency is so very important and where the electricity bill for two or three months can equal the price of the pump.

Shaft stress and rigidity problems require very careful study especially on applications where continuous operation at part load is required.

Vertical Operation

In the majority of cases, these pumps are operated with horizontal spindle but on the larger sizes, particularly on circulation duties, a vertical spindle arrangement will offer the advantage of a motor which is above flood level. Since the 'bottom' half of the casing is now at one side of the shaft, adequate support at the other side for the motor base is essential. (See figure 27.4).

Standard Single Stage Pumps

Fig 27.5 shows the British Standard for pump configuration. Duties and dimensions are given in British specifications and international Standards. BS4082 shows the in-line pumps; Fig 27.5 (a) l-type, (b) U-type. ISO 2858 and BS5257 show the end suction chemical pump (c).

Fig 27.6 shows in-line pump and motor.

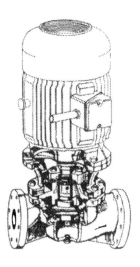

Figure 27.6 – Vertical in-line centrifugal pump motor unit.

MULTISTAGE SPLIT CASING PUMPS

It is usual to refer to pumps with casing split on the horizontal axis (ie the axis of the shaft) as split casing pumps and to refer to pumps having several stages, each separated from its neighbour by a vertical split normal to the shaft axis as sectional or cellular type.

Chapter 27 described horizontal split casing single stage pumps. We now consider the design and salient features of split casing pumps having two or more stages. In Britain and on the Continent, split casing pumps having up to two stages are manufactured, those with more stages being more generally of the sectional type. In America, on the other hand, horizontally split casing pumps have two, four, six or eight stages. (Fig 28.1).

Design of Two Stage Pumps (Fig 28.2)

In general, two stage pumps up to about 250 mm branches will be very similar to the single stage pumps mentioned in the preceding chapter, but will have single entry instead of double entry impellers. For larger sizes, double entry impellers may be used in order to provide more favourable suction conditions and to minimize bearing loadings by avoiding risk of differential thrust when working close to cavitation conditions.

The two single entry impellers are generally arranged back to back, that is to say, their inlets point away from each other thus minimizing gland pressure and providing economy of material as a pressure vessel. The exception to this rule is the two stage pump for condensate extraction duties on steam power stations where advantage is taken of having the suction inlet in the centre of the pump so that the glands where the shaft enters the pump are subject to delivery pressure. This is an important matter when operating on vacuum of up to 750 mm Hg, since the leakage of air into the feed water, which would be harmful to the boiler, is thereby minimized.

Mechanical Problems

Since the pump has two stages, it will be developing heads of the order of 400 m and therefore requires a stronger pressure vessel than that of the corresponding single stage pump. The increased strength of the pressure vessel will be obtained by the use of higher

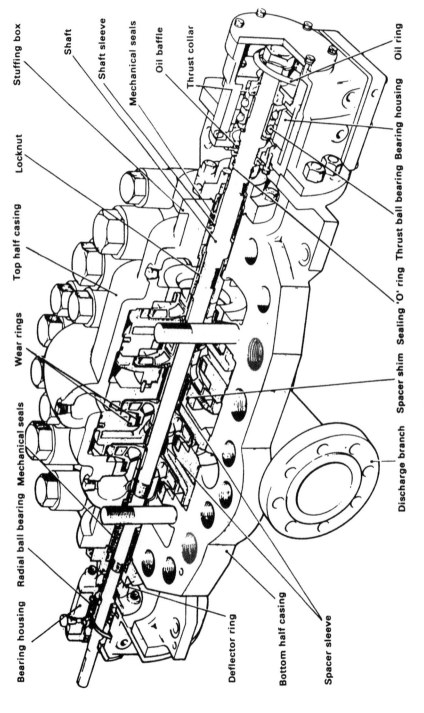

Figure 28.1 – 4 stage split casing pump

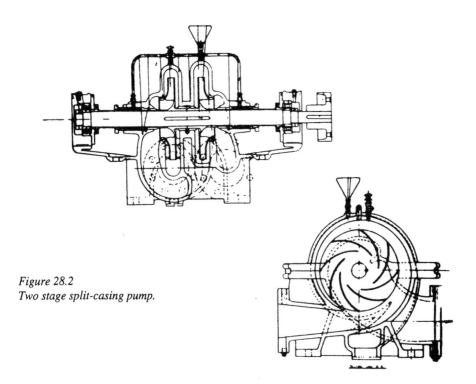

Figure 28.2
Two stage split-casing pump.

tensile materials and/or by the increase of metal thickness and provision of heavier bolting. It is usual in such cases to provide additional casing bolts between the impellers and near the suction passage, the bolts being mounted on long bosses to avoid prejudice to the water flow areas. The shaft similarly will be heavier, to transmit twice the power of a single stage pump, and will be provided with a central water lubricated sleeve bearing between the two impellers.

A double entry single stage pump, theoretically in perfect axial balance, will be provided with a thrust bearing to handle any departure from perfect theoretical balance. When a single stage double entry pump is operated at or near the cavitation point, the generated pressure will fall, but this will not involve any major departure from thrust balance.

On the other hand, in the case of a two stage single entry pump, operation at or near the cavitation point can give rise to axial thrust, since the first impeller can cavitate completely before the cavitation affects the second impellers, so that a thrust can arise, corresponding to one impeller totally unbalanced. This unbalanced thrust can readily occur in the case of a burst delivery pipe or faulty suction conditions giving rise to cavitation in the first impeller. It is therefore important that the thrust bearing for a two stage pump should have adequate margin to cover the possibility of these abnormal thrust conditions.

Rigidity of Two Stage Pump

In general, a two stage pump will have a more rigid shaft than that of a single stage because, although the span between bearings is greater, the shaft diameter is generally larger to transmit the greater power and, in addition, a certain amount of support is derived from the water lubricated central sleeve bearing. The gland conditions are more onerous on a two stage pump since the gland at the delivery stage is subjected to first stage pressure instead of inlet pressure. It is often convenient to provide a restriction bush to break down the pressure to a reasonable value for the type of packing or seal that is used, the discharge from this restriction bush being used to seal the first stage gland against air leakage.

Hydraulic Problems

In general, a single entry two stage pump could be compared to a double entry single stage pump of twice the quantity and half the head, a two stage pump representing two impellers in series whilst the single stage double entry pump represents two impellers in parallel. The

Figure 28.3 – Two stage pumps on water supply duties.

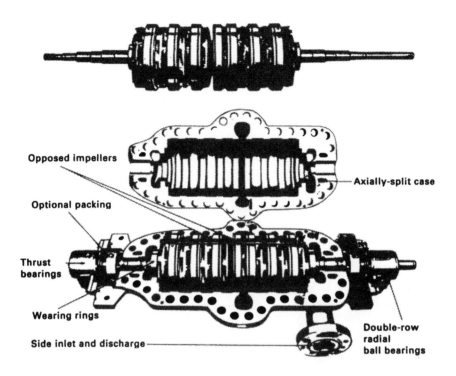

Figure 28.4 – Multistage split casing pump.

efficiency of the two stage pump is determined from the flow quantity and shape number curves in Chapter 7 by taking the generated head of one impeller. Some efficiency loss is incurred in the cross over passage, the design of which requires great care, but experience shows that this loss in efficiency is reasonably offset by the saving in mechanical losses which are somewhat less than twice the mechanical losses of a single stage pump corresponding to one stage of the two stage pump.

Position of Cross Over

Pump manufacturers in general favour a cross over between the stages which is arranged on the bottom half casing (although many pumps are seen with a cross over above). the cross over below gives somewhat greater rigidity to the pump and leaves the upper half lighter which facilitates opening up for maintenance. American multistage split casing pumps often have very long cross overs above and below the pump proper in order to minimize transfer losses between stages, but this practice has not found favour in Britain or on the Continent.

Double entry two stage pumps are occasionally seen but it would appear that they offer little advantage over two separate single stage pumps with common driver, since the standard patterns for the latter will probably be available.

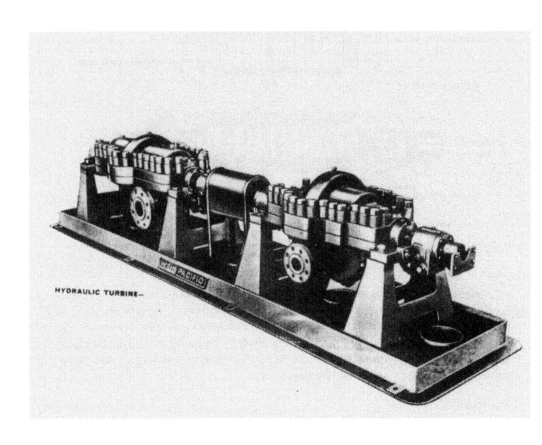

Figure 28.5 – Pump driven by similar pump operating as turbine.

Application of Two Stage Pumps

The two stage pump is the link between the single stage and the cellular multistage pumps. The end covers, shaft, balance disc and bearings of the cellular multistage pump are normally designed for duties up to ten or twelve stages; therefore, if such a pump were used for two or three stages it would be uneconomic in efficiency and in price. it is here that the two stage split casing pump fulfils a valuable role.

Two stage pumps are applied to the higher head duties of waterworks, to the medium head mine drainage duties and to many other chemical process duties between 100 and 300m.

A typical example is on the fixation of nitrogen from the atmosphere. Here two stage pumps delivery 350 l/s against a head of 270 m for scrubbing the carbon dioxide out of the hydrogen, nitrogen, carbon dioxide gas mixture. After passing through the scrubber towers the water, heavily charged with carbon dioxide, is passed through a turbine so as to recover approximately half the expended power. Each motor develops 1150 kW in order to handle the load of pump and turbine during the starting conditions, the turbine recovering 440 kW when the cycle is in full operation. After passing through the turbine the water and carbon dioxide are discharged to a degasifying tower, the water then returning to the two stage inlet - when the cycle begins again. Operation is quite stable and there is no need for a clutch in the drive.

MULTISTAGE CELLULAR PUMPS

General

Multistage pumps are required where the total head exceeds that which can be conveniently generated in one impeller. Depending on the size, speed, shape number and the nature of the drive the head per stage can vary between about 10 m and 1000 m, the pros and cons of these very low heads per stage being discussed in Chapter 12.

For economy and simplicity the individual stages of a multistage pump are arranged in cellular form so that they can be assembled on one common shaft with its associated glands and bearings and provided with inlet and outlet passages in the end covers which clamp the stages by means of long throughbolts. Heavier shafts to transmit the torque of more than one stage, single entry impellers and hydraulic balancing discs are features of cellular multistage pumps.

Construction

Cellular multistage pumps are constructed by building several stages together on the same shaft. Each stage contains impeller, diffusers, return passage to lead the liquid into the next stage, together with renewable wearing parts on which the impeller hubs and becks run. Each set of stage parts is contained within a pressure chamber generally comprising an outer ring and a closing or stage separation plate. The outer ring on the last stage withstands full pressure, but for convenience all stages are made equally strong. The closing plate of each stage with stands the generated stage pressure.

The neckring, which is fixed to the closing plate, has a pressure difference across it of almost one stage pressure whilst the chamber bush, fixed to the return water passages, withstands only the pressure rise through the diffuser, generally about 10-15% of stage pressure.

This method of construction affords, in general, circular machining, considerable strength with respect to internal pressure, straight forward manufacture and flexibility of application to various duties. The end covers are made of steel above about 35 bar and of cast iron below that. Forged steel covers are used for the highest pressures.

The covers carry the inlet and outlet passages and branches together with glands and bearings. The delivery cover also contains the balance disc. In order to provide ready access to the balance disc for maintenance, the multistage pump is generally driven by a shaft extension at the inlet side.

The support feet on the pump are provided on the end covers in the case of heavy duty units, or on the stages for the easier duties, since this permits the inlet and outlet covers to be taken from stock and rotated to any convenient position to suit site requirements.

Flexibility

The great advantage of cellular construction is the fact that all stage parts, ie chambers, impellers, diffusers, return passages, neckrings, chamber bushes, can be stocked in large numbers and, on receipt of order, assembled between stock end covers to the appropriate number of stages. It is only necessary, therefore, to manufacture the shaft, the coverbolts and the bedplate at the time of the order, thus giving very rapid deliveries.

Rotor

The impellers are of the single entry type and are unbalanced hydraulically, since to provide double neckring balanced impellers would be an unnecessary complication and would increase the overall length of the pump. The total thrust of all the impellers is therefore carried on a balance disc. This is a mushroom shaped valve which runs on a film of water escaping from the high pressure end of the pump. The annular space between the inner diameter of the waring face and the hub diameter of the balance disc is made appreciably larger than the annular area between the impeller neck diameter and the hub diameter. This ensures that the force lifting the balance disc exceeds any possible end thrust from the impellers. As soon as the disc is lifted from its face the resulting escape of

Figure 29.1 – Multistage waterworks pump with engine drive.

water across the fact acts as a pressure regulation so that in operation the balance disc runs on a film of water whose thickness is of the order of 0.05 mm. Balance discs are entirely self regulating and give a satisfactory means of supporting the thrust. A further advantage of the balance disc is that a high pressure gland at the delivery end of the pump is thereby avoided.

Balance discs were introduced to centrifugal pumps some 70 years ago and have given entirely satisfactory service, often running up to thirty years on clean water before maintenance is necessary.

On grit-laden waters more rapid replacement is required but the provision of harder materials and, in extreme cases, of external oil lubricated thrust bearings (see later), is practicable.

Mechanical Thrust Bearings

On vertical multistage pumps Michell or ball/roller thrust hearings are provided to take the static weight of the rotating element and any hydraulic thrust. On horizontal multistage pumps a Michell thrust bearing is provided to assist the balance disc, if any of the following conditions apply:

(i) violent pressure changes, such as may occur on hydraulic presses, cranes, etc.

(ii) operation in parallel with reciprocating pumps, which gives rise to severe pressure fluctuations.

(iii) the pumping of abrasive liquids, which cause severe balance disc wear.

(iv) frequent stopping and starting, such as may occur on automatic accumulator service.

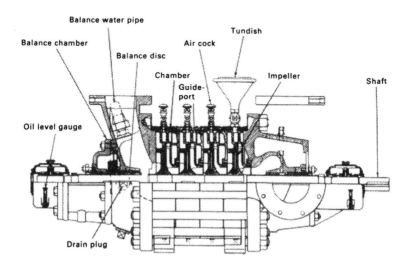

Figure 29.2 – Typical arrangement of multistage pump.

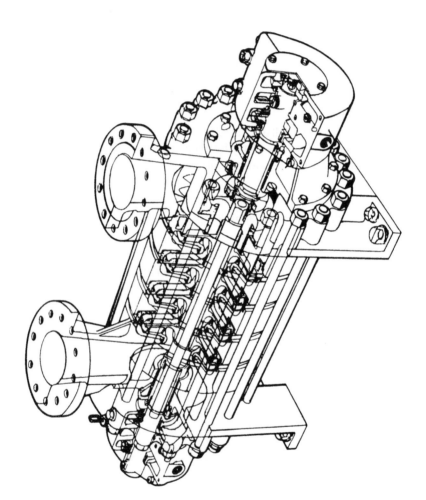

Figure 29.3 – Multistage general duty pump.

It is essential that the Michell bearing be rated at the full hydraulic thrust. In these cases the balance disc serves only as a pressure reducing device to protect the glands at the delivery end of the pump.

Disposal of Balance Water on Multistage Pumps

On normal cold water pumps, where a sump is available, balance leakage may be returned to atmosphere. In a pipeline booster station where the balance water is returned to the pump inlet, a water seal at the balance disc end of the pump is provided to avoid air drawing if there is a possibility of operating on suction lift.

Section Drawings

Figs 29.2 and 29.3 show low pressure multistage pumps (up to 35 bar), the type which would be available largely from stock, whilst Fig 33.2 shows a high pressure pump.

Materials of Multistage Pumps

Multistage pumps up to 35 bar are fabricated from similar materials to those used in the single stage pumps, namely, a cast iron pressure shell with a steel shaft and throughbolts, bronze, impellers, diffusers, balance disc and renewable wearing parts. Materials for heavy duty pumps are discussed in Chapter 33, Boiler Feed Pumps, and Chapter 37, Mine Pumps.

BOREHOLE PUMPS

General

Water is gathered in hilly country by building reservoirs and in the flatter country by pumping from rivers or from below the ground level. Underground water is pumped from deep wells or from boreholes. Historically, the first development would be the digging of a well which would naturally have a diameter large enough for a man to use a spade, after which the technical advance of machinery would permit the drilling of boreholes relatively smaller than the deep wells. The distinction is important in pumping practice since the economy of a borehole as opposed to a deep well has fostered the development of borehole pumps, which have a smaller diameter and a greater length than the conventional horizontal shaft multistage pumps. The smaller diameter and the lower rotational speed in consequence of long vertical shaft drive have led to a lower head per stage so that bronze renewable diffusers of the multistage pump are not required on certain duties, thus permitting the development of the bowl design in parallel with the diffuser design.

Description of Borehole Pump

A typical borehole pump comprises a vertical spindle motor at the ground level which rests on a stool containing Michell bearing housing for the vertical drive shaft, the final discharge bend for the water flow and the gland. (Head gear and motor section — Fig 30.1). Suspended from the stool is a length of piping and shafting sufficient to reach the lowest water level. Below the water level the pump proper is attached to the lower end of the piping.

Details of Pump

The shaft of the multistage pump is provided with bronze or stainless steel sleeves at its journal, running in bearings of rubber, carbon or plastic material. Because of the length of a vertical stage, it is usual to provide distance sleeves between the impellers.

The impellers are naturally of the single entry type discharging (in the case of a medium shape number) into the bowl chamber of cast iron or bronze. This bowl chamber will have

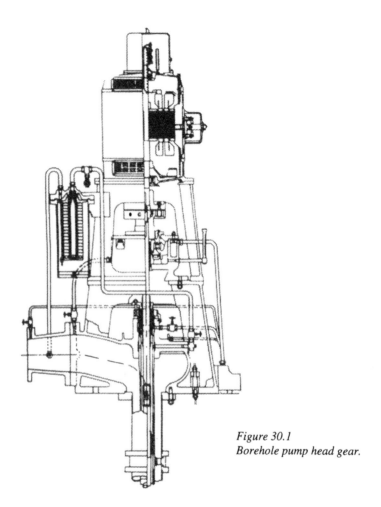

Figure 30.1
Borehole pump head gear.

a number of integral diffusers which decelerate the water and conduct it radially inwards to the entry of the next impeller. On the lower shape numbers the impeller will discharge the water into renewable bronze diffusers within a cast iron or bronze chamber which return the water to the inlet of the next impeller.

Since minimum diameter of chamber is desirable, the chambers are attached one to another by small interstage bolts, instead of the heavy covers and long throughbolts of the conventional horizontal multistage pump.

The inlet and outlet end covers contain the branches and bearings and are bolted respectively to the suction pipe and the rising main.

Details of Rising Main (Fig 30.1)

The rising main between the pump and the motor comprises, essentially, the drive shaft and the water flow pipe with spider bearings connecting the two. The bearings are

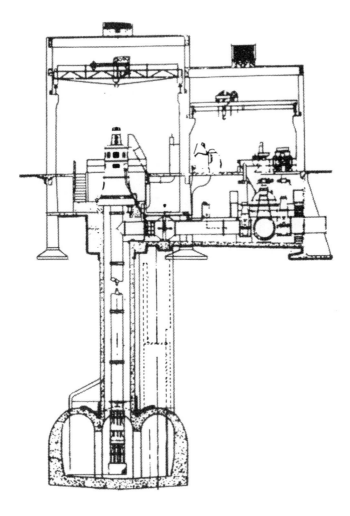

Figure 30.2 – Sectional elevation of large borehole pumps.

generally of lignum vitae, rubber or plastic and are usually enclosed in an internal tube which is supplied with clean filtered water. Where the pumped water is particularly clean, such internal tube and filtered water supply may be omitted, although it is necessary with rubber bearings to allow a continuous trickle of water down the shaft whilst the unit is standing so as to prevent the rubber adhering to the shaft, with consequent risk of tearing on start up. The lignum vitae bearings have the property of storing a certain amount of water during standing periods. Plastic possesses a similar property, but may suffer from growth over a period of months with consequent prejudice to the bearing clearances and the safety of operation. More generous clearances and the use of oil impregnated plastics have largely alleviated this problem.

Figure 30.3 – Thames Lea Tunnel, Metropolitan Water Board.
Cone flow pumps for 1.3 m³/s and 27 m head.

Efficiency Aspect

The pumping of water from below the ground involves additional losses in the frictional torque of the vertical shafting and in the power absorbed by the surface bearing on which this shafting is suspended. Additional head losses are incurred in the pipe friction between the pump at water level and the discharge branch at ground level. These additional losses are taken into account in any wire to water guarantee for the pumping system as a whole.

Comparison with Submersible Pumps

Submersible pumping plant (Chapter 36) having motors operating in water has captured a large proportion of the borehole pump field, but conventional shaft driven borehole pumps are still attractive for the large quantity low and medium head duties.

Suction Strainer

The suction strainer is a long tube, with small holes 6 to 12 mm in diameter, so as to exclude pebbles, having a total area of holes at least six times the pipe area. The incoming water may be from strata above or below the strainer, which explains the reason for the length of the strainer and the need to ensure that it diameter allows ample clearance between strainer and bore of well.

Suction Piping

It is usual to provide 10 to 12 m of suction pipe to avoid the risk of the pump overdrawing the well or borehole and bringing air into the pump impellers. With this long pipe, the water will be drawn down to the limiting suction lift capacity of the pump, which is of the order of 9 m at zero flow, after which partial cavitation prevents any further lowering of the well,

*Figure 30.4 – Borehole pump on water supply duties driven by one
MW 600 rev/min synchronous motor.*

so that the pump remains full of water and able to resume pumping when the well level rises.

Absence of Foot Valve

In the past valves have been fitted to borehole pumps, but so much difficulty was experienced with sticking of valve mushrooms or flaps that it is now usual to omit them. It is never possible to ensure that the water is perfectly clean and free from a tendency to corrode or leave deposits which can jam the moving parts of the valve. Even more serious is the risk that the valve may jam open until the reverse flow of water reaches a high value after which the valve may shut with a very severe shock to the pump, to the rising main and to the head gear.

The cost of stripping the rising main length by length and raising the pump so as to maintain the foot valve is quite prohibitive, so that at the present time, very few borehole pumps are fitted with foot valves. The water column within the vertical rising main is allowed to flow in reverse which, in turn, causes the pump shaft to reverse in direction of rotation until the water within the main falls to the level of the water within the well.

Investigations of inertia problems consequent upon the emptying of the rising main and the reversal of the pump are dealt with in Chapter 24.

The Pump Proper

The ratio between the flow velocity in the collecting chamber or diffuser and the flow velocity within the impeller passages is dependent upon shape numbers. The medium shape numbers have a relatively low flow velocity within the collection chamber or bowl and therefore this component can be made in cast iron, thus minimizing cost. In order to reduce the stage diameter as much as possible, to take advantage of smaller boreholes, the flow of liquid through the pump, as seen in a section through the shaft axis, approximates to a sine wave for each stage, so that the impellers and return passages are conical in form. The complete stage comprising stud ring, collecting chamber, return passages and bolt ring is one integral piece of cast iron or bronze. Each stage has impeller, distance sleeve, renewable neckring and chamber bush of bronze, with provision for rubber or plastic wearing surfaces on sandy waters. The shaft of high tensile steel is sleeved throughout its length.

In the case of the low shape number pumps, the flow velocity of the water in the collecting chamber is relatively high, so that renewable bronze diffusers are provided in the cast iron chamber, which also incorporates cast iron return passages.

Unlike the medium shape number described previously the flow pattern through the pump is not a sine wave, but is instead a series of relatively sharp curves and straight flow lines, the impeller shrouds being at right angles to the shaft, whilst the return passages have a very slight cone angle.

For constructional reasons, a closing plate is fitted between the chambers with a double ring of studs and nuts.

The suction and delivery covers of the pump carry the bearings of rubber or plastic and, at the delivery end, provision is made for filtered water lubrication where necessary from the internal tube.

30.5 – Dismantled view of borehold pump.

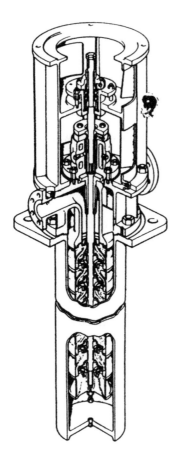

Figure 30.6
Borehole type
refinery pump.

Internal Tube and Shafting (Fig 30.1)

Co-axial with the rising main pipe is the internal tube containing clean filtered water to lubricate the rising main bearings.

Within this tube is the shaft of high tensile steel in lengths equal to the pipe lengths and joined together by a clamped cone or a screw coupling. The internal tube may be omitted if the pumped water is reasonably free from sand.

Head Gear (Fig 30.1)

At the surface is provided a circular head gear embodying the pipe bend between the vertical rising main and the horizontal discharge pipe, the lgland where the shaft enters the rising main and the provision for the filter with its pipes from the main discharge and to the internal tube.

Above the head gear is the motor stool containing a Michell or ball thrust bearing, which takes the weight of the shaft and impellers and the hydraulic thrust.

In the majority of cases, the impellers are unbalanced, so that the hydraulic end thrust is taken on the surface bearing, but on particularly severe duties, it is often convenient to balance each impeller by providing a neckring at the back (upper side) with holes for the back neckring leakage to return to the low pressure side.

Examples of Borehole Pumps

Borehole pumps are used for general waterworks supply, for industrial supplies for mine drainage.

Typical borehole pump duties are in the following ranges:

     up to 1 000 l/s

     up to 200 m head

     up to 1 000 kW at 1 500 or 1 000 rev/min driven by squirrel cage or wound rotor
          induction motors.

The above ranges cover the great bulk of borehole pumps but an example of 23 larger units, each at 1.2 m³/s against pressures varying between 55 and 64 m may be given. (See Figures 30.4 and 30.5.)

Borehole type pumps are also used on refinery duties, see Fig 30.6.

Ejector Pumps

An ejector at the bottom of a well, which is too bent or sandy to permit use of a borehole pump, can be fed by a conventional pump at the surface. this device is generally more efficient than an air lift and the submergence is less.

CONE FLOW AND AXIAL FLOW PUMPS

For low heads up to about 40 m per stage and for shape numbers above about 3000, cone flow and axial flow pumps may be used. In general, these pumps have vertical spindles and are arranged to draw water directly from a well or sump, for example, Fig 31.1, mixed flow pump.

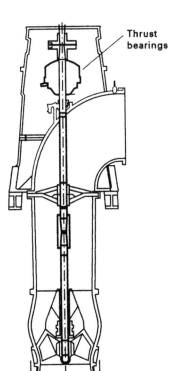

Thrust bearings

Figure 31.1
Suspended bowl mixed flow pump.

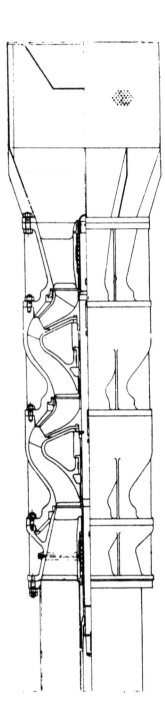

Figure 31.2 – Cone flow pump.

Cone Flow or Mixed Flow Pumps

A section of a two stage cone flow pump with a stage head of 30 m and specific speed of 2 000 appears in Fig 31.2 which, in conjunction with the description of a borehole pump in Chapter 21, is self-explanatory. The characteristic of a typical mixed flow pump appear in Fig 31.3 which shows the high rise of pressure from best efficiency point to zero flow and the consequential relatively high power at zero flow compared with a centrifugal pump the steep head characteristic gives the advantage of ability to operate against a varying head with relatively small efficiency change. In general, these pumps have a non-overloading power characteristic, that is to say, the maximum power occurs at or about best efficiency point.

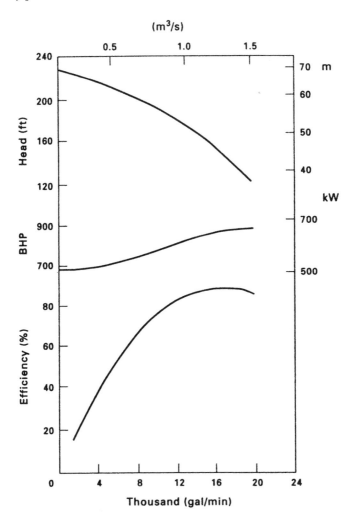

Figure 31.3 – Mixed flow pump characteristics.

Area ratios for the design of cone and axial flow pumps are shown in Fig 31.4, together
with usual type pumpers for which these area ratios may be used. Cone flow pumps are
illustrated in Figure 31.2.

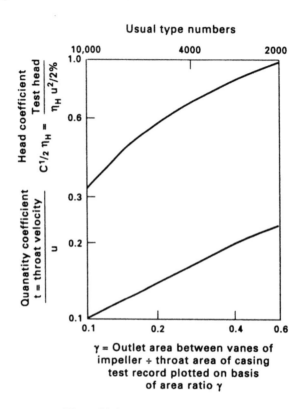

Figure 31.4
Mixed and axial flow pumps.

The Deriaz pump described in Chapter 35 is a special form of cone flow pump wherein
the impeller blades at 45° are adjustable in operation, so as to improve the efficiency flow
curve.

Axial Flow Pumps

For the lowest heads and for large quantities at shape numbers of 5000 to 10 000 axial flow
pumps having impellers resembling a ship's propeller are used (see Figure 31.6). These
pumps are usually fitted with a non-return valve but no sluice valve, since the head and
power at closed valve are considerably greater than at best efficiency (see Fig 31.5); over-
lading at start-up is thereby avoided. Stationary diffusers for axial flow discharge are
provided. On certain types, particularly the highest shape numbers (20 000) inlet guide
vanes with or without discharge vanes are used, the inlet vanes giving a whirl which is
removed by the impeller.

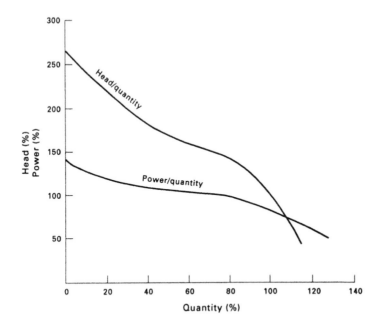

Figure 31.5 – Axial flow characteristics.

Figure 31.6 — Single Stage Axial Flow Pump.

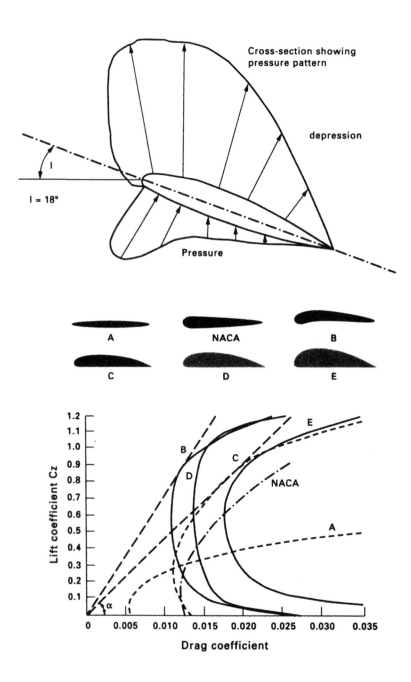

Figure 31.7.

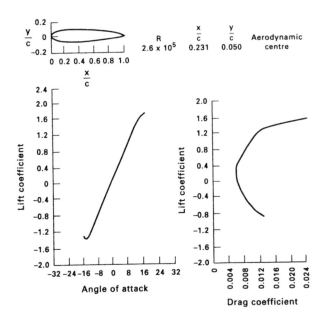

Figure 31.8 – Aerofoil Data NASA 23015.

Large axial flow pumps are generally made with movable impeller blades similar to a Kaplan turbine. The blades are controlled by level or head so as to give the best possible efficiency at each duty. The resulting quantity-efficiency curve is much fuller than that of the normal fixed blade type, being the envelope of curves for each angle of blade. The increase of blade angle results in a proportional increase of quantity for every point on the axial flow pump characteristics of head, power and efficiency; ie, a family of curves is produced showing different quantities at the same heads.

The discharge vanes for an axial flow pump occupy a conical space following the impeller in the usual pump application. For the largest flows, this cone can be replaced by a scroll similar to that of a Kaplan turbine, when a Kaplan pump results, having moving blades on the impeller and moving diffusers in the scroll or volute. Such a pump will give a further increase in the fullness of the efficiency flow curve and a similar family of head characteristic to that described previously. Flow variations of the order of 1 to 4 can then be attained against a given head with little variation of efficiency.

Cone flow and axial flow pumps are described by Wild in Ref 41.

Axial flow pumps can be designed on the author's area ratio system (Fig 31.4) or by treating each part of the impeller vane as an aerofoil and deriving lift and drag values as shown in Figs 31.7 and 31.8. For complete characteristics see Chapter 24, Figs 24.6 and 24.7.

Syphon Pipe Systems

Where the discharge point of a pipe system is below the highest point of the system some degree of syphon recovery of head may be possible.

With atmospheric pressure at the discharge point, the vacuum at the top point is equal to the static level from inlet level to the top of the system in order to start pumping. See Ref 85

References

41. WILD'L.O. 'Power Station Pumps', Pumping, May 1960 et seq.
85. Ryall, M.L., 'Development of a Self-Regulating Low Head Centrifugal Pump for Sewage and Land Drainage', (Vertical Free Vortex with Hollow Cone), British Pumping Manufacturers' Association, 7th Technical Conference, York, 1981.

SEVERE DUTY PUMPS AND SMALL HIGH SPEED PUMPS

Severe Duty Pumps

The foregoing has dealt with the more moderate duties which, for convenience, can be referred to as comprising stage heads up to 300 m. The large majority of small and medium pumps manufactured throughout the world fall in the category of small or moderate duty. When, however, we consider the more severe duties in respect of pressure, temperature or power, we find it essential to operate at very much greater stage heads so that, for a given expenditure of metal and space, a very much greater performance is attained. As can readily be imagined, many problems of stressing, vibration, expansion and metallurgy are encountered when designing such pumps.

These more severe duties are discussed in the chapters on Boiler Feed Pumps, Single Entry Pumps and Pumps for Storage Duties.

Head per Stage

Although the stage heads of the majority of pumps are relatively modest, investigations, into higher heads have been made over the years; eg Daugherty (Ref 1) 1915, recorded 300 m per stage, the author (Ref 36) 1951, suggested 600 m per stage as a maximum probable head at that stage of metallurgy. Karassik (Ref 42), recorded 25 boiler feed pumps running for several years at 500 to 600 m per stage, and present day makers of pumps for space rockets and jet aeroplanes are pumping kerosene at up to 1200 m per stage. Chapter 33 describes pumps operating at 1 200 - 1 400 m per stage.

The chief objection to a very high head per stage is the peripheral erosion which, in earlier tests, reduced the life of a stainless steel impeller at 1 200 m per stage on zero flow to an hour or so.

This is due to the fact that the impeller vane is passing the diffuser tips or the cut water of a volute pump at an extremely high speed with consequent local pressure and velocity fluctuations in the liquid which rapidly erode the metal of impeller and stationary parts.

Such wear is also a function of the departure of the duty from the best efficiency point since at that point the impeller vanes and the diffusers are deemed to be appropriate to the flow.

Figure 32.1 – Multistage hydraulic service pump 3 000 rev/min, 70 bar, on tappet control from bottle pressure. When the tappet control isolates the large motor (almost every minute), the smaller motor holds the set at 1 000 rev/min to avoid starting loads on the switchgear and motors.

Cavitation Problems

Where the high heads per stage are associated with relatively high shape numbers, ie approaching the optimum of 1400, a high submergence or inlet pressure may be required to avoid inlet cavitation.

Chemical Effects

Further complications arise where the liquid is corrosive. As described earlier, corrosion and erosion go together on high velocity turbulent duties, such as occur where high heads are associated with operation at duties far removed from the best efficiency point.

Local Factor

On the majority of these more severe duties the pump will not always be running at its best efficiency point. For example, the flow through a feed pump must be varied to suit the load on the power station; similar limitations arise on certain process duties and on storage

*Figure 32.2 – Large multistage descaling pumps in
South Wales steel works, 100 bar pressure.*

pumps. Even plant which runs continuously at best efficiency must have short periods of
operation at closed valve during start up and shut down when flow conditions within the
pump are far from ideal.

On the very large feed pumps the rotational speed is generally reduced for the low load
conditions. This considerably alleviates any tendency to attack the metal, and incidentally
saves power and wear and tear on the boiler control valve.

Increase of Size of Plant over the Years

Power plant, whether feed pumps, storage pumps, oil refinery or process plant, increases
in size to a greater or less degree over the years, thus presenting the pump maker with
increasingly severe problems. Thermal, nuclear and hydro power stations are advancing
in size and duty as quickly as present day metallurgical research will permit, and in
consequence, the severe duty pumps described in the following chapters represent today's
limits of what metals can do and review a totally different standard of performance with
respect to power for a given size from the normal run of everyday pumps.

Small High Speed Pumps

Geared Pump Sets (Fig 32.3)

For the lowest shape numbers (eg N_s equals 300), very large area ratios can be used. The resulting impeller flow velocity becomes so low that the impeller shape is not critical and it can be made open type with four or five radial blades and no shrouds. This design offers the advantages of the large clearances (thereby minimizing the risk of seizure) and of simplicity and strength. The casing can now be concentric with a tangential discharge cone, hence the description tangent pumps. By increasing the rotational speed to 10 000 - 20 000 rev/min the pump size can be reduced drastically. The cost, especially on duties demanding a stainless steel or titanium pump, is correspondingly reduced.

The pump set now comprises a two pole motor, a step up gear box and a small simple pump capable of duties up to 20 l/s at 170 bar. Gear box, seals and pump modules are easily accessible. This pump compete with the conventional multistage pump offering the advantages of low first cost and size, minimum risk of seizure if accidentally run dry, and a stable non-overloading characteristic.

The velocity in the casing throat at best efficiency point approaches the peripheral velocity of the impeller and consequently the head and the power characteristics fall rapidly at a point slightly beyond the best efficiency flow. This fall is caused by the fact that the throat velocity can never exceed the impeller peripheral velocity (except in the case of turbine operation). Since the flow can never exceed a predetermined value, a non-overloading characteristic is ensured.

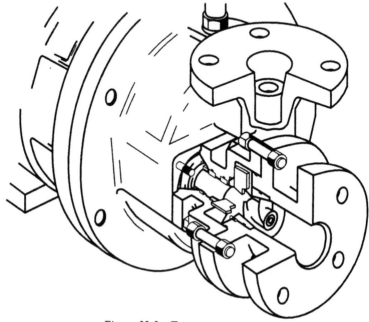

Figure 32.3 – Tangent pump.

The high speed tangent pump is slightly less efficient that the two pole multistage pump for the same duty, but the small size, the low cost, the simple robust construction and the aforementioned advantages render it an economic proposition especially for small refinery, process and spray cleaning duties.

Rocket Pumps

Aero and space craft must necessarily reduce mass to an absolute minimum. The fuel pumps are therefore designed to run at the highest possible speeds with relatively low NPSH values, which involves the use of inducers.

Figure 32.4 – High speed geared in-line pump set.

Inlet shape numbers up to 50 000 in l/s rev/min and metres are involved, the inducers being predominantly of the increasing pitch helix type. These high inlet shape numbers are acceptable in view of the short operating life required of the space travel pumps compared to the continuous duty of land or marine installations.

Rotating Casing Pumps

These embody a stationary nozzle receiving the rotating water and are suitable for duties with shape numbers in the region of N_s 100.

References

1.   DAUGHERTY, R.L., 'Centrifugal Pumps' (McGraw-Hill).
36. ANDERSON, H.H., 'Kempe's Engineers Year Book', 1951 et seq. Chapter on 'Liquid Pumps' (Morgan Bros. Ltd, London).
42. KARASSIK, I.J., 'Progress Report on High Speed Boiler Feed Pump', American Power Conference, March, 1957.

SECTION 8

Application Of Larger Power

BOILER FEED PUMPS

LARGE SINGLE ENTRY PUMPS

PUMPS FOR STORAGE AND ENERGY AND
OTHER LARGE POWER DUTIES

BOILER FEED PUMPS

This chapter illustrates the vast increase in capacity pressure and power of feed pumps by describing firstly, investigations and developments by the author which represented the state of the art in 1961, and secondly investigations and developments by Leith, McColl and Ryall which represented the state of the art in 1971.

The predominant changes over this decade are the doubling of main generator powers from the range of 350/500 MW to 700/1300 MW and the reduction of number of stages, this involving higher stage heads and higher speeds and calling for more stringent erosion, critical speed and vibration investigations. The consequent reduction of size and adoption of higher stresses permit thinner walled vessels which are less susceptible to thermal transients.

As shown in the second part of this chapter the development work was greatly assisted by the introduction and general use of computers for stress and vibration analyses.

SI Units

The thermal shock tests and calculations described were carried out prior to the introductions of SI units, which have subsequently been inserted on the charts.

Part 1 up to 1961

The increase in the size of power plant has been so rapid that the boiler feed pump of today is as powerful as a main turbo-alternator of a generation ago, operating pressures having increased tenfold. A typical duty for a 350 MW feed pump is 1 200 tonnes/h, 200 bar delivery pressure, 244°C feed-water temperature, 4500 rev/min for which a 8.5 MW turbine is required.

It is usual to provide one running pump and one or two standby pumps which come into service automatically if the running pump stops, the change-over being initiated by reduction of feed main pressure. The lower temperature of the standing pump causes a severe thermal shock on change- over.

Figure 33.1 – Ring section feed pumps for Nuclear Power Station.

For feed temperatures of the order of 230–260°C it is essential to provide booster feed pumps which receive water at, say, 4 bar 130°C from the deaerator and increase its pressure to 45-65 bar so as to ensure that when passing through the feed-water heater it can be raised to 230–260°C without flashing into steam. The feed pump then raises the water pressure from 45–65 bar to 200 bar or more in super-critical steam stations.

Booster pumps and feed pumps for feed-water temperatures up to about 150°C are generally of the throughbolt sectional state type (Fig 33.2). The pumps for feed water at 150-260°C (or higher) have a barrel enclosing the pump proper in order to minimize thermal stresses and to ensure uniform water flow, and, in consequence, uniform heat flow and expansion about the shaft axis during rapid temperature changes. These pumps are illustrated in Fig 33.2 and in Fig 33.6.

The major factor in the mechanical design of the feed pump is the thermal shock to which the pump may be subjected.

In addition to the stresses imposed by pressure and by operation, consideration must also be given to the stresses due to differential thermal expansion during rapid changes of feed-water temperature.

The bolts holding the casing elements together to form a pressure vessel are partially exposed to the air, and consequently will change temperature less rapidly than the casing wall which is in contact with the feed water. When the pump is heated suddenly the bolt stress is increased by the differential expansion between the hot casing and the cool bolts. During this heating period, the stress in the bolts must not exceed a safe proportion of the yield stress of the materials from which they are made.

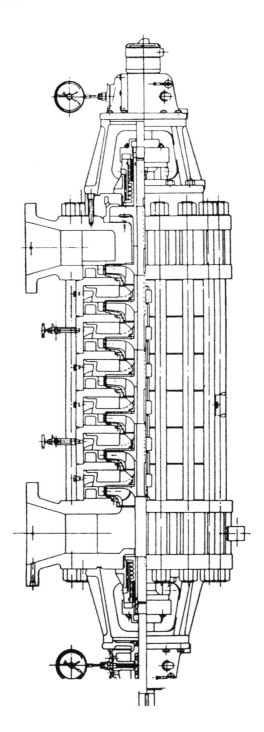

Figure 33.2 – Sectional arrangement of throughbolt boiler feed pump.

Figure 33.3 – Barrel feed pump for 500 MW generator.

Conversely, when the casing is cooled suddenly to a temperature below that of the bolts, the resulting differential expansion will cause a reduction of bolt tension.

It is essential that at their minimum stress conditions, the bolts contain a sufficient margin of tension to hold the main joints against risk of leakage.

The shell of the barrel is exposed to severe temperature fluctuations on its inner wall, whilst its out wall exposed to atmosphere would tend to lag in temperature behind the inner walls during temperature changes.

This gives rise to a temperature stress within the metal due to the differential expansion between the inner and outer walls. The temperature stress is integrated with the stresses resulting from pressure and operation to give the total stress. Care must be taken to ensure that any local plastic yielding or autofrettaging under this total stress is under control and is perfectly symmetrical.

The thermal shock will also cause risk of misalignment of bearings and bushes internally, and of pump and driver externally, if the elements of the pump are allowed to expand unequally with respect to the axis of the shaft. Stresses also cause risk to alignment if they are unsymmetrical around the shaft axis. To ensure correct alignment within a few

Figure 33.4 – Ring section turbine driven feed pump on large steam ship.

Figure 33.5 – Steam turbine driven feed pump.

hundredths of a millimetre under all operating conditions at present and future feed temperatures, it is therefore essential that all thermal flow during transient conditions, and therefore all water flow, all pressure vessel sections and all stresses, should be as nearly as possible perfectly symmetrical about the shaft axis.

The boiler drum has a similar general form and duty to that of the feed-pump barrel, but has a lower pressure and a higher temperature. Considerable instrumentation and control equipment are provided to ensure that the boiler is free from rapid temperature changes. Minor deflections, however, are not nearly so critical in the boiler as in the high-speed pump, which must be designed for rigidity and for perfect alignment under all conditions.

For safe operation, the boiler must have an uninterrupted supply of feed water which is within 50 deg C of its own temperature to avoid thermal shock and possible catastrophic failure of so vital a unit (Ref 44).

It is therefore essential that a standby pump be available at all times. The standby pump must be capable of accepting within a few seconds the full flow of full-temperature water irrespective of its standing temperature, which may be 100–160 deg C lower than that of the running pump, involving temperature changes at the rate of 500 deg C a minute or more.

In this manner, the feed pump shields the boiler from dangerous thermal shocks. furthermore, in an emergency due to failure of a heater, the feed water entering the pump may suddenly drop in temperature by 100 deg C in two minutes. Here, the pump takes the first shock as the residual heat of the pump, pipes and economizers raises the water temperature before it enters the boiler.

Certain makers ensure that the feed pump is kept warm by leaking back full-pressure water from the feed main through the standing pump, suggesting that, in the absence of such warming there is risk of seizure on start-up (the pumps are asymmetrical). Such warming procedure, however, involves power loss, control equipment, orifices etc, with consequent erosion on the full pressure difference and, moreover, does not take care of the thermal shocks of sudden reduction in temperature.

It is considered that a wiser plan is to provide symmetrical pumps capable of withstanding the most severe temperature fluctuation likely to be met in the power station.

The feed-water treatment is designed to give minimum corrosion at the boiler. Such treatment may, however, result in a liquid which is strongly corrosive and/or erosive at the high flow speeds and pick-up speeds associated with high-pressure pumping.

Erosion is minimized and performance improved by complete mathematical and geometrical control (in three dimensions) of the acceleration given to the water by the impeller vane, by correct transition from impeller to diffuser, by controlled deceleration in the diffusers and return passages and by the use of suitable grades of stainless steel in the pump proper.

Symmetrical Feed Pump

A thermally symmetrical feed pump was suggested in 1945 and manufacture commenced on fourteen pumps, each equivalent to a 60 MW turbo-generator.

*Figure 33.6 – Symmetrical barrel feed pump for 500 MW
generator with single stage booster pump.*

60 MW Thermal Shock Test: 1948

A further unit was built and tested as a prototype on thermal shock conditions (Ref 45).

This pump was run on closed discharge valve until a temperature of 260°C was reached by churning, then held steady at 260–265°C until component temperatures were stabilized. The discharge valve was then opened to cool the pump suddenly at 170 deg C/min in order to simulate the worst thermal shock anticipated in the operation of the pump at the power station. The pump was then stripped and examined to ensure that no distortion or internal leakage had occurred.

Barrel Feed Pump (Fig 33.7)

The pump proper, almost entirely of stainless steel and floating freely, is enclosed within a barrel in order to provide circular thermal symmetry and to minimize the number of joints carrying thermal loading in addition to pressure load. Each stage comprises impeller, diffuser, return flow passage and renewable bushings within a forged chamber. Water

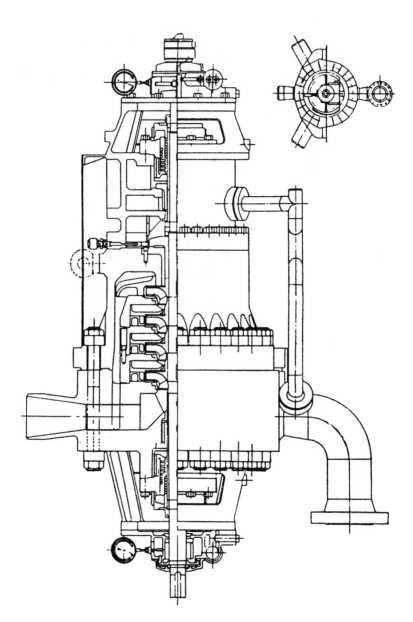

Figure 33.7 – Sectional arrangement of barrel type boiler feed pump.

Figure 33.8 – Barrel feed pump for 600 MW generator.

enters the end cover, passes through the stages, discharges into the annulus between the pump proper and the barrel and finally returns to the discharge branches in the end cover. The symmetrical enveloping annulus flow ensures that differential thermal stress is virtually absent in the internal stage joints. The total thrust of the single-entry impellers is carried on a hydraulic balance disc, the balance leakage being returned to the pump inlet. The pump proper and the barrel can expand freely and independently without prejudice to internal and external joints.

100 MW Thermal Shock Test: 1957

A later thermal shock test on a 100 MW feed pump was carried out in 1957 in order to ensure that the design principles still formed a satisfactory basis for the more severe conditions associated with duties from 100-800 MW and over. The record of this test is shown in 33.10.

The chart shows the gradual increase in temperatures as the pump, running on zero flow, converts the power of churning into heat.

Test Problems

Considerable difficulty and risk arise in simulating power station conditions by churning. With booster and gland circuits, a local flashing may occur in spite of careful instrumentation, since a feed pump is a steam accumulator and may become a steam generator. Several abortive tests occurred in 1948 and 1956 before a satisfactory layout was designed.

In order to ensure lubrication of internal bushes a seven-stage booster pump was used

Figure 33.9 – Barrel feed pump with self seal main joint for 660 MW generator driven by 18 MW turbine.

with sixth-stage discharge to feeder inlet and seventh-stage supply to internal bushes. This was for works tests only; the feed pump on site relied on heater pressure drop to lubricate these bushes as the additional sealing impeller was found to introduce risks of a slug of hot water reaching the booster.

For example, during the 1948 preliminary tests, the generator supplying the booster motor fell out of step, permitted the booster pressure to fall and caused cavitation in the feeder, which, instead of seizing, continued to run and thereby acted as a steam generator and approached dangerous pressures on consequent closure of the booster delivery non-return valve.

In order to avoid such a danger, the 1957 tests replaced this non-return valve by a sluice valve and provided an outlet to keep the booster cool and free from cavitation. However,

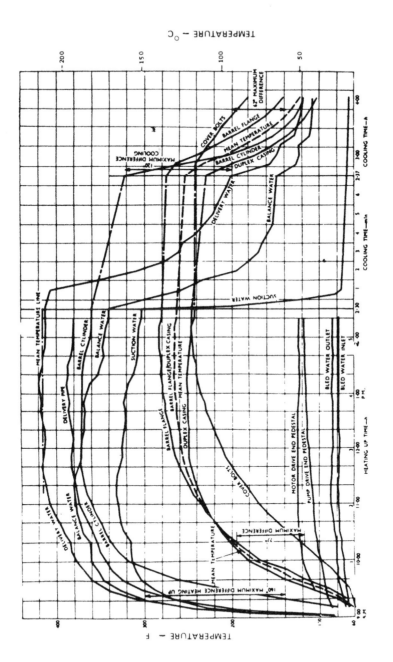

Figure 33.10 – Feed pump thermal shock test.

the accidental complete closure of this cooling outlet during temperature regulation caused a slug of hot water to re-enter the booster, cavitating both pumps. The feeder ran steam-logged two minutes before shut- down, but was undamaged.

The feeder, now a steam accumulator, was isolated by hand closure of the booster delivery sluice valve, but the water flow across the gland bushing to the booster discharge provided sufficient steam to heat the booster to 200°C. It was felt imprudent to switch off the booster motor as this steam might reverse the booster set at dangerous speed. The booster pump was therefore allowed to run steam-logged until, after 12 minutes, it was stopped by seizure of the first impeller, no other damage resulting. Further valves were fitted in the gland lines for the final tests.

It will be appreciated that the feed pump at 200–260°C is a steam accumulator at 16–46 bar, capable, when the motor is switched off, of reversing the feed pump at dangerous speed if an outlet for the steam is provided. A low pressure safety valve is impracticable, as the normal running pressure is 150–170 bar.

There is also the risk of the feed pump, when cavitating, acting as a steam generator to produce dangerous pressures if an outlet is not provided.

The booster delivery valve must therefore be open during running but quickly closed if the feed pump is stopped whilst at high temperature. Careful hand control of the whole test is therefore essential.

These experiences are described in order to show how such tests, necessary once or twice to prove design principles, require, to say the least, great care. The experience gained in these unique tests suggested that a symmetrical feed pump can run without damage for at least one minute under steam-logged or cavitation conditions, and this was confirmed several times during site commissioning.

Test Results

The result of the final 1957 thermal shock test are shown in Fig 33.10 where the abscissa shows hours during the heating-up period, then minutes for the sudden cooling period and afterwards hours for the final run.

It will be seen that after approximately five hours, all temperatures had stabilized and the delivery valve of the pump was then opened suddenly to pass a generous flow of cold water through the pump, thus giving a thermal shock of 550 deg C/min to the internals of the pump.

Main Bolt Stressing

The first consideration was the possibility of leakage, and for this reason the set was run after cooling under the maximum difference between main bolts and casing had occurred. During heating up, the bolt temperature is less than the temperature of the casing, but during the cooling period the bolt temperature is greater than that of the casing. The difference of temperature during heating up gives rise to an increase in bolt stress due to differential expansion of bolt and casing. During the cooling down, the difference of temperature gives rise to a reduction in bolt stress.

The purpose of the thermal shock test, as far as bolts and casing were concerned, was the ensure that no leakage took place and that no over-stressing occurred. This point was satisfactorily confirmed, showing that the original design principles rested on a sound basis, in that the heating-up stress on the bolts did not exceed a safe proportion of the yield point of the material, and that during cooling down, the alleviation of bolt stress was such as to leave an adequate amount of tension to maintain the joint.

Thermal Stress in the Shell of the Barrel

The next consideration was the question of thermal stress in the shell of the pump barrel. To this end, temperature indicators were placed at various points in the casing, so that a graph of temperature differential from outer to inner walls of various parts of the casing shell could be prepared.

The graph is shown in Fig 33.11, where wall thickness is plotted against temperature differences for operation at 212°C and extrapolated for 244°C.

From this curve it is possible to deduce the corresponding thermal stress due to differential expansion and the total stress, and to compare that with the yield point of the material.

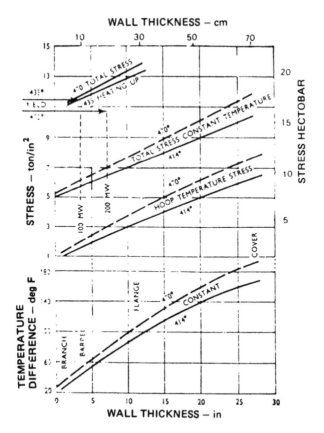

Figure 33.11 – Thermal behaviour of feed pump at 212°C
extrapolated to 244°C.

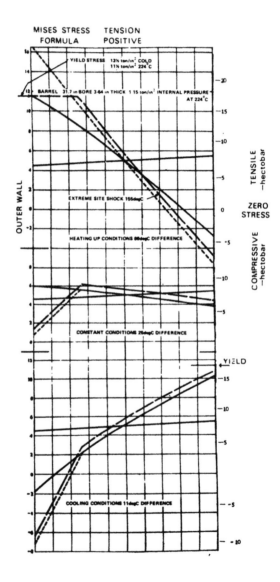

Figure 33.12 – Thermal shock test.

A graph of corresponding barrel wall stress during heating up, constant operation and cooling down is shown in Fig 33.12. It is seen that the 100 MW pump barrel of mild steel, 9.2 cm thick, will be just on the yield point at its outer surface during the slow heating up, and conversely, will approach the yield point at the inner surface during the cooling shock of 550 deg C.

Site Tests on 100 MW Units

The works' tests involved slow heating up and rapid cooling, but at the power station the conditions are reversed. The station designers stated that a severe cooling shock at the power station would not be likely to exceed 100 deg C in two minutes.

The warming-up period, however, was very much more severe in the power station than in the test that could be simulated at the manufacturer's works where the warming-up was done very slowly by the power taken to run the pump at closed valve.

At the power station when the running pump (at full load, full temperature) was tripped, the standby pump, relatively cold, accepted full temperature, full load, within the eight or ten seconds taken by the motor to reach full speed.

This involved a temperature difference across casing walls of 156 deg C, the thermal shock being of the order of 550 deg C/min, raising the stress figures to the dotted line shown in Fig 33.12.

As a result the outer 25% of the casing thickness was autofrettaged, (ie, locally plastically yielded) so that the stress lines for the heating-up, constant temperature and the cooling-off conditions turned sharply downwards, as shown at the left-hand end.

There is no danger in this, since the frequency of such shock is relatively small with respect to the fatigue range of the material, and since the reversal shock of cooling is very much less severe in the power station, so that autofrettaging of the inner surface does not occur.

Barrel Material

In the light of this information on the behaviour of thick walls of metal under the works' thermal shock test and the site operation of the 100 MW pumps, it was considered that to proceed from a 100 MW unit to a 200 MW unit where a 18.5 cm thick mild-steel barrel would be involved, would not be practical since the resulting total stress, under constant temperature conditions, would be 11 hectobar, and the shock conditions would cause autofrettage which was excessive in extent and on both sides, inner and outer, of the barrel wall.

It was known that certain feed pumps had unsymmetrical barrels with wall thickness of 28 cm, since their design was not so economical in respect of carcase diameter to impeller diameter, but in these cases the standby feed pumps are kept within 10 deg C of the temperature of the running pumps.

For the 200 MW feed-pump barrels, therefore, a high- tensile stainless steel was used. (FV 520). The resulting reduction of thickness to 5 cm had the advantage of the very much lower temperature differential between walls and since the material had a higher yield point, autofrettaging would not occur.

An improvement of corrosion resistance was also obtained with the higher tensile stainless steel and a big reduction in weight and cost resulted.

Thermal and Stress Symmetry

The thermal shock test also demonstrated that under these severe conditions the correct alignment of the shaft, bearings, and motor was maintained so that seizure was avoided. The whole of the structure, comprising barrel, end cover and bearing housings, was completely symmetrical about the shaft axis. The barrel contained no holes, so that its stress was uniform about the circle and was accurately determined. The end cover was symmetrical, expanded radially in a plane when heated, and was robust enough to permit the holes for the branches to pass between the main bolts.

The flow of hot or cold incoming liquid in the 200 MW feed pump was from the branches in the end cover, thence through the pump to the annulus between the pump proper and barrel in a guided symmetrical manner. In this way, during thermal transients, the whole circumference of the barrel received an equal heating or cooling effect. The hearing housings were fixed to the barrel and to the end cover by a conical member at 360°, thus affording symmetry of the whole unit.

The autofrettaging of the barrel would be symmetrical in this arrangement.

In the absence of such stress symmetry and guided flow symmetry, operation would be prejudiced in the following manner:

(i) The incoming flow of hot or cold water during temperature transients would tend to be on that part of the barrel circumference which is near the branch, thus causing uneven heating.

(ii) The presence of a hole for a branch in the barrel wall would be a local weakness in the longitudinal strength of the barrel. Such a weakness in a barrel whose thickness and stressing will result in autofrettaging, (ie, in local plastic yielding), is almost certain to cause the barrel to bend, since it is quite impossible to reinforce the barrel near the hole to make it as strong as the rest of the circumference, and bending of even a tenth of a millimetre would prejudice alignment.

At the first heating-up shock, an unsymmetrical barrel with a hole in its length subjected to autofrettaging would tend to distort, with prejudice to the safety of fine running clearance parts, especially if cavitation is approached.

End Cover

On the 200 MW feeder, a further approach to perfect thermal symmetry was attained by arranging the inlet and outlet branches in the end cover in cruciform, the two suction passages being top and bottom, with the two delivery passages at the horizontal centre line.

End Cover Stresses

The flow passages in one typical end cover correspond to a 15 cm pipe with varying wall thickness. Fig 33.13 is a chart of total wall stresses, determined by the integration of pressure and thermal stresses and plotted against wall thickness.

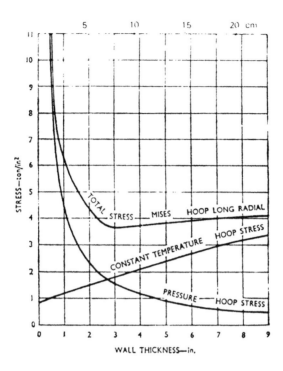

Figure 33.13 – 150 mm pipe stress at 230 bar closed valve pressure 244°C.

The hyperbolic curve is the pressure stress, whilst the rising curve is the thermal stress for constant temperature operation based upon the test recorded in Fig 33.10.

It will be seen that for all thicknesses from 4 cm upwards, the total stress is safely within the limit of 7½ hectobar.

Stress Calculations

The three dimensional stresses are integrated by the Mises-Hencky formula to give the total stress corresponding to the proof stress of the material.

The error is assuming the stress lines on Fig 33.12 to be straight lines instead of curves is reasonably offset by the assumption that the inner metal wall has the same temperature as the water.

Feed Range Characteristics

On full flow, the pressure required at the pump discharge is equal to the boiler pressure, plus the static lift from pump to boiler, plus the friction in pipes, heaters, etc.

At minimum flow the pump pressure is increased by approximately 15% but the pressure required is only the boiler pressure plus the static lift. There is, therefore, a surplus pressure of the order of 30% of the available pump pressure at zero flow.

In order to minimize the wear and tear on the boiler regulating valve and to save power,

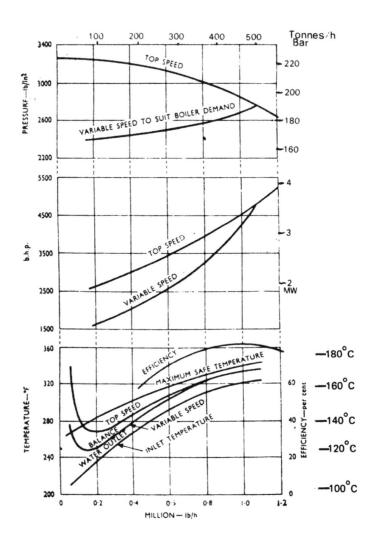

Figure 33.14 – 300 MW feed pump characteristics.

variable-speed drives, which reduce the pump speed as the load on the turbo-generator reduces, are provided. These are described in greater detail later.

Pump Characteristics

Typical characteristics of quantity, pressure, power, and efficiency are shown in Fig 33.14. The upper pressure curve shows full-speed operation, whilst the lower curve shows operation corresponding to boiler demand at reduced pump speed.

Pump Operation on Zero Boiler Demand

On zero flow, excessive heating of the pump occurs with risk of seizure on prolonged running. To obviate this, an automatic valve leaks a sufficient flow from pump discharge back to the deaerator. The operation of this valve is initiated by a signal from the flow-measuring device, with a standby signal, in case of failure, from a temperature switch at the balance flow. Fig 33.14 shows the temperature corresponding to full and reduced speed operation plotted against flow. The lowest curve is the inlet temperature of the feed water. The next two curves give the temperature at the balance flow outlet for full and reduced speed operation.

The top temperature curve gives the maximum safe balance water temperature to avoid flashing into steam. This is deduced by taking the saturation pressure corresponding to the inlet temperature for various flows, adding the pressure corresponding to the static height of deaerator water level above the pump centre line to give a total inlet pressure. The saturation temperature corresponding to this total inlet pressure is the maximum safe temperature.

The minimum leak-off values for safe operation are seen to be 27–54 tonnes/h, namely the points where the full or reduced speed temperature lines respectively cross the maximum safe temperature line.

Where the full-speed operation can occur and the excess pressure between upper pressure curve and boiler demand is broken down by the boiler regulator valve, the flow of 54 tonnes/h corresponding to full speed must be taken as minimum safe leak-off.

If, however, the pump speed is controlled by the boiler demand, that is, a computed integral of steam flow, water flow, and boiler water level, and there is no major pressure drop across the boiler regulator valve, then the lower flow of 27 tonnes/h, corresponding to reduced speed, will suffice. In the latter instance, if the feed pump operates on top speed to pressure test the boiler and steam pipes, such test will be on cold water, and 27 tonnes/h leak-off will still be safe.

Transient Temperature Conditions

During rapid load rejections the deaerator pressure may be less than that corresponding to the saturation pressure of the hotter water in the supply pipe between the deaerator and the pump itself. Some cavitation will, therefore, occur in the pump, but no damage will result provided the pump is symmetrical and provided flow is maintained. A careful arithmetical integration of these transient conditions is made to ensure that the extend of the cavitation is within safe limits.

Hydraulic Design

The design of impellers, diffusers, etc in respect of shape number, head characteristic, efficiency and cavitation is given in earlier chapters.

Efficiency change with temperature and speed is given later.

Determination of Impeller Vane and Passage Shape

Kaplan's method of error triangle vane development is stated by Stepanoff (Ref 8) to be responsible for recent high pump efficiencies.

It has, however, been the author's experience that the error triangle method is indeterminate and haphazard, particularly in the low pressure portion of the impeller or turbine runner. Moreover, it fails to give true mathematical and geometric knowledge of the position of the blade at any point of the flow areas involved, and is rather an inefficient method in the pattern shop.

An improved method, which has now been in use for a number of years, produces (by numerically controlled generation) the optimum blade shape in three dimensions for the hydraulic duty involved. This procedure, described in Chapter 11, gives true mathematical knowledge of every portion of the vane and of the designed acceleration given to the water by the vane in three dimensions. The impeller is rotated as in the pump whilst a milling cutter representing a wave of water cuts the required surface. The resulting smoother flow avoids erosion and improves performance on the very high heads (above 500 m per stage) involved in boiler feeding.

Chemical Aspects

The generation of high heads per stage of 500–1500 involves correspondingly high flow velocities of the water.

As already mentioned, a liquid which is comparatively inert at low velocities can be highly corrosive and/or erosive at high velocities, since the protective film of the salt or oxide of the metal, normally found in static corrosion, is eroded away by the high-velocity particles of water.

This corrosion and/or erosion occurs in spite of the fact that there may be no abrasive particles in boiler feed water. Obviously, on abrasive duties lower heads per stage must be used. the high stage heads of boiler feed pumps are only practicable since abrasive particles are virtually absent and since stainless steel is largely used.

It is fortunate that we have stainless steel available, since cavitational attack on stainless steels in a boiler feed pump is virtually unknown except for one or two isolated cases where thermal stations have been installed in an area predominantly served by hydro stations, so that, at certain seasons, in order to avoid spilling of reservoirs, the steam stations have had to run on 10% load almost continuously. A simple modification enabled the impellers to handle such duties without cavitational attack.

Providing these conditions are known beforehand, therefore, the internal parts of the pump can be designed to cope with them.

Metallurgical Details

The effect of high flow speeds on the corrosion and/or erosion of metals and the vulnerability of mild steel and cast iron to these conditions have already been mentioned.

Austenitic stainless steel, BS1631 containing 18% chromium and 8% nickel, has an extremely high resistance to corrosion, and is readily weldable, but is coefficient of expansion is approximately 16 millionths per degC, compared with mild steel which is approximately 11 millionths, thus presenting difficulties of slack fitting impellers at higher temperatures.

A further disadvantage of BS1631 is the relatively low yield point at 200°C and above, and it is therefore little used in boiler feed pumps above 120°C.

The martensitic stainless steels, BS1630 and En 57, containing 13% and 18% chromium respectively, have high strength, the same coefficient of expansion as mild steel, but the disadvantage of poor welding characteristics and rather difficult casting characteristics.

Corrosion resistance of BS1630 and En 57 is inferior to BS1631 but quite adequate for feed-water duties. BS1630, En 57, FV 520 and 17.4 are at present the main stainless steels in normal use on heavy duty boiler feed pumps, the BS1630 being used for impellers, diffusers, chambers, etc, and the En 57 for shafts. FV 520 and 17.4 may be used for any or all of the pump parts (see Chapter 15).

The problem arises of running two types of stainless steel together without seizure in the moving and stationary replaceable parts of the pump. Satisfactory operation of two stainless steel surfaces has been obtained by arranging for a sufficient difference of Brinell hardness number between the two faces in question and by correct surface finish.

Other Mechanical Design Aspects

The uniformity and symmetry of the pressure vessel has already been described, together with the stress considerations of thermal shock.

Rotor

The pump shaft is made of En 57 stainless steel and is subject to a maximum stress at the first impeller, the stress here being the combination of the total torque and the end thrust of the first impeller.

Since each impeller has, for simplicity of fitting, its own diameter on the shaft, the end thrust reaches a maximum at the balance disc where the torque stress is nil. This gives the advantage of economic use of the shaft material.

The operating speed of the pump is between the first and second critical speed, since it would be uneconomic to run below the first critical speed.

Analysis of general data from earlier designs gives the shaft size and critical speed which is checked by the following analysis:

(i) the usual graphical integration in the drawing office to determine deflection of shaft.

(ii) a checking of this deflection by mounting the built-up rotor on a surface table and measuring its deflection.

(iii) an oscilloscope test of the natural period of vibration of the shaft in air by hitting the centre with a light hammer and holding an oscilloscope pick-up at this point.

(iv) a similar check of the second critical speed at the $\frac{1}{4}$ and $\frac{3}{4}$ points along the length of the shaft.

(v) an oscilloscope test during the running up of the pump from rest to full speed, and during the running down after power is switched off.

All these tests give reasonably comparable results, the last test on the finished pump being, of course, the most important.

Main Bolts

The pump bolts are of 120 hectobar steel and the straining is based upon the temperature differentials observed in the two thermal shock tests. Since the greater part of the bolts is in open air, they are generally of the order of 55 deg C below the pump temperature which, whilst heating up and cooling down, serves to minimize the thermal stress.

These bolts are drilled for three purposes:

(i)   to permit electrical heating for assembly and for dismantling.

(ii) to provide accurate micrometer measurement of length and extension.

(iii) to help to maintain them at an intermediate temperature between hottest and coldest
      conditions of the pump.

The bolts are nominally tightened with a 60 cm spanner and a light hammer, then electrically heated; the nuts are rotated a specified number of degrees and the bolts are then allowed to cool. This gives the required amount of tightness which is checked by micrometer measurement through the bore. Unfastening of the bolts is carried out in a similar manner.

Alignment

The conical end-cover bearing housings are designed to have as much rigidity as possible in their 360° anchorage to the main casing, which is suspended by a cruciform system of keys permitting free expansion in all directions without loss of alignment with the motor. The pump is suspended at its centre line and a check is made on the high-temperature test to ensure that the temperature rise of the stool supporting the pump is comparable in magnitude to the temperature rise of the motor carcase. This maintains very close alignment, but in order to render the unit less sensitive to minor alignment changes of the order of 5 to 8 hundredths of a millimetre, a spacer coupling is used, which also has the advantage of assisting access to glands.

Glands

The feed pump glands for the first thermal shock test were subjected to 55 bar and 260°C, which is a very severe condition for the packing.

For later designs, the packing conditions were reduced to 2–4 bar, 125°C, by interposing the booster temperature and pressure across a bush between the high-pressure high-temperature inlet water and the packing.

A flow of bled water from the booster discharge leaks across this bush to the 2–4 bar region. A considerable improvement of packing life naturally resulted (Fig 28.1).

At the present time, parallel investigations are in hand on mechanical seals and on packingless glands.

Pressure Joints

The pressure joints, stage and main, are all metal, since soft joints are useless for thermal shock duties. Corrugated rings of stainless metal are inserted and crushed almost flat so that virtually no elasticity remains.

The rings provide a replaceable member thus giving a more satisfactory joint than scraped metal-to-metal faces or corrugated faces which would require re-machining.

Each joint has its own ring of bolts completely independent of other joints. This avoids the risk of leakage attendant upon the simultaneous holding by one external bolt ring of internal and external joints on separate components, each of which has its independent expansion and contraction during thermal transients.

Minimizing of High-Frequency Vibration and Noise

As each impeller blade approaches a diffuser, a small hydraulic shock occurs which is of relatively high frequency, for example, seven or eight times the running frequency, but of relatively small amplitude. In order to reduce this shock and the consequent alternating stress on the casing to a minimum, the number of impeller blades is prime to the number of diffusers and the impeller blades individually and as a group are of helical form, giving the whole rotor a skewed effect so as to reduce this high- frequency hydraulic shock and noise to a minimum.

General Chart of Mechanical Design

The total stress loading on pressure shell and bolts is compounded of the pressure stress and the thermal stress. It therefore follows that for a given design an increase of temperature involves a reduction in permissible pressure. Furthermore, as pressure and temperature increase, the materials change from cast iron, bronze fitted on cold water low-pressure pumps to high-tensile steel shell with stainless steel inner pump on the most severe conditions.

Fig 33.15 shows, in very general form, the typical pressure and temperature ranges for the various materials and constructional types. It will be seen that the thermal advantages of the barrel type permit it to deal with the higher temperature duties leaving the lower temperature duties to the through-bolt pumps.

Thermal Comparison of Pump Casing and Boiler Drum

Topley and Nicholson (Ref 44) describe how a 100 MW boiler drum is instrumented and guarded with the greatest care to avoid exceeding a temperature difference between inner

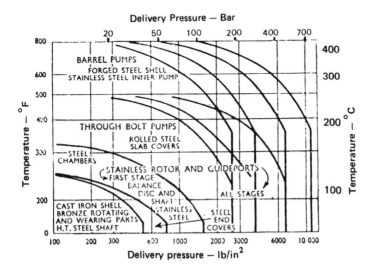

Figure 33.15 – Mechanical design chart.

and outer wall of 55 deg C when increasing load, and 40 deg C when reducing load.

In contrast to this, the pump casing must accept 155 degC temperature difference between walls and a rate of liquid temperature change of the order of 550 deg C per min. Table I gives a dimension, temperature and stress comparison of the boiler drum and pump casing.

TABLE 1 – DIMENSION, TEMPERATURE AND STRESS COMPARISON OF BOILER DRUM AND PUMP CASING

100 MW	Boiler Drum	Feed Pump Barrel
Pressure, bar	110	140–190
Temperature, °C	320	236–71
Bore, cm	170	81
Wall thickness, cm	14	9.1
Maximum temperature difference across walls, deg C	55	155
Maximum thermal shock, deg C min	27 estimated	550
Cooler wall hoop stress on heating up, hectobar	6	7
Cooler wall thermal stress on heating up, hectobar	7	19
Total stress, hectobar	13	17.5 Autofrettaged at yield stress for 186°C

It is reasonable to compare these two cylindrical members since the combinations of their respective pressures and temperatures involve similar wall proportions.

There is, however, a considerable difference in the rigidity requirements since the boiler, having no high-speed shaft, is not critical to a few hundredths of a millimetre distortion.

Comparison with Steam and Gas Turbines

Turbine design is based upon symmetry about the shaft axis in order to allow rapid heating. 125 MW turbines now in operation permit very rapid load acceptance indeed, but virtue of sectional stage and barrel design, comparable to the feed pump just described.

Feed Pump Drives

Normally, feed pumps are driven by two-pole a.c. motors and this has proved suitable for most duties.

Where variable speed is considered desirable, (that is to say, on duties of 100 MW and over in Britain, and 60 MW and over elsewhere), a hydraulic coupling between pump and motor is often provided. Alternatively, the motor may have a wound rotor with continuously rated slip-rings. Such reduction of speed by slip-ring or hydraulic coupling reduces to one half the power loss consequent on operation at other than full speed and, of course, almost eliminates the erosion of the boiler regulating valve.

For units below 500 MW it is generally practicable to run at higher speeds than 3000 rev/min for which a gear drive or turbine drive would be provided.

A reduction of efficiency of $1\frac{1}{2}$ to 2% occurs in the gearbox, but this may be offset by the overall saving of capital cost, particularly as the introduction of a gearbox to permit speeds higher than those corresponding to a two-pole motor would, at no extra cost, permit a ratio of gears suitable for a slower motor which, in certain circumstances, (eg with wound rotor) may be desirable.

Recent large turbo-generators have raised the consideration of a feed pump directly attached to the generator shaft, running at a generator speed or, by gear box, at a higher speed.

It would appear that the generator sizes are approaching the limit, at 50 cycles, that can be manufactured and transported in Britain, whilst the turbine limit has not yet been reached.

The use of electric feed pumps of, say, 20 MW would therefore subtract 20 MW from the total output of the generators, but a mechanical drive from the generator would only involve power in the steam turbine leaving the full generated output available for the grid. Here hydraulic couplings have the further advantage of isolating the pump during the rundown time of the turbine.

Alternatively the feed pump may be driven by a bled steam turbine which gives speed variation very roughly comparable to load demand without the need for a hydraulic coupling and unlimited choice of a suitable operating speed for the pump duty.

Variation of Pump Efficiency with Temperature of Feed Water and with Rotational Speed

Effect of Viscosity Change

Impeller disc friction is reduced with increase of temperature, resulting in higher efficiency. Leakage losses are increased with increase of temperature, resulting in lower efficiency. Hydraulic skin friction loss is reduced with increase of temperature resulting in higher efficiency. Impact and eddy losses, being dynamic, are assumed unaltered.

Make-up of Losses in Multistage Pumps (Ref 30 and Chapter 11)

Impact and eddy loss		8%	
Disc friction impellers and balance disc		3%	Typical case of
Neck leakage	2%		20% loss,
Balance leakage	2%	4%	80% efficiency
Hydraulic skin friction		5%	

Variation of Losses with Viscosity and Density

Disc Friction varies as

$$\frac{1}{\sqrt[5]{\text{Reynolds number}}} \quad \text{x} \quad \text{density}$$

Reynolds number varies as $\quad \dfrac{1}{\text{kinematic viscosity}}$

Hence disc friction varies as $\quad \sqrt[5]{\text{viscosity} \quad \text{x} \quad \text{density}}$

Leakage Loss

Quantity of leakage varies as $\quad \dfrac{1}{\sqrt{f}} \quad$ in formula $h \quad = \quad \dfrac{flv^2}{2gd}$

f varies as $\quad \dfrac{1}{\sqrt[4]{\text{Reynolds number}}}$

Hence leakage varies as

$$\sqrt[8]{\text{Reynolds number}} \quad \text{or as} \quad \frac{1}{\sqrt[8]{\text{kinematic viscosity}}}$$

when   h is head loss

f is friction coefficient

l is pipe length

v is velocity of flow

d is diameter of pipe

g is the acceleration due to gravity

Hydraulic skin friction loss varies as f which varies as

$$\sqrt[4]{\text{kinematic viscosity}}$$

Effect of Clearance Change

(i)   where all metals have like coefficients of expansion no permanent efficiency change occurs, but clearance must be increased by a margin to cover the most severe transient expected during heating up.

(ii)   where impeller material has greater coefficient than casing, an increase of efficiency occurs due to reduction of clearance and therefore of leakage at higher temperature.

Pump metals fall generally into two groups:

(i)   those having coefficients between 10 and 12 x 10^{-6}; for example, cast iron, mild steel (cast or fabricated); 13% chrome stainless steel (BS1630); 18% chrome stainless steel (En 57), FV 520 and 17.4.

(ii) those having coefficients between 15 and 17 x 10^{-6}; for example, bronze and austenitic stainless steel (18/8 chrome nickel) BS1631.

The difference is approximately 5 millionths per deg C. Where bronze or 18/8 impellers etc, are fitted to steel or cast iron casings, increase of efficiency occurs at higher temperature owing to reduction of clearance and in consequence reduction of leakage.

Normal diametral clearance is one thousandth of a cm per cm of diameter. If impeller neck increases for each cm of diameter by 5 millionths x T (where T is temperature difference), the clearance falls by a similar amount and the leakage will be reduced very approximately in proportion to the area change.

The efficiency temperature chart appears in Fig 33.16.

Effect of rotational speed on efficiency is deduced by change of flow quantity which serves as an artificial Reynolds number. A chart of this efficiency change appears in Fig 7.3.

Determination of Leak-off Quantity

The leak-off quantity is to be that amount which will prevent steaming of balance water

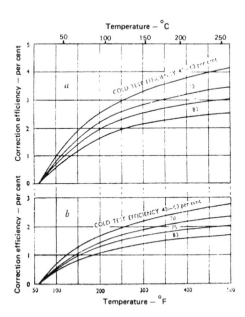

Figure 33.16 – Efficiency corrections due to viscosity and clearance on high temperature pumps. Head increases by 60% of efficiency increase
(a) Clearances change, for example, 18.8 or bronze impellers/cast iron or steel casing.
(b) No permanent clearance change, for example, 13 chrome impellers/cast iron or steel casing.

flow. The NPSH converted via steam tables is regarded as a safe temperature rise for the balance water.

Procedure for Approximate Determination of Temperature Changes

Add NPSH in bar to the saturation pressure corresponding to the pumped water temperature. This gives the absolute inlet branch pressure. Read from steam tables the saturation temperature corresponding to this pressure. This is then the temperature to which the balance water may be allowed to rise. The neglecting of radiation losses provides a safety margin.

From this balance water temperature rise determine the efficiency at leak-off quantity as follows, which from the pump characteristic gives the actual leak-off quantity. (Specific heat can be regarded as unity except above 150°C where a correction is made by noting the change in liquid enthalpy).

$$\text{e equals} \quad \text{Efficiency at leak-off quantity} \quad \text{equals} \quad \frac{\text{head in metres at leak-off quantity}}{427 \ \text{x} \ \text{specific heat} \ \text{x} \ \text{rise in deg C}}$$

Determination of Other Temperatures

$$\text{Temperature rise from inlet to discharge branch} \quad \text{equals} \quad \frac{\text{head in metres}}{427 \ \text{x} \ \text{specific heat}} \quad \frac{(1-e)}{(e)}$$

$$\text{Temperature rise balance only} \quad \text{equals} \quad \frac{\text{head in metres}}{427 \ \text{x} \ \text{specific heat}}$$

$$\text{Total temperature rise inlet to balance outlet} \quad \text{equals} \quad \frac{\text{head in metres}}{427 \ \text{x} \ \text{specific heat} \ \text{x} \ e}$$

Temperature Rise of Mixture

$$\text{of balance flow and main flow entering first impeller} \quad \text{equals} \quad \frac{\text{head in metres}}{427 \ \text{x} \ e \ \text{x specific heat}} \quad \text{x} \quad \frac{\text{Balance water flow}}{\text{Main flow}}$$

pH and Material Chart

Fig 33.17 Karassik shows material selection chart for various pH values and temperatures. The higher pH values show iron, steel or stainless steel but stainless steel may be required at even lower pH values due to high flow speeds as described previously.

Conclusion

Some 200 feed pumps of the thermally symmetrical barrel design are in operation on duties up to 500 MW and a number for larger duties are under construction.

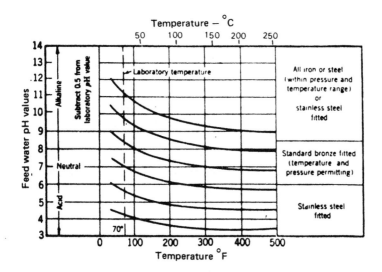

Figure 33.17 – Boiler feed pump materials based on pH values of feed water.

Part 2: The 1970's by Leith McColl & Ryall (Ref 46)

The need for totally reliable boiler feed pumps for large modern generating sets has led to the specification of pumps capable of surviving dry run and thermal shock conditions.

Special design features, introduced with the object of improving reliability with minimum effect on performance, are described. The development work leading to the adoption of these features and the reasons for the choice of materials and methods of manufacture are reviewed.

Details of construction to accommodate rapid dismantling and replacement of the internal pump assembly are given.

A comparison is made between glandless and geared suction booster pumps and the relative advantages are discussed.

Early 1970 developments are described which include the use of a suction inducer to improved the performance of the suction stage pump and the possibility of an integrated turbo feed pump running at high speed without a separate booster pump.

Design Philosophy

It is a prime requirement for modern large electrical generating sets that outage should be minimal. In common with other major ancillary equipment, boiler feed pumps must therefore be extremely reliable, and a new generation of boiler feed pumps has been developed with this specific objective kept clearly in view.

These pumps, illustrated in section in Fig 33.18, are designed so that contact cannot occur between rotating and stationary parts under any condition of operation, even if the pump should be run vapour-bound or dry. The pumps run at high speed and have only two stages with glands which have been designed to be of minimum axial length so that the shaft can be kept short and stiff with a deflection that is very small in relation to the relatively large internal running clearances provided.

In addition to the high pressure stages, there is a suction stage in a separate casing, and running at an appreciably slower speed, comprising the 'suction pump' which generates the NPSH required by the high pressure pump(Ref 47).

The device used for balancing axial hydraulic thrust in the high pressure pump is of a type which will not be damaged if the pump runs dry. The axial hydraulic thrust is opposed and largely compensated by a balance drum and the residual unbalanced thrust is carried by an external oil-lubricated tilting pad thrust bearing.

If a pump does have a breakdown, or has to be overhauled during a scheduled outage, it is a specified requirement to be able to get the pump back into service within the space of a single shift of eight hours. For this reason the working parts are designed as a removable cartridge assembly complete with bearings, and equipment is provided to facilitate mechanical handling for quick changeover.

A prototype boiler feed pump embodying these features was designed and constructed under a development contract from the Central Electricity Generating Board, Lond, and this pump, after undergoing successful shop tests, was installed at Ferrybridge Power Station for endurance trials under service conditions as the main feed pump for a 500 MW set.

Main Pump Construction

The working parts of the pump form a complete cartridge inside the pump casing shown

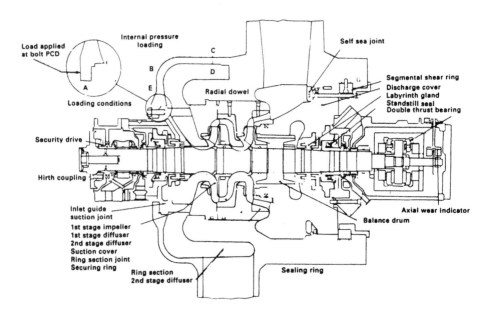

Figure 33.18 – Sectional arrangement of main pump.

in Fig 33.18. The pressure containment portion of the casing carries twin suction and discharge branches arranged to give maximum symmetry with minimum compensation for penetration, while uniformity of suction and discharge flow minimize unsymmetrical thermal distortion. Supporting feet are arranged on the centre line to give positive restraint to movement in both upward and downward directions and yet allow free thermal expansion. Lubricated sliding keys in cruciform configuration control direction of thermal movement and, together with the self lubricated supporting feet, maintain alignment with the driver even with pipe loadings of the order of 50 000 lb/ft (70 000 Newtons/metre) applied in each of three mutually perpendicular planes at the casing centroid.

The forged suction casing is welded to the main barrel forging, forming a suction annulus around the pump barrel guiding the water from milled ports in the barrel into the eye of the first stage impeller. The suction casing acts as a spring, which on final assembly is elastically deflected so that the stainless steel spiral wound suction cover join and the mild-steel copper-coated inner barrel joint are made simultaneously, a substantial preload being applied to both joints. During transient conditions of thermal shock the suction casing and inlet guide expand or contract more rapidly than the heavier section main barrel. The initial deflection of the suction casing compensates for differing reaction rates and maintains an adequate seating load on the joints to prevent leakage.

The suction casing is thus designed for three conditions of loading:

(i)   to give a substantial preload on the ring section joint.
(ii) for an internal hydraulic pressure loading of 500 lb/in^2 (35 bar).
(iii) to sustain a thermal shock, with the maximum temperature differential between suction casing and main barrel not exceeding 110 deg C.

Fig 33.18 gives details of the loading system used to establish the preload on the inner barrel joint. A stress analysis was carried out by finite element analysis techniques, using computer facilities. The structure was considered as excastre along C D (Fig 33.18), but free to move otherwise. This analysis indicated a maximum tensile radial stress occurring in the region of Point B, the maximum tangential stress being in the bore.

In the analysis for the effect of hydraulic pressure, it was assumed that the loading acted over the annular area from Point F to Point D. Stress analysis confirmed a general picture of low stress levels. The maximum stress in this case was again a radial stress occurring at Point B.

From thermal shock considerations, the design is such that differential expansion between the suction casing and main pump barrel can be catered for through the suction casing flexibility. It is necessary to ensure that preload will be maintained at the inner casing joint during a cold shock and that the suction casing will not be subjected to excessive stresses during a hot shock. Joint preload is unlikely to be lost during a cold shock since hydraulic pressure in the suction casing will assist in making the joint. During a hot thermal shock the deflection due to differential temperature is in addition to the suction casing initial deflection. This increases the levels of stresses under that loading, the positions of maximum stress remaining the same.

The final theoretical stress levels, obtained from a combination of the aforementioned three loading conditions, were confirmed by strain-gauge tests on a one-third size scale model. The basic design stress level of 20 000 lbf/in² (13.4 hectobar) was not exceeded.

Stress analysis using finite element techniques was also applied to other areas of the pump casing, including the region around the main high-pressure joint, which is of the self-sealing type described in Ref 35.

Since the pump is likely to be cycled about 15 000 times during its life the problem of low-cycle fatigue was considered. A quarter-size scale model casing was cycled to hydrostatic test pressure 25 000 times after which it was sectioned for examination and found to be free from cracks or deformation.

A precise procedure has been laid down for hydrostatic test with a view to limiting the number of applications of maximum hydrostatic pressure to which the casing is subjected.

During manufacture of these pumps, special care is taken to ensure that areas of stress concentration are free from surface defects.

Internal Cartridge Assembly

If a central generating station has, say, three 660 MW sets, internal cartridges for all three main feed pumps must be interchangeable. So also must the internal cartridges of the six starting-and-standby pumps.

For this reason, the cartridge is centred in the barrel casing at two locations only, at the suction end and at the discharge cover.

There are two internal ring sections, one of which is integral with the suction guide. They are centred to each other by radial dowels, near the periphery, which permit relative

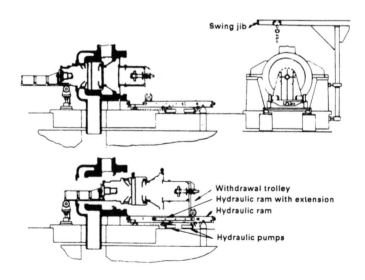

Figure 33 19 – Cartridge withdrawal equipment.

expansion while ensuring concentricity. The ring sections are held together by steel rings — which might be regarded as hollow radial dowels — on the circumference.

The second stage diffuser is located in the second ring section and is centred on the discharge cover where it is sealed by a special piston ring which allows relative movements due to expansion. The pressure differential to be sealed by the piston ring is modest, being equal to the pressure recovery across the second stage diffuser which amounts to about 280-300 lb/in² (19–21 bar).

Replacement of a pump cartridge assembly complete with integral bearings has been demonstrated as achievable within an eight hour shift, using specially designed hydraulically operated withdrawal gear as shown in Fig 33.19. The conception of the single shift changeover presupposes that a trained team of five men is available with the necessary equipment gathered together. Once it has been decided to replace a damaged cartridge and all permits to work have been cleared, work immediately starts on the removal of specially designed quick release cladding and assembly of withdrawal cradle. during the initial thirty minute period the pump is drained of hot water and force-cooled by flushing through with cold condensate until manual handling is possible using asbestos gloves. The pump is provided with swing-jibs and electric hoists of one ton capacity to assist in lifting pipework, covers and tools, etc. The spare cartridge is supplied in a cocooned and sealed condition to prevent ingress of moisture or foreign matter and is located on a combined maintenance and storage cradle for easy handling.

Pump Rotating Element

The pump shaft is made from forged 13% chrome stainless steel, heat treated in the vertical position to give an ultimate tensile strength of 45 ton f/in² (67 hectobar), and minimum impact resistance of 25 ft lb Izod (3.4 metre kg). It is ground all over to a surface finish of 32 micro inches CLA (0.8 micron) excepting at the bearing journal positions where the surface is chromium plated, as tests on plated and unplated journals have shown that chrome plating makes the journal less prone to damage and gives longer life in service (Reference 48). To minimize stress concentration effects, there is a minimum radius of 3/32 inches (3 mm) at any change of section. Torque is transmitted to the shaft through Hirth type radial serrations as shown in Fig 33.20 (a). This type of coupling has given very satisfactory service for many years on small turbo-feed pumps running at speeds up to 10 000 rev/min and is easily removable with minimum overhung weight. Satisfactory static tests were carried out on a full size coupling at 150% full load torque for a 660 MW main feed pump. The coupling was then uprated to transmit 210% full load torque in view of the live steam power capabilities of some of the driving turbines. One radial tooth is left uncut ensuring that the coupling can only be assembled in one position thus maintaining dynamic balance on re-assembly after removal. The central retaining bolt is tightened by means of an electric bolt heater. To give maximum resistance to fatigue, the bolt is made from high tensile stainless steel, waisted and ground all over, including the threads.

Should the bolt fail for any reason, torque can be transmitted by a screwed security sleeve, the screw threads being opposite to direction of rotation. An alternative drive

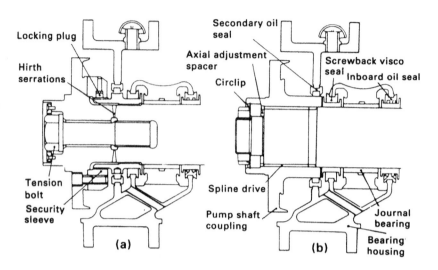

Figure 33.20 – Alternative designs of driving couplings.

which is being adopted for later machines is shown in Fig 33.20 (b), in which torque is transmitted by splines.

The pump impellers and gland rotors are all driven by splines because these give maximum symmetry of flexural stiffness and also ensure consistency of assembly and dynamic balance. These are to BS3550/1963, 10/20 diametral pitch with 20° involute angle and major diameter fit, this being the simplest spline to manufacture and most suitable where maximum centring accuracy is required. One tooth on each spline is chamfered off at the end to act as a reference and ensure that the rotors are fitted in the same position on every assembly. The splines on the shaft are flash chrome plated 0.0001 inch (0.0025 mm) thick to obviate any tendency to galling during assembly and dismantling. Radial integrity of the impellers and gland rotors and prevention of leakage along the shaft is achieved by making a portion of the impeller and gland rotor hubs an interference fit on the shaft. The interference region is removed as far as possible from the shroud, and a thermal barrier introduced in the form of an annular recess in the bore. A special gas ring has been developed to allow particularly rapid heat input to the interference region of the hub, so allowing the rotor component to be removed before much heat can soak past the thermal barrier into the impeller and gland rotor bodies or into the shaft. The balance drum is the only major component on the shaft which is not driven by splines. The drum is heated in an oil bath and shrunk on during assembly and is removed when necessary by high pressure oil injection.

Oil Seals and Coupling Guards

At high rotational speeds, it has been found that the air pumping effects caused by couplings, etc, adjacent to the inboard journal bearing can aggravated sealing problems. A double seal arrangement as shown in Fig 33.20 (a), has been found to be effective. Here the area between the two seals is kept at atmospheric pressure with a large capacity vent.

Any oil which gains access to this locality is drained back to the oil sump. The primary oil seal is of the 'visco seal' screw type which pumps any oil leakage back into the oil sump by viscous pumping action.

Because of high rotational speeds, disc friction in air in excess of 5 hp (3 kW) at the flexible coupling can be anticipated. To prevent overheating, the coupling guard cannot be totally enclosed and special attention has to be paid to the air circulation arrangement to avoid creating a local depression pressure which can induce oil leaks at the oil seals.

Rotor Dynamics

In the past few years much progress has been made in the calculation of critical speed, particularly with the ready availability of computer facilities. The critical speed of a shaft, especially of a shaft running in liquid, is however, a very complex subject.

For a pump of the class under discussion, operation at or near a critical speed must be avoided when the shaft is running in air or in vapour, as well as when it is running in liquid. It was a design aim that the first transverse critical speed in water should exceed, by a margin of at least 20%, the maximum running speed of the pump, and that the first and second transverse critical speeds in air should be well clear on either side of the minimum and maximum running speeds.

Existing computer programmes take into account, not only the many changes of section along the length of a shaft, but also the lateral stiffness of journal bearings and angular stiffness of thrust bearings. Account can also be taken of the stiffening effect of leakage flow through running clearances at the impeller eye and balance drum, but little account has been taken so far of the damping effect of the liquid in these clearances.

Early in the 1970s the authors' company had already made a start on a simplified analysis to determine the complete response characteristics of a shaft running under the particular conditions existing in large high speed pumps. It has been shown that good qualitative agreement exists between theory and experiment and the work has now been extended to investigate the aspects relating to different bearing types and the influence on critical speed of leakage flow through clearances. A general purpose computer programme has been developed to assess shaft response and stability data for use early in the design stage.

Hydraulic Design

Advanced class boiler feed pumps are designed as two stage units, each stage developing 3750 ft (1150 m) head at duty point and 4500 ft (1380 m) head at no-load — roughly double that currently in general use for boiler feed pump service. The very high head generated per stage involves very high fluid velocities in the impellers and diffusers and precautions must be taken to avoid erosion at the impeller tip and cavitation-erosion at the first stage impeller inlet.

High speed erosion tests carried out on rigs utilizing aluminium and stainless steel impellers and diffusers showed that heads of 4000 ft (1220 m)/stage could be achieved without damage, provided that a substantial radial gap existed between the impeller and diffuser and that sufficient NPSH was available to suppress cavitation in the eye of the first stage impeller. When the radial gap is small, the impeller periphery can be damaged by fan tail erosion, which is believed to be caused by cavitation due to localized high velocity and is largely independent of the NPSH at pump inlet. Increased radial gap reduces the local velocity and erosion does not occur.

The axial entry diffuser, with is inherently large radial clearance, has therefore a significant advantage where high stage heads are to be developed. Axial entry diffusers also enable the pump casing diameter to be reduced by about 25% when compared with radial diffuser construction, which corresponds to a weight reduction of about 40%, and there are secondary benefits in improved behaviour under thermal shock conditions by the use of thinner wall sections.

Reduction in general noise level has been a feature of pumps employing axial entry diffusers.

Normally, the shut-valve power with radial diffusers is about 60% of full load power, but with axial entry diffusers the power absorbed at zero flow is about 40% of full load power which permits some reduction in leak-off quantity.

Impellers and diffusers are cast in 17% chrome, 4% nickel precipitation hardening stainless steel which has very good mechanical properties and excellent corrosion/erosion resistance.

To obtain maximum mechanical integrity and ensure repeatability of hydraulic performance, the impellers are precision castings produced by the lost wax process employing a metal die. The blade profile is defined mathematically in tabular form and thus can be translated directly into metal cutting action without intermediate marking off which might introduce inaccuracy. It was the original intention to manufacture the diffusers as

Figure 33.21 – Impeller inspection machine.

lost wax precision castings but because of size and weight limitations, these were produced as semi-precision castings by the Avnet-Shaw process. It is of interest to record that the final impeller and diffuser castings were found to be comparable in passage accuracy, although the surface finish of the impellers at 80/120 micro inches (2–3 microns CLA) was slightly better than that of the diffusers at 100/125 micro inches (2.5–3.2 microns CLA).

An investigation was carried out subsequently to compare the physical properties of impellers cast under vacuum against those cast in air. The significant difference was that the Izod value of the vacuum poured impeller castings was 64 ft lb (8.8 metre kg), compared with 30 ft lb (4.1 metre kg) for air poured castings. A commercial sand cast impeller in the same material has an Izod value of about 15 ff lb (2 metre kg). Further development work is proceeding with a view to establishing improved methods of producing impeller castings which enable advantage to be taken of the fully developed properties of the material without introducing undue complication in casting procedures.

A special-purpose impeller inspection machine shown in Fig 33.21 was developed in conjunction with PERA.* This is used to check that impeller passage dimensions, blade thickness and profile accuracy lie within close tolerances.

Hydraulic Thrust

For reasons of hydraulic performance, single inlet shrouded impellers were selected, and as a direct consequence an unbalanced cumulative axial thrust of the order of 100 tonnes is present acting towards the suction end of the pump. complete and automatic compensation of this inherent axial thrust can be achieved by means of a balance disc mounted on the shaft which controls leakage through two orifices in series, one of variable area depending on the axial position of the shaft, to establish an intermediate pressure over its own area which exactly counter-balances the axial hydraulic thrust generated by the impellers. This well proven device unfortunately becomes completely inoperative if the pump is run dry or in the vapour locked condition.

Partial hydraulic balance can be achieved by an opposed impeller configuration or by means of a symmetrical balance face on the rear shroud of each individual impeller, or alternatively, as in the present case, by means of a separate balance drum. With each of these systems an external oil lubricated thrust bearing is provided to absorb any residual unbalanced thrust which may be present under normal operating conditions, and, if desired, the thrust bearing can be sized to cater for the maximum thrust which might occur on dry run or with disturbed suction conditions.

The balance drum is sized to compensate for about 85% of the total hydraulic thrust from the impellers under normal conditions, leaving about 15% to be carried by the external oil lubricated tilting pad thrust bearing, and it is, of course, essential to have reasonably accurate information on the magnitude of the hydraulic thrusts involved.

Investigations were carried out experimentally on the single stage pump rig shown in Fig 33.22. In this overhung pump the impeller attitude on the shaft was reversed, so that direct measurement of radial flow outward or inward along the back shroud could be undertaken. The magnitude and direction of this flow was controlled by varying the pressure in the chamber at the non-drive end of the shaft, and the effect on the pressure distribution across the back shroud was assessed from a series of static pressure tapping

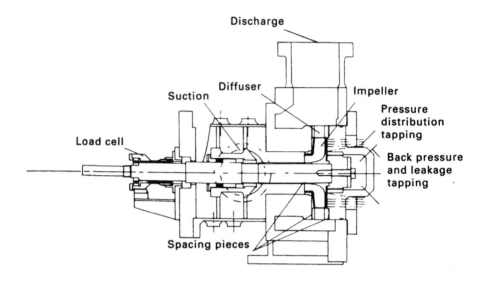

Figure 33.22 – Single stage pump test rig.

points drilled in the pump casing. The total thrust generated was measured by a calibrated strain-gauge load cell.

Tests were conducted with differing axial clearances between impeller and casing and with a selected number of sealing ring running clearances. It was concluded that the most important single factor affecting axial thrust is the impeller tip pressure acting between the casing and the front and back shrouds. The effect of variations in radial flow on the shroud pressure distribution was also of some significance, and its effect must be taken into account when assessing the hydraulic thrust.

The thrust bearing housing of the prototype pump is designed to accept a load cell so that the thrust bearing load can be measured on test over the whole of the operating range. The balance drum diameter is selected to ensure that the shaft is always in tension and that the residual thrust is unidirectional to avoid any possibility of axial shuttling of the shaft. In the prototype pump the thrust bearing was designed to accept a normal continuous thrust of 10 tons (100 kilo-newtons) in either direction, corresponding to a specific loading of 500 lbf/in² (35 bar) on the pads and is capable of absorbing a transient thrust of 25 tons (250 kilo-newtons) in either direction for short periods. Thrust load data measured in the course of shop trials is plotted in Fig 33.23.

Impeller Seal and Balance Drum geometry

The incorporation of a balance drum, subject to a pressure drop of up to 2500 lbf/in² (172 bar) and having a diameter of approximately 250 mm, has highlighted the need for a fuller understanding of the factors controlling leakage flow, and of developing seal labyrinth geometries to achieve minimum leakage.

Speed has a significant effect on leakage, and this effect was examined theoretically for

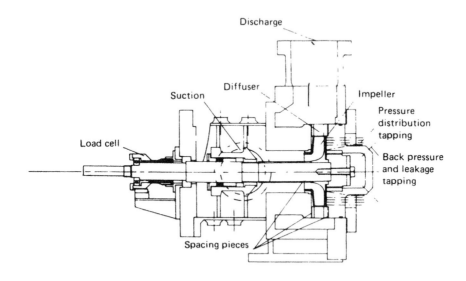

Figure 33.23 – Axial load on thrust bearing

plain drums; ie those with no serrations. Leakage rates over a range of speeds, clearances and pressure drops were measured experimentally, initially for plain drums, and the comparison, giving good agreement with theory, is shown in Fig 33.24 (a), illustrating the marked effect of speed in reducing leakage. In the case of high speed advanced class feed pumps, the leakage rate through the rotating drum clearance is 50% lower than it would be if the drum were stationary. This is regarded as a further small but significant advantage to be gained from the use of high rotational speeds.

Alternative types of drum and seal surface geometry, including opposed screws and circumferential serrations were tested on a high speed rig, and a typical comparison between the leakage flow through a plain annulus and that through an opposed screw configuration is shown in Fig 33.24 (b). A plain drum, in association with circumferential serrations on the static bush, was adopted for the production pump as a compromise between manufacturing simplicity and optimum hydraulic performance.

External Gland Design

To obtain acceptable periods of maintenance free operation, it has been customary to transform the intractable high temperature, high pressure shaft sealing problem into a relatively simple low temperature, low pressure gland by means of high pressure unloading connections and cold condensate injection systems. This has normally been accomplished at the expense of axial length, and it is true that in many conventional

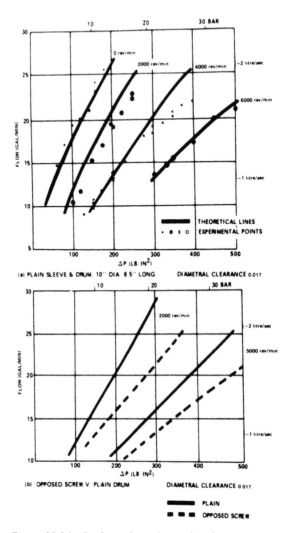

Figure 33.24 – Leakage through annular clearances.

designs of boiler feed pump the length of the two external glands dominates the distance between bearing centres and that the effective pumping section is comparatively short. Glands for modern large boiler feed pumps are predominantly of the fine clearance floating ring type which depend for their satisfactory operation on the presence of a hydrodynamic water film between the stationary ring and the rotating shaft. These floating ring glands also have the advantage of being able to accommodate shaft movements during run up to speed, and in many cases the gland housing itself is angled during manufacture to match the slope of the very flexible pump shaft of those earlier pumps, where the shaft was designed to run between its first and second critical speeds which involved a maximum deflection of between 0.02–0.04 inches (0.5–1 mm). However, gland designs of this type were not acceptable in view of the requirement to survive dry running, since continuity of the gland sealing water supply could not be guaranteed.

Instead, fixed labyrinth bushes having a minimum diametral running clearance of 0.02 inches (0.5 mm) were adopted with multiple flow restrictions stacked radially to minimize axial length. Bearing in mind that the maximum central deflection of the very stiff pump shaft is only 0.003 inches (0.075 mm), there is no possibility of radial contact occurring at the gland positions.

These fixed labyrinth bushes are supplied with cold condensate injection under normal conditions of operation, and centrifugally operated standstill seals of the spring loaded segmental type are incorporated to minimize gland leakage when on standby duty.

Under dry run conditions, the fixed bushes will operate satisfactorily without an external supply of sealing water, and no significant erosive damage will occur when accepting flashing flow for short periods of time.

Compatibility of Materials

The proper selection of materials capable of surviving high speed rubbing contact without serious damage was considered to be of prime importance and a materials' development programme was initiated in which different material combinations were rubbed together in a water environment with varying contact pressures up to 250 lbf/in² (10 bar) and at surface speeds up to 200 ft/sec (60 m.s). The principal conclusion drawn from these tests was that differential hardness between the two materials in contact was not the prime factor, but that absolute hardness of the stationary material was the most important feature. Plated and coated materials, non-ferrous alloys and austenitic cast irons were found to be totally unsatisfactory where a combination of high peripheral speed, in association with high specific loads, could be encountered. Fully hardened stainless steels of the 17% chrome 4% nickel (precipitation hardening) and 13% chrome 1% nickel types were found to be acceptably compatible under these severe conditions, and these combinations were adopted into the final design. A detailed report on these tests is given in Ref 48.

Test Programme

The prototype pump was designed to meet the requirements of the main boiler feed pump specification for a 660 MW sub-critical set, but since endurance testing would take place in a 500 MW installation one of the two impellers was reduced in diameter by machining the blades between the shrouds. Thus the pump as a whole would match the system resistance line of a 500 MW boiler at its designed running speed of 7500 rev/min and ensure that one stage was generating its full designed head of 3 730 ft (1145 m). A spare set of impellers was provided so that at a later date the full sized impeller could be cut (or renewed if necessary) and its neighbour replaced and tested at full diameter, so providing endurance test data for both stages.

In the Works a full sized test rig was available but driver power limitations dictated that shop trials should be conducted at 3600 rev/min — just below 50% of the designed running speed of 7500 rev/min. After completion of satisfactory performance tests with hot water at 186°C under simulated site conditions, during which the predicted efficiency was achieved at duty point, a cavitation test was conducted with water at 100°C boiling off to atmosphere. The onset of cavitation was observed with a suction head of 29 ft (9 m) corresponding to an inlet shape number of 8400 (5600) in litres/sec and metres). Fully developed cavitation occurred when the suction head was reduced to 21 ft (6.4m) corresponding to an inlet shape number of 10 700 (7100). On completion of the cavitation test the pump continued to run under fully cavitating conditions for a further period of ten minutes, during which time the suction isolating valve was progressively closed until completely shut. The power absorbed by the pump fell to a very low value (some 10% of the normal value) and the pump remained in service for a further period of ten minutes. Both suction and discharge pressure gauges indicated zero readings, and the pump ran smoothly, emitting a mild and not unpleasant siren-like note. The pump was then stopped, still in the 'dry run' condition. After re-opening the suction isolating valve the pump was restarted and run for ten minutes at rated flow. Measurement of performance indicated no deterioration and subsequent examination of internal components revealed no damage.

Following installation as the turbine driven main boiler feed pump in a 500 MW set at Ferrybridge Power Station, similar units have run for several years at loads up to 660 MW in an entirely satisfactory manner.

Choice of Suction Pump

The stiff shaft, two stage concept defined for the main high pressure pump, coupled with design constraints on hydraulic proportions and physical dimensions, implies high rotational speed which necessitates a nett positive suction head well in excess of that available from the high level deaerator storage vessel. It is therefore necessary to provide a suction booster pump running at a lower speed driven by an independent driver, or through gears from the main pump driver. For accessibility reasons, it was considered undesirable to have a reduction gear unit driven from an extension of the main pump shaft, and at that time it was also felt to be undesirable to take a secondary power offtake from the non-drive end of the boiler feed pump turbine. Bearing in mind that the 'dry run' capability also applies to the suction pump, the ready availability of suitable glandless electric motor/pump designs led to their adoption for suction booster duty for which they were considered to be eminently suitable because of the absence of an external gland. Three 50% duty glandless pumps were selected, two working and one standby. Each pump was designed to be able to run out in load to match turbine MCR, whilst the standby suction pump starts and runs up to full speed in the event of failure for any reason of one of the working pumps.

The sectional arrangement of a 2 pole 3.3 kV 625 kW glandless suction pump forming part of a 660 MW 3 pump installation is illustrated in Fig 33.25. The basic design of these pumps has been fully described in Ref 47, although in the installation currently under consideration, it was possible to select a higher running speed of 2950 rev/min by incorporating a suction inducer to improve suction performance which resulted in appreciable cost savings.

Subsequently, it has become practicable to obtain a secondary power off-take from the high speed shaft at the non-drive end of the boiler feed pump turbine, and a glanded suction pump driven through gearing can be used without loss of accessibility for quick changeover of the main pump cartridge. The pump illustrated in Fig 33.26 is of the single

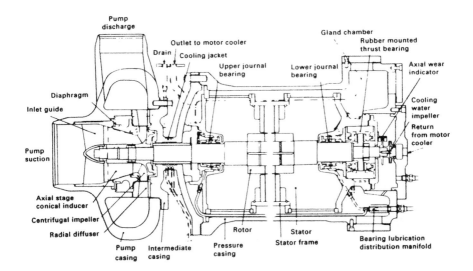

Figure 33.25 – Glandless suction booster pump.

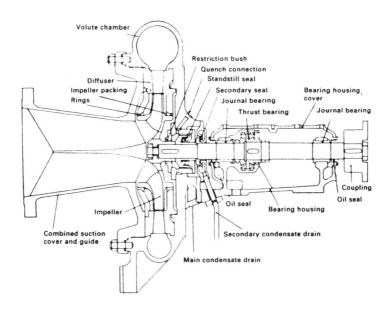

Figure 33.26 – Gear driven glanded suction booster pump.

stage end suction type, requiring only one external gland, similar in design to that employed on the main pump.

It is important to recognize that the gear driven suction pump is a variable speed unit, (as is the main pump), whereas the independent suction pumps are constant speed units. It is thus vital to ensure that the head developed by the gear driven glanded suction pump, when running at reduced speed, is adequate to meet the requirements of the main pump.

The design criterion adopted in the present case where independent suction pumps are employed, was that the pressure supplied by the main pump inlet branch by the suction pumps should be sufficient to provide at least twice the NPSH at which the onset of cavitation occurs at the main pump.

Future Developments

The design concepts described here have enabled users to reap the benefit of the improved reliability of boiler feed pumps by reducing the standby capacity which a prudent operator provides to minimize the possibility of main plant outage due to non-availability of an essential auxiliary. Thus for the 660 MW sets for CEGB there is only one 50% duty starting and standby pump in addition to the main turbo feed pump, and on certain projects, only one 15% duty starting pump is employed with no standby capacity.

It is equally natural that pump specialists should plan for further improvements in boiler feed pump design, and one promising line of investigation relates to a turbine driven pump in which the boiler feed pump and its driving turbine are integrated as a single unit. In the normal arrangement of turbine driven pump, both the pump and the turbine are provided with external oil lubricated thrust bearings. By solidly coupling the pump and turbine

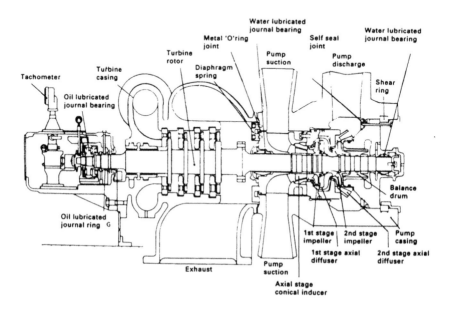

Figure 33.27 – Integrated turbo feed pump.

shafts, the pump thrust bearing can be eliminated, and with the satisfactory development of water lubricated journal bearings, it is possible to eliminate external glands on the pump. Thus, the only oil lubricated bearings in the composite machine would be the common thrust bearing, and the journal bearing at the non-drive end of the turbine.

The development of suction inducers to improve suction performance should make it possible to avoid the need for a separate suction booster pump whilst still permitting a high rotational speed to be maintained.

The resulting integrated turbo feed pump is illustrated in Fig 33.27 which embodies the essential design features for reliability and quick changeover of the internal cartridge. The dry running capability requires continuity of the water supply to the water lubricated bearings, but this can be provided for a limited but acceptable period of time by storage.

References

8. STEPANOFF, A.J; 1948 'Centrifugal and Axial Flow Pumps' (Wiley, New York; Chapman & Hall, London)

30. WORSTER, R.C., 'Flow in the volute of a Centrifugal Pump and Radial Forces on the Impeller, BHRA RR 543, 1956.

35. ANDERSON, H.H., 'Self Seal Joint for Very High Pressures' ASME 67 WA, PVP 5.

43. ARKLESS, G.F. 'Development of High Pressure Boiler Feed Pumps in Britain During the last Decade' Proc. IME 1966-67, Vol 181, Part 30.

44. TOPLEY, H. and NICHOLSON, G. 1958 Elect. Rev., London, Vol 163, page 63, 'Unit Generating Plant'.

45. ANDERSON, H.H., 1961 'Design of Modern Boiler Feed Pumps', Proc IME Vol 175, No 12.

46. LEITH, T.O., McCOLL, J.R., RYALL, M.L. Advanced Class Boiler Feed Pumps for 660 MW Generators. IME September 1970. Symposium on Advanced Class Boiler Feed Pumps.

47. LEITH, T.O., 'Glandless Pump-Motor Units for Feed Heating Plant in Central Power Stations'. I Mech E Technical Conference on Glandless Pumps for Power Plant - April, 1970.

48. McCOLL, J.R. 'Some Aspects of Boiler Feed Pump Design and Application' and Bocking, et al 'High Speed Centrifugal Pumps; Investigation of Some Material Problems'. BPMA Technical Conference, December, 1968 and 1969.

LARGE SINGLE ENTRY PUMPS

Review of Developments

The most popular centrifugal pump for low and medium heads is the double entry split casing type described in Chapter 27.

This type of pump has enjoyed favour in Britain and the United States for about fifty years. During this period, however, heads, powers, and speeds have increased rapidly and the need has arisen for greater rigidity to meet the upper range of these increased duties without excessive wear or maintenance. Research into the existence of radial forces acting on the impeller at duties other than the best efficiency point has demonstrated the need for heavier shafts, particularly on high heads.

Where shaft rigidity is inadequate, neckring contacts may occur, thus wasting power and shortening life. In addition, shaft deflection may cause the opening of shaft sleeve-joints, permitting air drawing and/or access of liquid to the shaft, and, in extreme cases, causing shaft breakage by corrosion-fatigue. The problem of withstanding a pressure difference, where the shaft enters the casing, is rendered more difficult if shaft deflection occurs.

The double entry feature involves a split and several bends in the flow of liquid from the inlet branch to the impeller. This causes friction loss and poor distribution of flow in the impeller inlet, which prejudices performance. The split of the casing is a further cause of hydraulic loss since exact matching of cast surfaces is difficult.

The double entry impeller with two narrow passages is more difficult to cast and has a greater wetted surface than a single entry impeller for the same duty. These facts again prejudice efficiency. The horizontally split casing presents a difficult shape for a pressure vessel.

It should be emphasized that the faults mentioned above are mainly associated with the upper ranges of head and with fairly severe duties, and that the split casing type of pump, within its field, gives excellent service. Nevertheless, if a heavier and more efficient design can be developed there is every incentive to adopt it, particularly on the higher heads and for the more difficult duties.

Pumps Designed for Rigidity

When an attempt is made to increase the diameter of the shaft on the split casing pump it is found that it can be done only at the expense of water passages, thus prejudicing hydraulic performance. By making the impeller of the single entry type, however, the flow of water can enter at one side and the shaft can enter the impeller at the other, with no restriction to diameter.

At the same time the water flow to and through the impeller is simplified, thus giving an improvement in performance. The large increase of shaft diameter that can therefore take place without prejudice to water-flow passages permits a very generous margin of safety in respect of rigidity. The actual increase in rigidity is about tenfold and is proved by running the pump dry.

A single entry impeller can readily be made 'blind' and the hub of the impeller can extend out to the atmosphere, thus preventing any air-drawing. The question of the opening of the sleeve joint under radial loading does not therefore arise and, as shown above, shaft deflection is reduced to a negligible figure.

Axial balance of the pump impeller can be readily obtained on the single entry design. The annulus between the eye diameter and the shaft diameter is balanced by providing a back neckring and piping its leakage to the inlet branch. The area of the shaft and sleeve diameter is balanced when the inlet pressure equals atmospheric pressure, but with a suction lift or a positive pressure, an end thrust occurs which, however, is relatively small in comparison with the capacity of bearing that would be associated with the diameter of shaft employed.

Water Turbines

A parallel development has already occurred in water turbines which are now entirely of the single exit type.

Unit Construction; Single Entry Pumps

The change from split casing double entry pump to end-cover type of single entry pump facilitates the construction of a combined pump and motor unit, since the motor shaft is ideally suited to an overhung impeller. This development, is, however, so intimate a union of pump and motor that very close collaboration between electrical, hydraulic, and mechanical designers is essential.

The firm with which the author is associated has developed single entry combined pump and motor units for large powers. Since pumps and motors are made in the same factory it is an advantage to make a combined machine, avoiding shaft and bearings in the pump proper and saving power, cost, and space. The impeller is, therefore, fitted directly to the motor shaft which, together with its bearings, is designed to carry the additional loads involved. A saving in efficiency accrues from the elimination of pump bearings and couplings and from the reduction of shaft ship at the gland.

In the vertical arrangement, the pump casing becomes the motor stool, giving a very compact unit occupying a minimum of space. Examination of the pump is carried out by lifting the motor and impeller, thus avoiding the disturbance of pipe joints.

A section of a combined unit is shown in Fig 34.1. Attention is drawn to the particularly

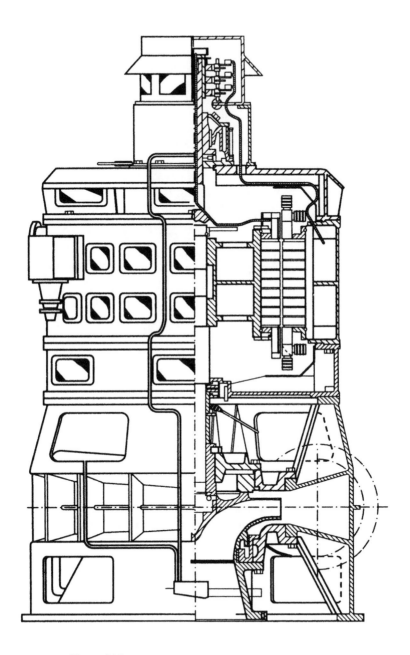

Figure 34.1
Section of combined motor pump unit for vertical operation.
Fabricated construction is employed throughout, all plate
in contact with water being of clad stainless steel.

Figure 34.2 – Vertical cooling water pumps

heavy shaft, the single entry features with the blind impeller, and the manner in which the opening in the casing for the entry of the impeller is closed by an end cover carrying the stuffing box. When opening up the pump for examination the stuffing-box bolts and the holding-down bolts for the motor are removed, thus permitting the motor and impeller to be lifted as a unit.

This combined unit can also be arranged with a horizontal shaft, but the advantages of accessibility and space economy naturally decrease somewhat with horizontal application.

A considerable number of these units are in operation in waterworks, chemical plant, oil refineries, etc. See also Ref 17.

Construction and Application Details

The pump casings are fabricated from mild steel plate, which involves some modification of hydraulic practice. Careful collaboration, however, between designers and fabrication experts has produced a casing design simple enough or fabrication. Here any loss due to the unorthodox shape is offset by the smoothness of plate as opposed to cast surfaces.

The fabrication construction avoids pattern costs and permits greater freedom of design, so that units can be tailor made for each installation instead of being fitted to the nearest existing pattern.

On one waterworks application the pumps are controlled by a rotary plug valve on the

Figure 34.3 – Eight combined diesel engine pumps in a British oil refinery (Harland). Units were tested at makers works. Care was taken to ensure that the impeller was positively locked against cyclic variation of engine, and that the imeller blades did not cause harmonic vibration with the engine. The imeller has damping effect on torsional oscillation.

discharge branch. These valves replace sluice and non-return valve and, since they provide a smooth bore, the friction loss through them is negligible. They are designed for easy opening under full pressure and have joint faces of monel metal as protection from scour. The valves open and close automatically to avoid water hammer or collapse of the pipe from sudden power failure. They are operated by oil servomotor, water servomotor, hand or electric drive. Basically, this valve comprises a rotary plug which is lifted from its conical seating before turning.

For general industrial application it is often practicable to operate pump units in the open air without a pump house. This provides a very simple arrangement with open culvert and ejector priming.

Stainless Casings

For the handling of aggressive waters the pump casings are sometimes fabricated from stainless-steel-clad plate. All surfaces exposed to flow are, in view of the nature of the water, made from austenitic stainless steel. This involves a further radical departure from orthodox pump construction in that all spigots, joints, tappings, etc must have full stainless continuity. After fabrication and before stress relieving, the casing interior is sponged with dilute acid to prove this continuity, afterwards being scrubbed free of acid.

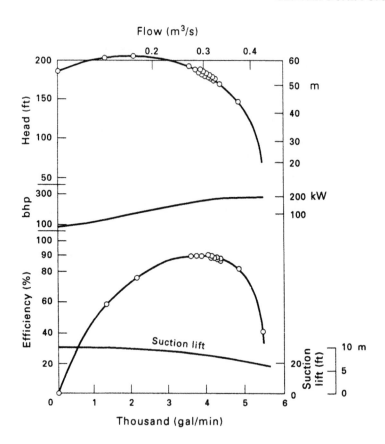

Figure 34.4
Test characteristics of 5 million gallon waterworks pump.

Diesel Drive

The unit construction principle has also been applied to a group of eight horizontal cooling-water pumps (Fig 34.3). In this case, however, the impeller instead of being fitted to the shaft of a motor, is fitted to the stub shaft of a diesel engine. The diameter of the shaft is relatively large in comparison with that of the impeller because of the necessity for rigidity and the need to withstand torsional oscillation.

Characteristics

Typical single entry characteristics are shown in Figs 34.4 and 34.5. The wire-to-water efficiencies of these units were confirmed on site tests and bonus money was paid for the efficiency exceeding the no tolerance guarantee.

(a) Characteristics

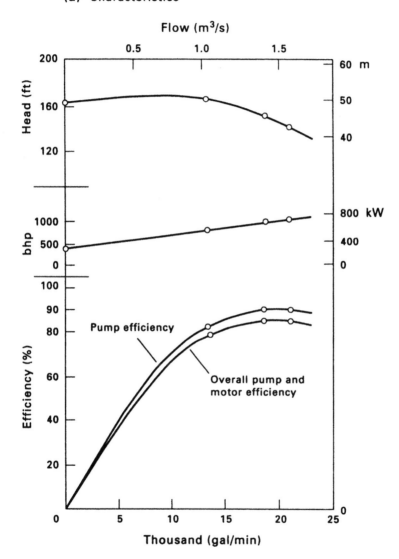

Figure 34.5
Test on 600/750 mm refinery cooling-water pumps and 800 kW motors.
(Pump Test Sheet on next page)

PUMP TEST SHEET

To the order of

						Serial No ZX2236	W No 60577/1
18500 gal/min	145 ft/head	495 rev/min				TYPE 55VA 24/30	Ph 3 Cycles 50

MOTOR A.C./D.C. Frame or make AV 35   Serial No. ZX2239   W No. 60577/308   HP 1050   Volts 3,300   FL Amps/ 65   rev/min 495   (Thrust brg loss - HP)

GAL WATER	QUANTITY			PUMP HEAD							WATER HP	rev/min or slip	cycles	MOTOR								Pump BHP	PUMP EFFY	Comb Effy
	Reading	gal/min	Percentage of normal gal	Dia charge ft	Suction ft	Distance between guages	Vel head corr	TOTAL HEAD	Percentage of normal head					Volts(K)	Amps(K)	KW1(K)	KW2(K)	Input E HP	Effy	Motor BHP				
36.8	21.6	18600		13.8	8.1	3.0	2.4	151.5		863		91		3196	132.4 / 154	0.410	0.214	1000	95.4		964	89.4	85.3
37.7	23.3	20860		12.8	9.0	3.0	3.2	143.2		908		103	50.5	3150	152 / 164	0.426	0.227	1050	95.4		1002	90.6	86.4
37.7	24.18	12200		12.0	9.8	3.0	3.6	136.4		917		166	50.5	3150	167 / 169	0.431	0.229	1062	95.4		1014	90.6	86.4
57.7	17.20	13080		16.9	6.7	3.0	12.4	168.94		669		614	50.5	3210	133 / 135	0.351	0.180	853	95.3		813	82.3	78.4
68.6	12.84	12360		1.61	2.8	3.0	0.51	167.11		423		613	51.0	3270	108 / 108	0.290	0.144	497	96.0		662	63.9	60.8
71.5	-	94		1.61	0	3.0	-	164.00		-		36	51.0	3280	76 / 72	0.194	0.064	416	93.9		390	-	-

K = 1200

Water measured by 72"WEIR	Cycles 50	Supply 2000 KVB TRANS	Balance water 36.8	gal/min at duty		Tested by W BOGGRA A CC
Time started up	Time shut down		Test No. OFFICIAL TEST			Date 14 - 12 - 50
Remarks:	PUMP MECHANICALLY O.K.			W.O. No 60577-1		Approved by

Figure 34.6

Figure 34.7
Single entry single stage booster pump.

Storage Pumps

Hydroelectric generating plant is of particular interest at the present time since a large amount of development is in hand. In situations where a large reservoir can readily be made but a relatively small continuous flow of water exists, it is practicable to pump water up to the reservoir during the night using off-peak thermal power and to run the water back through a turbine during peak hours. In this way valuable power at peak times is obtained at the cost of cheap surplus power during periods of very low demand. The thermal units are, moreover, permitted to operate nearer to continuous full load, with consequent improvement in efficiency and life. Pumped storage may also be used to avoid wasteful spillage.

A single entry pump is in operation for a duty of 10 m^3/s against a head of 46 m driven by a 5 MW motor at a speed of 300 rev/min. The pump is also capable of operating over the head range of 30–50 m.

Because of its large size this pump is constructed more on the lines of a water turbine casing, having stay vanes to prevent the opening of the horse-shoe section of the volute under pressure. The impeller is of stainless steel and is fitted to the heavy stub shaft of an induction generator/motor which carries a turbine runner on its other end. Owing to the rigid construction the pump can run in air whilst idling, thus avoid the need for a hydraulic coupling. Previous power storage installations required a coupling to disengage the pump (of the split casing type) to avoid risk of seizure.

Figure 34.8
Combined motor pumps operating in open air on water supply duties
at a large chemical works.

The pump casing is made of mild steel and comprises a cast-steel speed-ring with plate-steel scroll sections. The casing is made in two parts to facilitate transport and maintenance. (See also Chapter 35).

Model Tests

In order to obtain confirmation of the design data on which the large machine at Sronmor was based, a model to one fifth scale running at five times the speed was made and tested. With this combination of size and speed, heads, velocities, and stresses in the model represent those of the full scale machine and, by applying a suitable correction for efficiency based upon change of Reynolds number, an accurate forecast of the full scale performance was obtained. The model and full size pump are illustrated in Figs 34.11, 34.12 and 34.13 which show how the scroll construction of the full size unit was faithfully reproduced.

The full scale pump requires very special characteristics to enable it to operate over the wide head range of 30-50 m with good efficiency and without risk of cavitation.

Figure 34.9
Combined motor pumps for waterworks. On site tests with no tolerance, efficiency
is financially guaranteed as the weighted average over the variable duties between
1.1 m³/s 100 m head and 1.5 m³/s delivery at 82 m head. The pumps operate
against variable duties, the whole group covering duties nominally rated at 3 m³/s
against heads of 30, 60 and 100 m driven by motors of 200 kW to 2MW.
Alternatively the 30 m pumps can run in seies with the 60 m pumps for high head
areas. The inset shows heavy shaft.

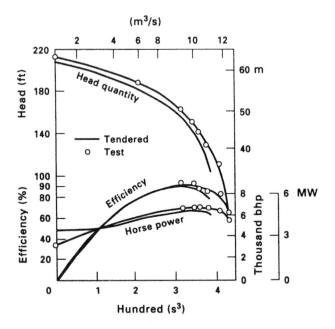

Figure 34.10
Characteristics of 5MW energy storage pump.

Fig 34.10 shows how the test characteristics of the model (when stepped up to the full scale performance) agreed very closely with the intended characteristics.

The increase in efficiency from model to prototype was calculated from both Moody formula and Reynolds number analysis as follows:

$$\frac{1-E}{1-e} = \left[\frac{d}{D} \right]^{1/5}$$

where d is the diameter and e the efficiency of the model; and D is the diameter and E the efficiency of the full scale prototype.

The efficiency increase of a complete range of machines is represented by a Reynolds curve having a varying index, but it is often more convenient on a particular pair of model and prototype to express the efficiency change as the straight line joining two points on a Reynolds curve, ie, by a single constant index which in this case is 1/5. This index was agreed in the contract. (See also Chapter 7).

Transient Conditions (See also Chapter 24)

On an emergency failure of power the pump would lose speed, the water column would stop and reverse, thus acting as a brake on the pump, stopping it, and finally driving it in reverse as a turbine. It is necessary, therefore, to determine the performance of the unit running forward as an energy dissipator; acting as a restriction to flow at zero speed; and

Figure 34.11
Scale model of storage pump.

as a turbine when running in reverse. The turbine characteristics must include operation at no load other than the windage and bearing losses of the induction generator, so as to give data to ensure that no mechanical damage or risk of bursting will occur at the reversed runaway speed.

The energy dissipation phase was tested by driving the pump in its normal direction of rotation, and forcing water into what is the pump delivery branch by applying a higher pressure than the pump itself is capable of generating. (The term 'dissipation' is used since no useful work is done: strictly speaking the energy is converted into heat).

In this way the pump received torque energy from its shaft and water energy from the supply to its delivery branch and was able to dissipate both these forms of energy. The test therefore involved measuring the power input and the quantity and head input.

For the transitional point between forward rotation as an energy dissipator and reverse rotation as a turbine, a test was taken with the rotor locked under the control of a spring balance. Water was again fed into the delivery branch under pressure, and the static torque and the flow quantity were measured.

Figure 34.12
Impeller and casing of 5 MW storage pump

The next stage was to test as a turbine, feeding water into the pump delivery branch and permitting the shaft to rotate, driving its motor as an induction generator.

The water power input was measured by determining quantity and head entering the pump as a turbine, and the power output at the pump shaft was measured by feeding the electrical energy of the motor as induction generator into a suitable measuring system.

Water Turbo-Alternators on the Unit Construction Principle

The unit construction principle has already been applied to water turbo-alternators. The alternator shaft has two bearings, the turbine runner being overhung at one end.
In a similar type the horizontal alternator shaft is carried in two bearings, and overhanging each end of the shaft is an impulse runner driven by two water jets - making four jets in all. The output is 6000 kW at a head of 230 m, the speed being 428.6 rev/min.

Small Combined Motor Pumps

Small pumps/motors on the unit construction principle have been developed for duties of

Figure 34.13
5 MW storage pump installed in Scottish site.

10–100 kW. Several sizes of pumps and motors have common matching dimensions so that a range of duties can readily be supplied from stock. These small units with cast iron casings embody the features of the larger machines, described above.

Application of Single Entry Combined Motor Pumps

It would appear that this development is suitable for heads in the range of 50–500 m in one stage, which would involve peripheral and flow speeds similar to those in comparable Francis turbines on heads from 60–600m.

In addition to the units already in operation on heads up to 120 m, good results have been obtained in research on a 600 kW model for ahead of 250 m in one stage.

This model is intended to forecast performance of a 13 MW two-unit series pump for a typical high head pumped storage scheme. The test characteristics are given in Fig 34.14.

Whilst the new development provides ample rigidity for the higher range of head, the high peripheral and flow speeds involved suggest that duties of, say, 400-600 m should be limited to applications where the water or other liquid is reasonably free from abrasive particles, at least with the metals at present in use. The purpose of the limitation is to avoid erosion and pitting that might otherwise occur at high speed points of flow transition.

Mechanical Design

When a pump and a motor are so intimately united that they have a common shaft and bearings, these parts must be designed adequately to carry the double duty without undue deflection or risk of failure.

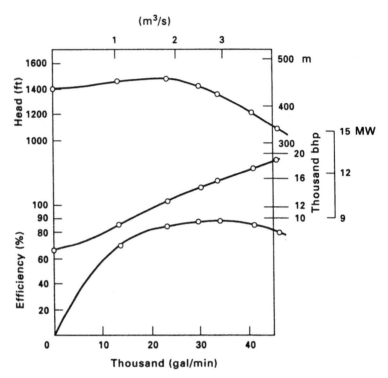

Figure 34.14
Stoprage Pump – Two pumps in series. Characteristics
based on quarter size model at four times prototype
speed. Model heads equals full head.

Efficiency correction

$$\frac{1-E}{1-e} = \left[\frac{d}{D} \right]^{1/5}$$

it is in this respect that the single entry impeller is of considerable value since it allows unlimited increase of shaft diameter.

An overhung single entry pump must necessarily have two bearings to support the shaft. For economy in space, power, and expenditure of metal, the next logical step is to place the rotor and stator of the electrical unit between these two bearings. Mechanical calculations for this combined unit are given in Chapters 16 and 17.

Hydraulic Design

The hydraulic design of single entry pumps is generally similar to that of the double entry type with, however, the following differences:

The direct axial entry permits a higher axial flow velocity so that the single eye is little greater in diameter than the double eye. Depending on the method of attachment of the impeller to the shaft, the inlet blades can be brought well down towards the centre of the impeller, which, together with the high rate of flow, results in relatively larger inlet blade angles.

The higher flow velocity and the absence of a shaft on the inlet side of the impeller permits the single entry type to handle almost as much liquid as the double entry type for a given motor level and speed, particularly on large vertical units where the whole circumference of the impeller eye can be at uniform low level. (See Chapters 9 and 10). NPSH can be minimized by the provision of an inducer preceding the impeller.

Conclusion

A large number of the unit construction single entry pumps have been supplied and have been operated for sufficiently long to indicate their capabilities. High efficiencies have been realized. The greater rigidity avoids mechanical losses, air drawing and shaft damage. Maintenance has, therefore, been considerably reduced. The simpler flow pattern on the inlet side avoids hydraulic losses, thus giving improved distribution of liquid to the impeller. This type of pump is, therefore, inherently more efficient than the split casing type. The improved flow permits the single entry type to handle almost the same quantity as the double entry type for a given suction lift and speed.

The greater rigidity has been proved in each case by running the pump dry — a very drastic test indeed, which could not be applied to the split casing type.

It has also been possible to operate the unit in some circumstances without stuffing-box or packing — a simplification and an advantage where high inlet pressures are concerned.

These pumps are particularly applicable to the high pressure range of single stage duties, as the shaft diameter may be made very large. The cover closing the casing is circular in shape and the whole design is simple, providing uniformity and strength.

The saving in space and the simplification of layout results in a considerable reduction in size of pump house. As will be seen from the illustration, this is one of the most important advantages in the new development. In many cases a vertical pump operates in the open air with short suction pipe and ejector priming. (See Fig 34.9).

The above mechanical principles are followed in the 100 MW 340 m reversible single stage pump turbines for Cruachan, Chapter 35.

References

17. ANDERSON, H.H. 'Modern Developments in the use of Large Single Entry Centrifugal Pumps'. 1955 Proc IME, London, Vol 169, No 6.

49. DAUGHERTY, R.L. 'Centrifugal Pumps for the Colorado River Aqueduct' ASME Mech Eng April 1938.

50. LUPTON, H.R. 'Pumping Machinery', Dugold Clerk Lecture, 1948, Inst CE.

51. LUPTON, H.R. 'Pumping Stations International Water Supply Assoc Congress, July, 1955'. Report published by Inst of Water Engineers.

52. WOOD, F. 'Ashford Common Works, with Particular Reference to the Pumping Plant'. Journal of Inst of Water Engineers, Vol 9, No 4, July 1955.

PUMPS FOR STORAGE OF ENERGY AND OTHER LARGE POWER DUTIES

With our modern alternating current power systems, it is still not possible to store energy other than by pumping water in large volumes to an elevated reservoir. The large alternators in central power stations must have a generating capacity to meet the highest peak demand, which means that during periods of lower demand, potential generating capacity is available. There is, therefore, economic advantage in storing energy in the form of water raised to a higher level during off peak periods, so that hit may be used to generate power in water turbines at peak periods.

Economically, if, during the night, electricity is worth half the peak period rate, this difference of cost may justify the capital expenditure of a pumped storage scheme.

In general, our power demands have a diurnal variation: eg, a peak demand of power between 8.30 am and 9 am followed by a larger peak between 4.30pm and 5.30pm. Optimum operating conditions for a fossil fuel steam power station or a nuclear steam power station, would be continuous operation at full load, whilst a tidal power scheme would generate power with a cyclic variation following the phases of the moon.

Incidence

During the last 50 years, several storage schemes have been commissioned on the Continent with total heads between 10 m and 1000 m, the stage heads ranging up to 300 m and powers up to 100 MW.

The Continental schemes in general have so far combined separate pumps and water turbines with an electrical machine which can operate as a motor or an alternator, according to whether the unit is pumping or generating. In America, several reversible units are in operation with heads between 70 and 300 m. Here the electrical and hydraulic machines operate as motor and pump during off peak periods and as alternator and turbine during peak load periods.

In Britain there is one 5 MW storage pump in operation against heads of 30-50 m with separate turbine and motor alternator, four 75 MW pumps at Ffestiniog with separate

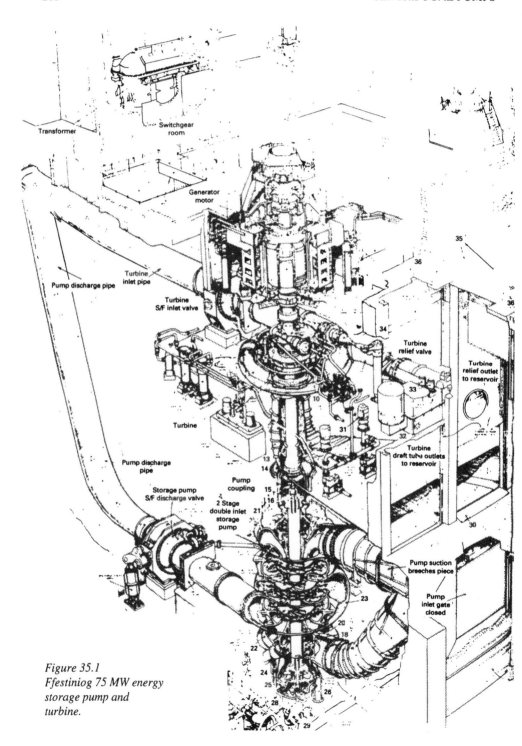

Transformer

Switchgear room

Generator motor

Pump discharge pipe

Turbine inlet pipe

Turbine S/F inlet valve

Turbine

Pump discharge pipe

Storage pump S/F discharge valve

Pump coupling

2 Stage double inlet storage pump

Turbine relief valve

Turbine relief outlet to reservoir

Turbine draft tube outlets to reservoir

Pump suction breeches piece

Pump inlet gate closed

35

36

36

34

33

31

32

30

10

13

14

15

16

21

23

20

18

22

24

25

26

28

29

Figure 35.1
Ffestiniog 75 MW energy
storage pump and
turbine.

turbines and motor alternators and four 100 MW reversible units operating in Cruachan at 340 m head.

The Deriaz machine is a British development embodying a schroll with movable gates containing a cone flow runner, also with movable blades, giving the efficiency advantages of a Kaplan turbine or pump at considerably higher heads by virtue of the 45° flow through the impeller or runner. This machine can be used as a pump or a turbine and is in operation at Niagara as a reversible pump turbine against a head of 23 m with a power of 34 MW.

A continental development is the built turbine wherein the rotor of the electrical unit surrounds the hydraulic runner or impeller, thus affording simplification of civil engineering work. These units are placed in a dam thrown across a river and operate as reversible pump turbines for storage work. They are particularly valuable for combined tidal power and storage schemes.

Foyers (Scotland) 150 MW reversible units operate at 178 metres head. Dinorwic 200 MW units are under construction. Reversible pump turbines up to 500 metres in one stage are being designed. (See also Ref 53).

The Ffestiniog Scheme (Ref 54)

The four 75 MW pumps and separate turbines at Ffestiniog are comparable in their design to the many Continental schemes.

Plate 35.1 shows the general view of the power plant which incorporates generator motor, Francis turbine disengaging pump coupling and two stage double inlet storage pump embodying three impellers, two single entry impellers for the first stage delivering into a double entry impeller in the centre of the pump for the second stage. The pump is shown in cross section in Fig 35.2 which is self explanatory. The weight of the pump rotor and any departure from theoretical axial balance is carried on a thrust bearing at the bottom of the pump, where provision is also made for the servo motor to operate the disengaging coupling. The purpose of this disengaging coupling is to eliminate the power losses of rotating the pump whilst turbining. The turbine is filled with compressed air so as to permit it to rotate with minimum power loss when pumping.

Inlet conditions are more favourable for a turbine than for a pump and therefore it is convenient to put the pump below the turbine, so that it receives the greater submergence from the inlet reservoir.

The Ffestiniog pump operates against heads of 296–324 m and in view of the fact that boiler feed pumps can operate on stage heads much greater than this, the question may be asked as to why Ffestiniog has been made a two stage unit. The answer lies in the fact that progress in the increase of speed, head and power must be gradual and must be dependent upon metallurgical investigations. A further point arises in connection with very large pumps and that is the relationship of the stationary metal to the mass of the rotating impeller. For example, a high pressure boiler feed pump has relatively small impellers with very heavy static masses to withstand the total pressure; in contrast, a storage pump has a heavy impeller with casing which is relatively thin since it withstands the pressure of only one or two stages. The shell is generally buried in concrete to give additional rigidity. Stage heads have since been increased to figures of the order of 400–500 m upwards (following the water turbines which now range up to 500–700 m) (See Ref 55).

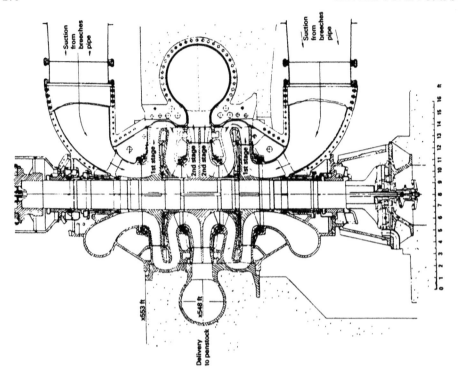

Figure 35.2
Cross section through pump.

Reversible Pump Turbine

American Francis Pump Turbines

Typical units are the single speed 7 MW pump turbines in Brazil, the 55 m Hiwassee pump turbines rated at 75 MW pumping and 100 MW turbining and the two speed flat-iron pump turbines rated at 7 MW 76 m 257 rev/min when turbining and 10 m³/s 73 m 300 rev/min when pumping, (Ref 56). The two speed motor alternator involves a double winding, (Fig 35.3). The Taum Sauk unit operates at 200 MW 250 m head with split rotor for transport. Grand Coulee 700 MW turbines are in operation.

Deriaz Units

The Niagara pumps, Figs 35.4 and 35.5, are rated at 27 MW 23 m when turbining, 130 m³/s 34 MW 23 m when pumping, both duties being obtained at a speed of 92.3 rev/min. A 23 MW 55 m Deriaz turbine is in operation on reversible duties in Scotland, and a 76 m unit in Spain, (Ref 57). With rotor blades closed, the Deriaz unit acts as a valve and when started requires minimum torque.

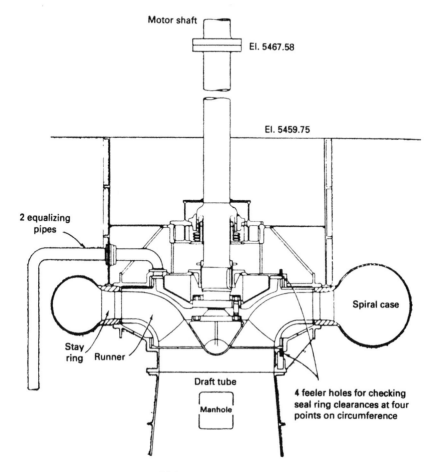

Figure 35.3
Flat-iron pump turbine. 13 m3/s 80 m head,
300 rev/min, 10 MW

Higher Head Deriaz Units

A 50 MW 260 m head, single stage reversible Francis unit of American design is in operation at Provvidenza in Italy.

There is also a Canadian reversible pump turbine in service at heads of 17-28 m 200 MW.

Problems of High Head Reversible Application

There is considerable saving in civil engineering costs by adopting the reversible pump turbine instead of separate pumps and turbines, as Fig 35.6, illustrating an investigation

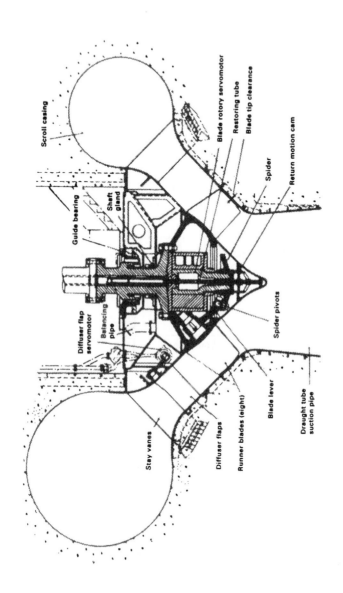

Figure 35.4
Deriaz unit at Sir Adam Beck, Niagara.

Figure 35.5
Above: Deriaz pump turbine, Niagara,
showing rotor blades closed.

Below: Showing rotor blades open.

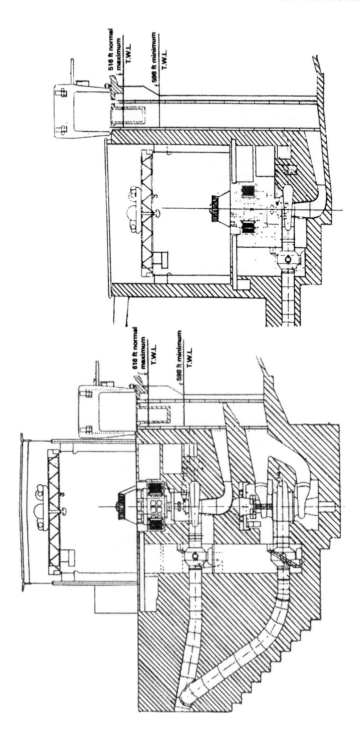

Figure 35.6
Alternative station arrangements for Ffestiniog: (left) separate pumps and turbines and (right reversible machines.

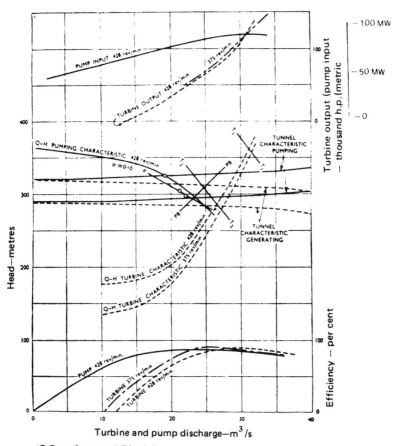

Figure 35.7
Approximate pump turbine characteristics for
Ffestiniog.

prior to the ordering of the Ffestiniog scheme, shows quite clearly. However, there are very severe problems arising in the reversible unit, as is shown in Fig 35.7. A major difficulty is the fact that, when turbining and pumping are running at the same speed and operating on the same reservoir and piping system, the reversible unit cannot operate at the optimum efficiency point for both.

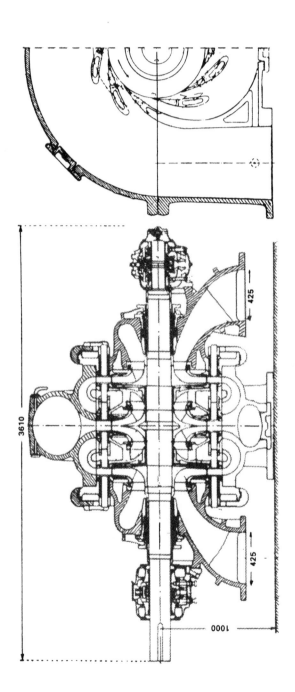

Figure 35.8
430 mm two stage pump 1 050 l/s, 230m, 1 000 rev/min, NS = 1 540. Storage pumping at Schwarzenbach works (Dimensions in millimetres).

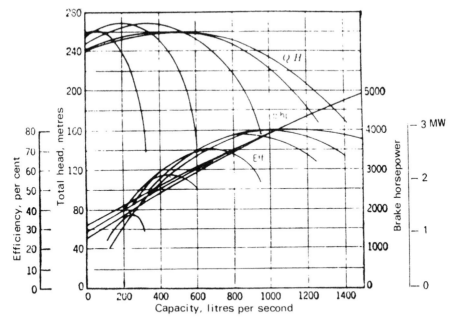

Figure 35.9 – Performance of pump shown in Figure 35.8.

This is illustrated in Fig 35.7 which shows the approximate pump turbine characteristics of the Ffestiniog alternative (not proceeded with). Some alleviation of this problem occurs in the provision of a two speed set, such as the Flat Iron pump turbine, but considerable increase in generator motor costs arises. Mechanical speed changing, for example a two speed epicyclic gear box, may help this problem in the future, since, at the same time, it permits entire freedom of choice of speed to the electrical designer and may thereby economize on civil engineering costs. Fig 35.7 also illustrates impeller and casing lines, as on Figs 6.3, 10.2 and 10.3.

The present tendency towards very large machines in order to justify the economics of storage schemes, involves turbine load control by movable diffusers. When pumping, pressure fluctuations impose a very great load on the controlling mechanism of the diffusers, necessitating positive hydraulic locks, especially during starting, to avoid vibration. There is little advantage in diffusers for pumping, but they do confer some slight extra efficiency, as well as the ability to vary load when turbining. (See Figs 35.8, 35.9 of 2 stage Continental pump with movable diffuser gates).

The mechanical problems of movable diffusers are liable, in the future, to impose a limit on permissible head per stage.

In order to cut down the electrical load when starting, the pump is often emptied by compressed air, the water being admitted when the pump is up to speed. Readmission of water involves a severe electrical load shock and very careful arrangement and design is needed in order to minimize vibration.

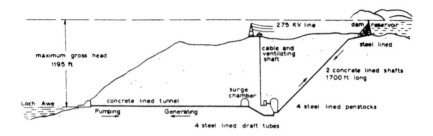

Figure 35.10 – Cruachan Scheme.

Turbine starting and synchronizing, either with diffusers or on a gateless unit with a valve by-pass, would appear to present no insuperable problems. Rapid changeover from pumping to turbine is practicable, since the hydraulic unit and pipeline can reverse almost in unison, the control valve or diffusers being closed to a point corresponding to the speed 'no load', and then opening when turbining, before the changeover takes place. When the power to the pump is switched off, the rotating element pump and motor will slow down and reverse up to approximately turbining speed, when the unit can be synchronized and switched in to the power line.

Cruachan 100 MW Reversible Units
(Figure 35.10, 35.11, 35.12 and Refs 53 and 55).

Here the operating duty is 358 metres 28.6 m³/s pumping and 34.3 m³/s turbining. For pumping, each unit is emptied of water by compressed air, then run up to speed by a pony motor of 7 MW. When water is admitted there is a sudden load of 14 MW thrown on to the electrical grid. However, after considering various hydraulic means of running the set up to speed in the pumping direction, this is accepted as the least objectionable method.

Figs 35.10 and 35.12 show one of the two English Electric Sulzer units running at 600 rev/min, Ref 55. Two Cruachan units, supplied by AEI Boving, run at 500 rev/min, (Fig 35.11).

Tehachapi (Fig 35.13)
This Californian irrigation scheme is a good example of large high pressure pumps on which 1% efficiency was assessed at $4 million in 1960.

The duty involved lifting 140 m³/s through a height of 600 metres and discharging the flow to a region 600 km away. Ref 58.

Several proposals were considered giving alternative duties and numbers of pumping stations and pump stages. These alternatives are shown in Fig 35.13. Ref 59 indicates test and estimated pump efficiencies for the model and prototype pumps in the various alternatives. The mean line of all the efficiencies recorded in Ref 58 lies within the highest of the four parallel bands defining the efficiency scatter of Fig 7.3. All pumps operate at the optimum shape number with respect to efficiency.

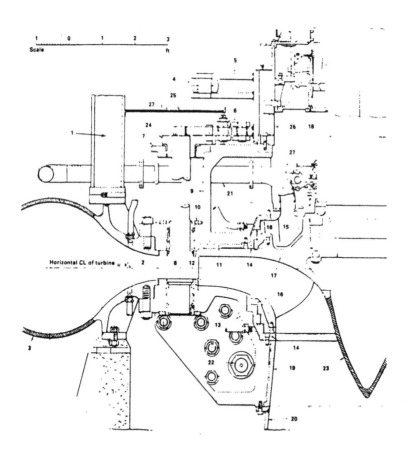

Figure 35.11
Section of 100 MW Reversible Pump turbine set.

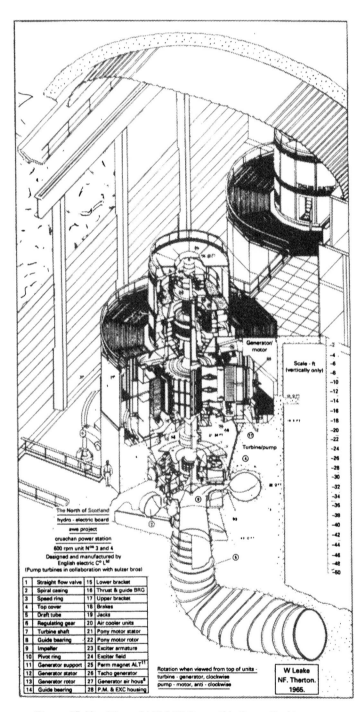

The North of Scotland
hydro - electric board
awe project
cruachan power station
600 rpm unit N⁰ˢ 3 and 4
Designed and manufactured by
English electric C⁰ L^td
(Pump turbines in collaboration with sulzer bros)

1	Straight flow valve	15	Lower bracket
2	Spiral casing	16	Thrust & guide BRG
3	Speed ring	17	Upper bracket
4	Top cover	18	Brakes
5	Draft tube	19	Jacks
6	Regulating gear	20	Air cooler units
7	Turbine shaft	21	Pony motor stator
8	Guide bearing	22	Pony motor rotor
9	Impeller	23	Exciter armature
10	Pivot ring	24	Exciter field
11	Generator support	25	Perm magnet ALT¹¹
12	Generator stator	26	Tacho generator
13	Generator rotor	27	Generator air hous⁴
14	Guide bearing	28	P.M. & EXC housing

Rotation when viewed from top of units -
turbine - generator, clockwise
pump - motor, anti - clockwise

W Leake
NF. Therton.
1965.

Figure 35.12 – View of 100 MW Reversible Pump Turbine.

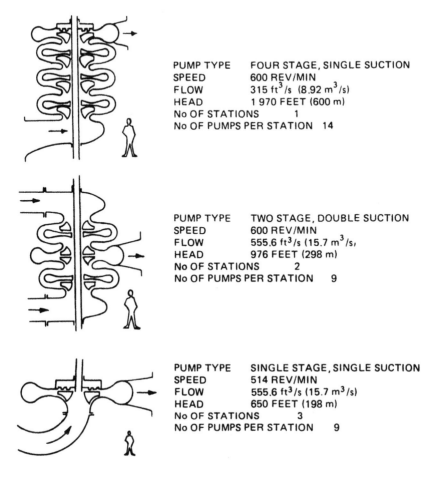

PUMP TYPE   FOUR STAGE, SINGLE SUCTION
SPEED   600 REV/MIN
FLOW   315 ft^3/s (8.92 m^3/s)
HEAD   1 970 FEET (600 m)
No OF STATIONS   1
No OF PUMPS PER STATION   14

PUMP TYPE   TWO STAGE, DOUBLE SUCTION
SPEED   600 REV/MIN
FLOW   555.6 ft^3/s (15.7 m^3/s,
HEAD   976 FEET (298 m)
No OF STATIONS   2
No OF PUMPS PER STATION   9

PUMP TYPE   SINGLE STAGE, SINGLE SUCTION
SPEED   514 REV/MIN
FLOW   555.6 ft^3/s (15.7 m^3/s)
HEAD   650 FEET (198 m)
No OF STATIONS   3
No OF PUMPS PER STATION   9

Figure 35.13
Tehachapi pumps, alternative proposals.

In order to ensure freedom from cavitation the limiting inlet shape number was determined by the consultants at 4250 rev/min l.s and metres.

Scale models were submitted by American and European makers to the National Engineering Laboratory, Ministry of Technology, Scotland, for testing. A comprehensive record of these tests, which had an overall accuracy better than 0.3% and a comparative accuracy of 0.15%, is given by Nixon in Ref 59.

Future Developments

The hydraulic limitation of size is dictated by transportation and indeed, in future it may be desirable to split the impeller/runner diametrically. This has been done on Taum Sauk turbines, but presents considerable mechanical problems.

It would appear that, although the motor generator is heavier than the pump turbine, it

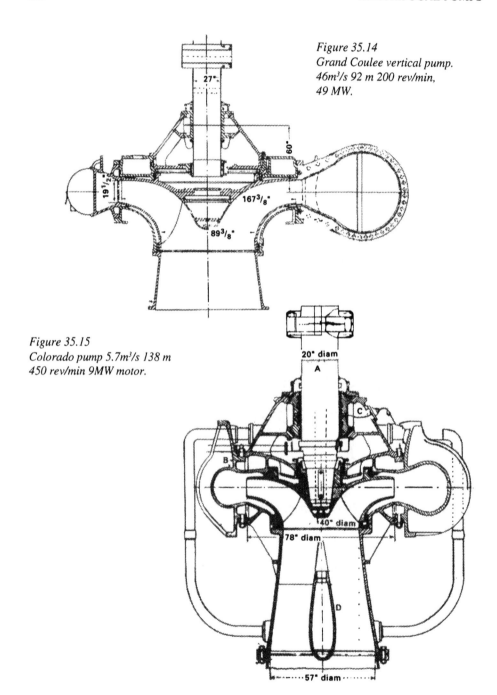

Figure 35.14
Grand Coulee vertical pump.
46m³/s 92 m 200 rev/min,
49 MW.

Figure 35.15
Colorado pump 5.7m³/s 138 m
450 rev/min 9MW motor.

can, however, often be assembled and wound on site. On future plant, overall efficiencies of the order of 70% would appear to justify the economics of storage schemes and such an efficiency is within present endeavour.

Diffusers may be fixed and governors may be omitted on some of the machines in a large station, thus saving costs (provided the electrical gird can accept block loading). Simplification of valve control etc, can ease the water hammer problem by permitting the pipeline to empty. For example, the 49 MW Grand Coulee pumps have no valves, but work on a syphon break. These are normally operated as pumps, but have on occasion been used for generation of power. (Refs 38 and 60).

Heavy casing walls and large diameter shafts are desirable with the higher stage heads of the future, since, (as mentioned above in the comparison of Ffestiniog and boiler feed pumps), large vibration forces can arise — particularly at low flows. Severe vibrations were experienced on the Grand Coulee installation (Ref 38) where the pumps, Fig 35.14, Refs 60 and 61, had very much higher shafts and casings than the Colorado pumps, Ref 49, Fig 35.15 or the single entry pumps described in Chapter 34.

See Appendix E

(See International Electrotechnic Commission Code 497 for tests of Storage Pumps and Turbines)

References

11. ANDERSON, H.H. 'Efficiency Majoration Formula for Fluid Machines' Int Assn Hyd Res 7th Symposium Vienna, 1974.

12. ANDERSON, H.H. 'Statistical Records of Pump and Water Turbine Efficiencies'. Conf 'Scaling for Performance Prediction in Rotodynamic Machines', Stirling, 1977 IMechE.

38. BLOM, C., 'Development of Hydraulic Design of the Grand Coulee Pumps', ASME, Paper 49/SA 8.

49. DAUGHERTY, R.L., 'Centrifugal Pumps for the Colorado River Aqueduct' ASME Mech Eng April 1938.

53. ARMSTRONG, N.A. 'On Site Efficiency Test on the Reversible Pump - Turbines at the Scottish Cruachan and Foyers Pumped Storage Plants.' Conference 'Scaling for Performance Prediction in Roto-dynamic Machines' Stirling 1977, IMechE.

54. HEADLAND, H. 'Blaenau Ffestiniog and other Medium-Head Pumped Storage Schemes in Great Britain', Proc IME, 1961.

55. Cruachan. 100 MW Reversible Units. Electrical Times, 21 October 1965.

56. PARMAKIAN, J. 'Flat-iron Power and Pumping Plant', Mech Eng, pp 677-680, August 1955 (ASME).

57. 'Deriaz Reversible Pump Turbine (Niagara)', Engineering, 10 February, 1961.

58. MILLER, D.R. 'Selection of Four Stage Pumps for Tehachapi', April, 1968 Proc West Water Power Sym ASME etc.

59. NIXON, R.A. and SPENCER, E.A. 'Model Testing of High Head Pumps' IMechE Symposium on Model Testing of Hydraulic Machinery, Cranfield April 1968.

60. MOSES, E.B. 'History and Development of the Grand Coulee Pumping Plant', Mechanical Engineering, September, 1948, (ASME).

61. PARMAKIAN, J. 'Vibration of Grand Coulee Pump Discharge Lines', Trans ASME Vol 76, 1954.

SECTION 9

General

SUBMERSIBLE PUMPS IN GENERAL

MINE PUMPS

SEWAGE AND SIMILAR DUTIES

OFFSHORE OIL DUTIES

DEEP SEA MINING

FLUID STORAGE CAVERNS

THERMAL POWER STATION PUMPS INCLUDING NUCLEAR

OFFSHORE, MARINE AND AEROSPACE PUMPS

REFINERY, CHEMICAL AND PROCESS PLANT – PROPERTIES OF LIQUIDS

VARIABLE FLOW

ELECTRIC MOTORS

VARIABLE SPEED FOR CENTRIFUGAL PUMPS

OIL, WATER, GAS AND STEAM TURBINES ON GLANDLESS DUTIES

PUMPS FOR VARIOUS SPECIFIC DUTIES

MISCELLANEOUS ASPECTS

THE PAST AND FUTURE

SUBMERSIBLE PUMPS IN GENERAL

The development of submersible pumps over the last few decades is entirely due to their success in performing their specified duties with remarkable reliability. This stems from rigidity, from permanently correct alignment and from the success of bearing designers in accepting lubrication by the pumped product.

Shaft rigidity and alignment has already been discussed so we now consider bearings. With a variety of materials from carbon, rubber, plastics, metals, ceramics to stellite and composites with almost the hardness of diamond, bearings can now operate in the most aggressive liquids with success, adequate monitoring forecasting the times for overhaul. Mechanical seals have developed to a standard where clean lubricating liquids are separated from pumped liquid and can feed the bearings thus excluding any harmful deposits.

Lubricating liquid pumps within the motor or other driver can ensure long, troublefree operation and the pumped product cools the stator winding and the bearings. With rigidity, alignment, cooling and liquid separation nothing can go wrong.

The advantages of submersible pumps when compared with horizontal pumps, motor and bedplate sets are as follows:

A. Mechanical

1. Better rigidity and alignment.
2.1. Bearings require no attention, there are no oil or grease problems of dirt, under or overfilling and no stuffing boxes to leak or be maladjusted. At ICI the author saw padlocks on stuffing boxes to prevent every passerby tightening or slackening them.
3. Mechanical seals suffer no malalignment.

B. Hydraulic

4. There are no NPSH problems or priming problems since the impellers are always below water level at start up. Air drawing cannot occur.
5. With no line shaft or malalignment higher rotational speeds are possible.
6. Symmetrical direct inlet gives efficiency improvement

C. Ambient

7. Impervious to heat, humidity, dust, acid or sea spray, etc.

D. Simplicity

8. No underground pump house required in a mine or elaborate pump house at ground level.
9. The pumpset is smaller and more compact. No leakages to be disposed of.
10. Easier to install and remove compared with shaft driven borehole pump or horizontal pump, motor, bedplate and switchgear which cannot descend the mineshaft fully assembled.
11. Suitable for remote automatic control with resultant saving of manning costs.
12. Easily removed for overhaul without access below ground.

E. Safety

13. No leaks or toxic, flammable or radioactive liquid. Virtually no noise. Flameproof.
14. Less danger of damage by seismic forces, thermal distortion or motor overheating.
15. Failure of power supply does not endanger the pumpset and cannot endanger the safety and freedom from flooding of the mine. The pumpset is unharmed by flooding.

F. Installation and Maintenance

16. Pumpset can be easily installed and removed for maintenance without access below ground. Discharge joint is a gravity wedge.
17. Overhaul in a clean workshop and re-assembly for despatch avoids contamination risks of re-assembly on site or at the pit bottom. Pumpset is truly portable.

G. Size

18. Because of unit construction and possible higher speed the pumpset is smaller and can be installed in a minimum of space, eg in a pipeline.

H. Cost

19. Absence of line shafting, bedplate inlet piping, motor stool, flexible coupling glands, inlet valves and reduced manning reduces cost.
20. Smaller size consequent upon unit construction and higher speed reduces cost.

Rigidity and alignment and the need for clean bearing liquid have been repeatedly emphasized here because these aspects are completely vital to safe pump operation.

General

The borehole pump, described in Chapter 30, comprises a motor at the ground level, a length of piping and shafting and a pump at the water surface down the well. The running speed of a borehole pump is limited by the shafting to a normal maximum of 1500 r/min in the general case. For high head duties at small flow quantities 3000 r/min is a very much more attractive speed providing a smaller, cheaper pump with, in the majority of cases, a higher efficiency.

Submersible pumps having motors running in the water permit an operating speed of 1500 or 3000 r/min up to duties of about 3 MW and have a lower first cost.

In consequence of this, the submersible pump has captured most of the borehole pump field and is firmly established on the score of reliability and efficiency on all duties apart from the large quantity low pressure duties.

Introduction of Submersible Pumps (Reference 10)

Prior to 1900, borehole and deep well pumps were of the reciprocating type. The introduction of the electric motor fostered the development of submersible pumps, particularly on borehole duties where a vertical spindle motor at ground level would drive a pump at the water level through a length of vertical shafting.

Some 50 years ago, insulated cables capable of withstanding water were made available and this permitted the development of the submersible pump. The factors obtained in a borehole favour the use of a submersible pump on the small quantity

high head duties, since the economic minimum of bore hole diameter will lead to a small diameter high speed pump with several stages to generate a high head and the omission of the vertical drive shafting will permit this high speed.

The shaft driven borehole pump with motor at the surface will suffer the losses of the thrust bearing, of the intermediate vertical shafting and of the lubrication system, etc, on the efficiency aspect and will involve relatively high cost in the provision of vertical piping, shafting, internal tubes, bearings, etc.

Since its motor runs in water the submersible pumping set will have a lower motor efficiency than the shaft driven pump, where the motor is running in air. In addition, the submersible pump set will have $1^2 R$ losses of the electric supply cable to the motor. Each type of pump therefore suffers losses over and above those found in a normal horizontal pumping application, the incidence of which determines the relative advantage. In practice, small quantity high head pumping duties (where high speed is an advantage) favour the submersible pumping plant, but the large quantity lower head duties appear to be more satisfactorily met by vertical shaft driven pumps and, in consequence, slow speed large power submersible motors (eg over 500 kW) are not generally used on higher duties and flows.

The cost and weight of the vertical cable has fostered the design of high tension motors for example 3 300 volts.

Mr F. Wood, Metropolitan Water Board, 1960, reported that, at a depth of 25 m (182 ft), the shaft driven pump and the submersible pump are approximately equal in capital cost, the greater depths favouring the submersible pump, whilst the lesser depths favour the shaft driven pump. He further stated that the running costs (excluding labour) for fuel and maintenance generally, electrical and mechanical, are slightly higher for the shaft driven pump. This is still true 25 years later.

Figure 36.1
Low shape number pump.

Pump Details

Figures 36.1 and 36.2 show respectively a low shape number pump and medium shape number pump with motor. The pump is placed above the motor, the water inlet being between the units, which arrangement keeps down the overall diameter of the set since the discharge water need not pass round the motor and since an additional pressure wall cable gland is avoided.

The suction end cover, however, contains a cylindrical strainer arranged between the pump and the motor so that the water can be drawn radially inwards from the well. Embodied in this suction cover is the housing for the bearing and for the flange fitting to the motor end bracket. The motor is connected to the pump by a claw coupling.

The delivery cover carries the top journal bearings and the small thrust pad to cope with the shaft uplift occurring on excessive flow quantities explained later.

The upper bearing is closed by the cover to this thrust pad, which prevents any sand or grit settling on to the bearing during standing conditions. A small umbrella attached to the shaft provides similar protection for the bearing at the lower suction end.

The pump shaft is supported in rubber carbon stellite or plastic bearings at the journals, but the weight of the pump shaft and some measure of the hydraulic thrust is carried through the claw coupling on to the motor shaft and to the motor shaft bearings.

In general, the pump impellers are balanced by a neckring at the upper side with holes through the driving shroud so as to return the neckring leakage to the lower pressure side, but recent developments in plastic bearing materials render such balancing unnecessary on small sizes.

Materials of Construction

The pump materials are very largely dictated by the quality of the water being handled, but, in general, the pump covers and chambers are made of cast iron or bronze and the impellers and wearing bushings of bronze. The shaft is normally of stainless steel or high tensile steel and carries stainless steel nuts and sleeves running as journals in water-lubricated guide bearings of carbon, plastic or rubber. The uplift thrust bearing on the top of the pump and the umbrella to the lower bearing are made of bronze.

Where a bronze pump is fitted to a mild steel motor and a mild steel rising main, discs of bakelized fabric are fitted between the flanges to provide insulation against electrolytic potential.

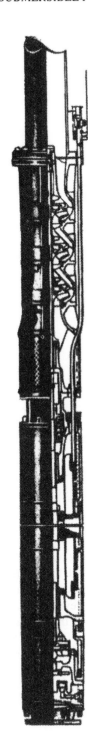

Shape Number

As mentioned earlier, the shape number of a borehole or submersible pump is dictated firstly by the hydraulic duty and secondly by the diameter of the borehole or well in which it is to be installed.

In general, submersible pumps for waterworks duties are inserted into boreholes, so that the smallest diameter pump for a given duty will be the most economical as far as the cost of drilling the borehole is concerned. Medium or high shape number pumps are more generally used, therefore, on waterworks duties.

On mine drainage duties up to 600 m (1968 ft) head, the submersible pumps are installed in the pit shaft where there is no limitation of diameter. The usual mining application therefore on submersible duties is likely to be a relatively high head and small quantity which is suitably met by the low shape number pump with renewable diffusers. (See Chapter 37).

Characteristics

Figure 36.3 shows the characteristics of quantity, head, efficiency and power of a six stage pump of low shape number in comparison with a ten stage pump of smaller dimensions and duty of medium shape number. It will be seen that the medium shape number has a non-overloading power characteristic, whilst the low shape number power characteristic continues to rise beyond the best efficiency point of the pump. This increase of power beyond best efficiency point will give rise to an overload during the first minute or so, whilst the rising main is being filled, but such an overload is generally accepted on a cold start.

Journal and Thrust-Bearing Loadings

Radial Loadings

Since the pump stages contain diffusers, the radial thrust on the pump bearings is very small, even in cases where the pumps are inserted in a horizontal pipeline for boosting duties on waterworks applications.

Axial Loadings

The pump axial loading downwards is carried on the motor thrust bearing, whilst the axial thrust upwards is carried on small uplift pad at the top of the pump discharge bearing.

Figure 36.2
Medium submersible pump and motor.

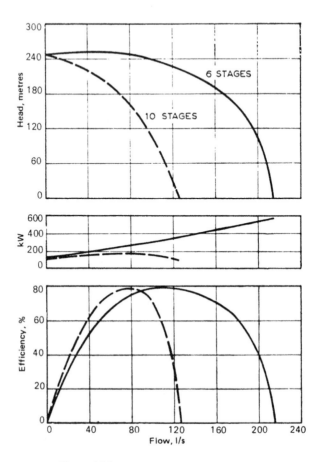

Figure 36.3
Pump characteristics for low and medium
shape number.

In general, it is normal to balance the impellers by providing a neckring on the upper side of the impeller in addition to one at the lower (inlet) side. Holes through the impeller driving shroud then return the back neckring leakage water to the lower pressure side of the impeller, thereby effecting balance. The top end of the pump shaft, however, is usually subject to delivery pressure whilst the lower end is subject to inlet pressure: this produces a vertical downward thrust proportional to the generated head.

Recent developments in plastic materials for the motor thrust bearings permit operation of small sets with unbalanced impellers, with consequent saving of cost.

Where the impellers are balanced, however, the axial thrust loadings are as shown in Figure 36.4, where the impeller thrust in Newtons is plotted against percentage flow. The downthrust on the end of the shaft will be proportional to the generated head so the fine

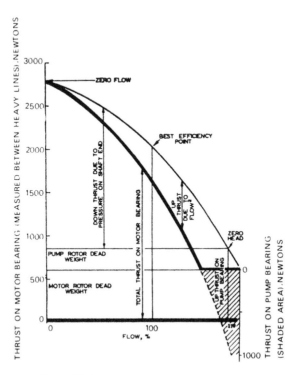

Figure 36.4
Axial thrust loadings on submersible pumpset.

upper line in the diagram is a representation of the pump head quantity characteristic in terms of downward thrust.

Each impeller turns the water through approximately 90° which gives rise, according to Newton's second law, to a dynamic upward thrust proportional to the square of the flow. The motor and pump rotors impose downthrust equal to their respective weights The summation of these thrusts is shown in Figure 36.4, plotted against flow.

At zero flow the weights of the pump and motor rotors and the downthrust due to differential pressure on the pump shaft end are carried by the motor thrust bearing. As the flow increases there is a dynamic upthrust proportional to the square of the flow, which ultimately exceeds the decreasing pressure thrust as zero generated head is approached The dynamic thrust at large flows produces an upthust on the pump shaft which is held by a small thrust bearing at the top of the pump. This removes some load from the motor thrust bearing, which, however, must always carry at least the motor rotor weight, since the coupling faces are free to slide apart for a short distance.

Figure 36.4 shows the downthrust on the motor bearing as the vertical distance between the heavy lines, the dynamic upthrust being measured as a reduction of the pressure thrust The pump upthrust is shown as a vertical height of the shaded area where the dynamic thrust line is below the pump-rotor weight line for the flow in question. The motor thrust bearing is of the Michell type water lubricated, and is continuously rated for maximum load. The pump uplift bearing is a small bronze pad against which a bronze cap on the shaft can run during uplift conditions. This bearing is generally short term rated at zero head, but is continuously rated at half normal head.

At unattended or automatic stations, where the water level in the well may be high at start up, the uplift bearing has an important duty to fulfil, since, dependent upon the well characteristics, uplift thrust may persist for a considerable time until normal well level is reached. (See Chapter 24).

Influence of Borehole Diameter on Type of Pump

A submersible pump stage is conveniently of multistage pattern, and advantage is taken of the vertical operation to reduce the casing diameter and increase the overall length without prejudice to mechanical stability. Beyond a certain point, however, this reduction of diameter is a prejudice to efficiency, and consequently this aspect must be investigated with respect to cost of the borehole. Some slight efficiency reduction has occasionally been justified where a large yield is required from a relatively small borehole or where the boring cost is high.

In the overall economic study of any scheme, the additional cost of a larger borehole must be balanced against the capitalized saving in efficiency.

When considering the relative merits of a normal borehole pump/surface motor installation and of a submersible pumpset, many of the aforementioned factors have to be taken into consideration. Normally, there is a substantial reduction on the capital cost when a submersible unit is installed. In many instances it is unnecessary to build a pump house, by operating instead in open air, or in a closed manhole. A pump house may, however, be required for other reasons, such as housing booster pumps, measuring equipment, control gear, etc, but it can frequently be made smaller and therefore cheaper.

The effect of depth must also be considered, since sacrifice of efficiency is involved where a high head per stage with a small quantity is required.

References

10. ANDERSON, H. H., and CRAWFORD, W. G., Submersible pumping plant, Proceedings Institution of Electrical Engineers, vol 107, part A, no 32, April 1960.
51. HEWMANN, A., The design and characteristic features of the submersible motor pump. World Pumps, May 1984.

MINE PUMPS

Mines are wholly dependent on pumps for safety in keeping them free from flooding and for many auxiliary services.

The chief duties are drainage, both main and coal face, coal washing, powerplant duties, mine sinking, etc. Almost all these pumps are of the centrifugal type, but for certain duties within its field the reciprocating pump still gives valuable service.

Duties

The drainage of mines involves raising water through heights up to about 600 m in Britain and up to 2 500-3 000 m abroad. For these duties centrifugal pumps are used almost entirely, with reciprocating pumps for the very low-quantity high-head duties. Permanent pump houses at the bottom of the shaft are installed, the plant being designed for continuous operation. Small portable pumps are used for local drainage at the coal face.

When sinking a mine shaft or dewatering an existing shaft, a pump may be used which can follow the work, pumping from a varying level to the surface. These pumps and motors are usually suspended on a wire cable so that they can be lowered as required, (see also Chapter 36, Submersible Pumps In General).

Coal washing requires the service of simple rugged pumps able to withstand liquids containing coal particles and grit.

Main Drainage Pumps

The primary pumps in a mine are situated at the pit bottom. They are generally of the multistage type as described in Chapter 28, with, however, a tendency to heavier design and superior metals in order to cope with the mine waters which may have acid content or grit in suspension.

Details are as follows: (See Figs 29.2, 29.3 and 33.2).

1. *Impeller.* The impeller must withstand the centrifugal and torque stresses for the peripheral speed and power involved. The hub and shroud must transmit the end thrust due to the hydraulic unbalance, with in turn is carried on the balance disc. The impellers generally are high-grade bronze, but cast iron, stainless steel, etc, are also used where the liquid to be handled demands other than standard metals.

2. *Casing*. Multistage pumps have cellular chambers fitted with diffusers to convert the velocity into pressure. The diffusers are of high-grade bronze. A bearing bronze bushing is fitted to the main wall of the chamber. After the impeller is inserted into it the chamber is closed by a plate which withstands the pressure difference of one stage. This plate is fitted with a neckring of bronze.

For low pressures the casing is made of cast iron, but for high pressures steel chambers with cast iron or nickel iron diaphragms are used. Since cast iron has a high compression strength, the combination of a shrunk steel chamber and a cast iron diaphragm gives an attractive design.

3. *End Covers*. The chambers are held together by suction and delivery end covers, which in turn are clamped by long through-bolts. These covers contain the suction and delivery passages and the branches. They also carry the bearing housing and the balance disc and chamber. The covers are made of cast iron for low pressures and cast or fabricated steel for high pressures.

4. *Shaft*. Increases of speed and head owing to modern demands have given rise to increases of shaft diameter. This results in a longer life, since the loading on the internal bushes is decreased.

5. *Sleeves*. The sleeves of the multistage pump fulfil two purposes. Firstly, they protect the shaft from corrosive attack and assist rigidity; secondly, they carry the hydraulic thrust of the impellers via the end nuts to the balance disc. Sleeves must therefore be machined to a high standard of accuracy so that the tightening-up process preserves straightness of the shaft. They must also have sufficient radial thickness to avoid overstressing on thrust load.

6. *Balance Disc*. The balance disc has proved to be one of the most reliable devices ever used in multistage pumps. Balance discs have increased considerably in thickness and in diameter as duties have increased, and renewable faces are now standard. On average mine water the balance disc gives a very satisfactory life, and it is only where excessive grit, pressures surges, rapid stopping and starting, etc are frequent that Michell bearings are fitted to take the thrust load.

7. *Glands*. The suction end gland is sealed from the first or second stage of the pump so as to permit operation on suction lift. The delivery end gland, owing to the pressure-reducing effect of the balance disc, is not subject to pressure. On the whole, therefore, mine-pump glands are not severely loaded except in the case of series sets, where the second pump is handling water at the first pump discharge pressure and thus requires a high-pressure gland.

Stuffing-Boxes

When a rotating shaft passes through a wall into a chamber containing fluid under pressure, the clearance between the diameter of the shaft and the bore of the chamber wall permits a leakage of fluid from the chamber. In the case of a pump this leakage of fluid is a direct loss of efficiency. The majority of centrifugal pumps work with vacuum or slight pressure at the stuffing-boxes, the high-pressure chamber of the pump being isolated from the lower-pressure chamber by neckrings at the sides of the impeller or, in the case of a multistage pump, by the thrust balancing device. There are, however, certain cases where the stuffing-boxes must withstand a considerable pressure.

1. Pumps on suction lift must be water-sealed to prevent air drawing at the stuffing-boxes. This sealing, taken from the first or second stages, may be as high as 100 m or more.

2. In series sets embodying two separate pumps, the low- pressure pump discharging into the inlet of the high- pressure pump, the stuffing-box at the inlet end of the high-pressure pump must withstand the discharge pressure of the low-pressure set. This pressure may be as high as 500 or 600 m in extreme cases.

3. In mines it is sometimes necessary to arrange a pump at a lower level than the sump. The difference in levels imposes a pressure on the water entering the pump suction.

4. Pumps can be arranged to draw simultaneously from different levels in the mine. Here again, the water from higher levels will enter the pump at some particular pressure.

5. Where the sump is above the pump house any leakage from the pump gland must be reduced to a minimum, as this leakage involves further pumping equipment.

Pressure at the pump inlet can be held by packing in a conventional stuffing-box or can be relieved before entering the stuffing-box. The pressure that can be imposed upon the packing in a stuffing-box depends upon the absence of abrasive impurities in the water and the truth of the pump shaft. The presence of abrasives considerably increases the sleeve wear, whilst eccentricity of the shaft will break up the structure of the packing. In general, it is recommended that stuffing-box pressures for mine drainage pumps should not exceed 100 m.

Sinking and Dewatering Pumps

When sinking a mine shaft (or dewatering an existing shaft) a pump may be used which can follow the work, pumping from varying levels to the surface. These pumps and motors are usually suspended on a wire cable so that they can be lowered as required.

Vertical arrangement is desirable in order to save space and give stability of suspension. The pump and motor are mounted in a girder framework, the motor being above the pump. Where the switchgear is simple it can be installed in the frame with the motor, but otherwise the top of the shaft is a suitable place for this switchgear. The whole pumping set and girder frame is suspended by means of a pulley and wire rope. Further lengths of rising main must be added as the set is lowered, and the electrical power cable must be lowered at the same time.

As the pump must work against heads varying from zero to maximum, a discharge throttle valve must be fitted to control the quantity pumped, so as to avoid overloading the motor or vaporizing in the first impeller, with consequent risk of seizure. As this throttling is rather wasteful or power, dummy stages can be fitted to the pump during the earlier stages of pumping to be replaced by live stages as the work of dewatering proceeds. Each lowering of the pump involves a certain amount of pipe fitting, so that it is economical to pump to as low a depth as possible for each setting of the pump. To this end the pumps are designed to give the greatest possible suction lift.

The weight of the rotating element and the hydraulic thrust of the impellers are carried by thrust bearing on the pump. In order to reduce gland pressure a balance disc is used, arranged at the delivery or upper end of the pump. Provision must be made for taking the reaction torque of the motor on all suspended sets.

Vertical sinking pumps can be so arranged that they may be used as permanent horizontal sets when the sinking is completed. When dewatering a sloping shaft the use of a horizontal pump arranged on a trolley adjustable to any angle of dip is the most convenient method.

Submersible sets are becoming more popular in mine work, where they offer considerable advantages of space saving and cost, and, moreover, in many cases avoid the need for underground supervision. For certain mine dewatering duties this type of pump is valuable, particularly in cases where a sudden rush of water must be dealt with. The ability of the submersible pump to operate in practically any position without the lining-up problems of shaft-driven pumps has on many occasions proved of value in preventing mine flooding.

Coal-Face Pumps

Small low-head pumps are required for local drainage at the coal face. These pumps must be of simple and robust construction to withstand accidental damage. It is generally of advantage to incorporate a self-priming device, as they frequently run on 'snore'. The

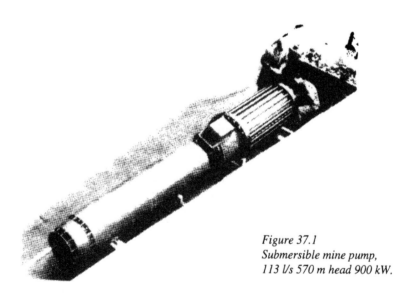

Figure 37.1
Submersible mine pump,
113 l/s 570 m head 900 kW.

water to be handled contains coal dust and grit, and therefore the wearing parts of the pumps must be cheaply and easily replaceable.

Coal-Washing

Coal-washing duties involve large pumps operating against fairly low heads. The chief problem is abrasion of the pumps by coal grit. As single stage pumps are used, it is possible to embody a simple construction with very thick metal walls to give long life. Wearing plates are fitted at vulnerable points.

The pumps run with flooded suction so as to obviate priming troubles and prevent any aggravation of erosion by vaporizing in the impeller. Careful hydraulic design reduces eddies, and consequently local abrasion, to a minimum. Nickel iron is used for the impellers, and rotation speeds are kept low. In this way it is possible to obtain reasonably long life when handling heavy slurry.

Sludge Pumping

The sediment from settling tanks must be raised to the surface by some means. One method is to use one of a battery of high-lift pumps for this purpose, the metals of this particular pump being especially resistant to abrasion. The sludge pump is then duplicate and standby, if necessary, to the clean-water drainage pumps.

In order to reduce deterioration of packing and wear of sleeves, the stuffing-boxes of the sludge pump are sealed from a clean water supply; for example, from the first stage of the main pumps.

Pump Characteristics

In contrast to waterworks duties where a lift of several hundred metres may involve a pipeline of several thousand metres in length, the mine drainage duties involve a pipeline which is generally vertical and therefore little different in length from the static head. this means that the pipe friction is a very small percentage of the total head.

Multistage pumps designed to give the maximum possible head for a given expenditure of metal may have unstable characteristics associated with the high area ratio. As far as possible every designer tries to make all his pumps stable, but this ideal may not be achieved every time.

As described in Chapter 22, it is essential that the zero flow head should materially exceed the duty head so that a second or later pump can be brought into operation in parallel with the No 1 pump.

The existence of an unstable pump and a very small amount of friction in a mine drainage system may, however, give rise to difficulty when a low frequency of electric supply occurs, such as could be the case when a mine has its own independent power supply. On public electricity supplies, low frequency is unlikely, but may occur.

Fig 37.2 illustrates this problem in an unstable pump which meets the duty of 152 m static lift, 8 m friction with a total head of 160 m at 50 cycle speed 1480 rev/min. It will be seen that the closed valve head is materially above the duty head of 160 m.

On 47½ cycles, 1 400 rev/min, however, it will be seen that the closed valve head is actually less than the static lift, which means that the pump will be unable to start pumping

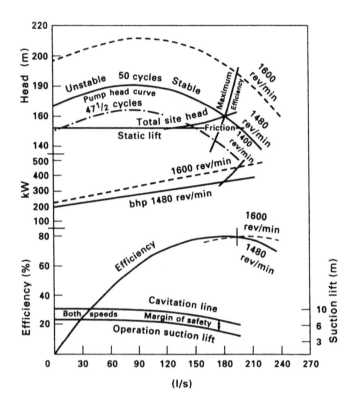

Figure 37.2

*Centrifugal Pump Characteristics. A hypothetical increase to 1 600
rev/min is illustrated to explain change of characteristics with speed.
A head characteristic at 47½ Hz is also shown to illustrate the fact
that on this low frequency the pump head at closed valve is less than
the static lift, and therefore pumping cannot commence.*

except by partial drainage of the vertical column, since the pressure holding the non-return valve shut is great than the pressure generated by the pump at no flow.

This chart also illustrates the increase in power consumption by the pump consequent upon an increase of frequency to give a higher speed of 1600 rev/min.

Maximum Speed

The suction pressure (see Cavitation, Chapter 9) is dependent upon operating speed in rev/min, since the velocity at which the impeller inlet vanes pick up the water increases with rev/min. There is therefore a limit of quantity at every speed — at 3000 rev/min this limit is in the neighbourhood of 80 l/s on suction lift.

It is found that centrifugal pumps will not give reasonable efficiencies at quantities below about 5 l/s at 800 m, and it is below this limit that reciprocating pumps are of value. (See Chapter 32).

The Choice of Speed

The majority of mine pumps run at 1500 rev/min or 3000 rev/min (less slip). Certain aspects of speed have been discussed above. For small quantities on very high lifts 3000 rev/min permits the use of a single unit where 1500 rev/min would require a series set and a considerable increase in space — always an important matter where every cubic metre has to be excavated and roofed. The low-speed set will be more expensive in first cost and spares, and involves left and right-hand patterns of impellers, chambers, etc, if the balance discs are to be accessible. The stuffing-box of the high-pressure unit is exposed to the discharge pressure of the low-pressure unit.

In respect of the choice of a high-speed set (3000 rev/min) the following points must be considered:

1. As the head per stage is greater, the water flow velocities within the pumps are greater, with correspondingly greater wear, especially on abrasive waters.

2. The clearances and passages of a 3000 rev/min pump will be smaller than those of a slower-speed set, and therefore more sensitive to wear or choking.

3. Where the head is insufficient to demand the maximum possible number of stages at 3000 rev/min, a slow- speed set of 1500 rev/min may give slightly better efficiency.

4. Two-pole squirrel-cage motors can now be designed to give low starting currents, especially on the low starting torque of the centrifugal pump. Slip-ring motors have been made at this speed up to quite large powers, but are not generally recommended for colliery work.

On the whole, therefore, it may be desirable to use a 3000 rev/min set on small-quantity high-head duties, where the water is reasonably pure and where electrical considerations permit this speed. The saving in first cost and space must be balanced against the rather greater wear but cheaper replacements of the 3000 rev/min set. The relative efficiency will depend upon the duty.

Standage

There are many reasons for recommending the largest possible standage of water:

(a) If the sump is too small the pump may overdraw it with risk of seizure. This is particularly the case on high-speed drives where a large amount of power and of rotor inertia are applied to a relatively small pump. When the pump is starved of water, flow ceases, and vaporization occurs in the first stage, with consequent lack of lubrication to the neckring and of sealing water to the suction end bush and stuffing-box. Seizure therefore may start from the suction end, and continue throughout the pump before overload trips cut out the motor — even then there is a large amount of energy in a 3000 rev/min rotor.

*Figure 37.3 – Ring Section pump for injecting
water into oil wells 800 l/s, 370 m head.*

(b) Electric power is generally charged at a unit rate plus maximum demand, and in
some cases may be cheaper at off-peak periods. The smallest possible pump
running 24 hours (or off-peak hours) would then have the smallest maximum
charge and be most economical in capital expenditure. Depending on the rate at
which the water passes through, a large sump may then be required. A further factor
is the incidence of shifts and supervision of the pump station — if pumping is
carried out only for eight hours a day the sump must be large enough to store at least
24 hours supply of water plus a safety margin.

(c) A large sump gives an opportunity for abrasive matter in the water to settle out and
also permits dosage to neutralize acids, etc, if required.

Against all this is the over-riding fact that every cubic metre in a mine is of vital
importance and cost. The standage should therefore be as large as this last factor will
permit.

Severe Mine Waters

The presence of grit, ochre, fine straw, and similar impurities in mine water may lead to
wear in the neckrings and thrust balancing devices in multistage pumps. The choice of
suitable metals for these parts is a problem for metallurgists and chemists, and each
particular mine water must be examined to determine the correct metals to use. In addition
to causing wear, ochre and similar deposits tend to fill up the passages of the pump, thus

reducing its output and efficiency so that frequent cleaning is required. It should be noted that this point has considerable bearing on the question of high speeds. The smaller high-speed set will be affected to a much greater extent by a given thickness of deposit than a larger slow-speed set for the same duty.

The percentage of solid impurities in the water can be reduced considerably by settling tanks arranged with low weirs such that the water velocity is a few cm per second.

The use of special metals to withstand the worst chemical corrosion involves considerable expense. Manufacturer therefore standardize on a high-quality phosphor bronze which will be suitable for the majority of mine duties. In the relatively small number of cases where this metal is unsuitable, more expensive materials can be used. Here again only a complete examination of the water to be pumped will determine the suitable metals. This question is complicated by the fact that corrosive, abrasive, erosive, and electrolytic tendencies may be present together in a given water, and these phenomena generally act in concert in attacking metal, (see also Chapter 33).

Motors (see Chapter 47)

The majority of mine duties permit the use of a squirrel- cage motor. This is because variable speed is almost never required, and because modern squirrel-cage motors can be designed to take reasonably low starting currents, particularly on the low starting torque of a centrifugal pump, (see Fig 24.1). Starting is usually direct on the line, star-delta or through auto transformer. A typical value of starting current for direct-on-line start is four times full-load current at a very low power factor.

Hence, in the case of most power systems, the power drawn from the main at starting is not likely to be serious, but kVa and voltage drop must be carefully considered in respect of effects on generator, cables, transformers, etc. Motor starting characteristics on full and reduced voltage are shown in Fig 37.4.

The squirrel-cage motor is a very robust machine, ideal for mine duties. For small powers ball-roller bearings are usual - larger powers have sleeve bearings. The motor power should exceed the pump-absorbed power at rated duty by at least 10%. This is to allow for variation between estimated and actual head and to allow for any frequency fluctuation.

Where installed cables are already heavily loaded, as may occur in the case of an old mine, starting current may be reduced by the provision of a wound rotor and slip-rings. whilst slip-rings are a perfectly reliable device, it is nevertheless true that this is a further point that calls for maintenance.

Submersible Motors

It is usual, to fit a mining-type junction box at the submersible motor so as to avoid the shipping and transporting difficulties of a motor with a long trailing cable. For small powers 500 V is used with three-core cable. For large powers the weight and size of the cable are important factors and therefore voltages of 3000 are used, for example at 400 kW. In determining the starting conditions it is important to allow for the voltage drop between the surface and the submerged motor, using a voltage drop figure that allows for the drop in the cable. Korndorfer auto-transformer starting is essential for submersible sets.

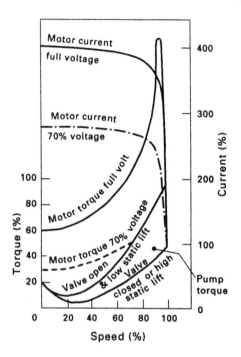

Figure 37.4 – Starting of electrically driven pump. For direct-on-line start (full voltage) no difficulty should arise provided that lines, transformers, and power supply are adequate. Switchgear should have inverse time lags on overload settings to deal with this.
·For auto-transformer starting an example of 70% tapping is given. Not that this would be critical on full-valve start, but satisfactory on high static lift such as occurs on mine drainage. Approximate run up time: 2T on valve open. 1.5T on valve closed.

A submersible motor is to all intents and purposes a flame-proof machine. Mining motors are generally more robust than the usual run of production machines, to allow for the difficulties of handling and transporting in restricted spaces with little crane facilities.

Fiery Mines

Flame-proofing of motors to Buxton certificate may be required for certain mines, but in the majority of cases the pump house is regarded as non-fiery.

Typical Examples of Drainage Pumps

In this country duties of individual pumps vary from about 10 to 150 l/s and from 60 m to about 700 m. The powers range from 50 to about 1000 kW. Abroad, higher heads may be involved: a 3000 m mine will use three 1 000 m pumps in series at different levels.

A few typical duties of British and South African mine pumps are given below:

120 l/s, 400 m, 1480 rev/min, 640 kW motor single unit. Britain.

40 l/s, 730 m, 1480 rev/min, 560 kW motor series set. South Africa.

9 l/s, 800 m, 2900 rev/min, 120 kW motor series set. Britain.

95 l/s, 1 060 m, 1480 rev/min, 1450 kW motor series set. South Africa.

60 l/s, 900 m, 1480 rev/min, 800 kW motor single unit. South Africa.

70 l/s, 380 m, 1480 rev/min, 410 kW motor single unit. Britain.

17 l/s, 200 m, 1460 rev/min, 50 kW motor single unit. Britain.

150 l/s, 1000 m, 1490 rev/min, 2100 kW, single unit. South Africa.

References

2. ANDERSON, H.H. 'Mine Pumps', Jl. Min Soc, Durham University, July 1938.
62. ANDERSON, H.H. 'Pumps for Mines with special reference to Coal Mines', Silver Jubilee, Indian Pump Manufacturers Association, Bombay 1976.
63. BRYSON, W, and DONALDSON, J. 'Pumping Plant for Colliery Dewatering'. Mining Technology. January 1975.

SEWAGE AND SIMILAR DUTIES

The development of the oil barrier air-filled motor has revolutionized sewage pumping practice. It is no longer necessary to provide inlet pipes or to descend into the well for the purpose of assembling pump and piping since the pumpset can now be lowered into its location and can automatically wedge into the delivery pipe joint. Removal is equally simple. Because the impeller is fitted directly on to the motor shaft the reliability is improved and the complication and cost of the previous surface motor and vertical line shafting is avoided.

A standard motor frame can embody several different impellers according to the type of sewage or other liquid to be handled. For example, the free flow or torque flow impeller in Figure 38.1 leaves a vane free volume on its inlet side which minimizes the risks of clogging especially on crude sewage which often contains fibrous material. The efficiency sacrifice that this design involves is justified by the reduced risk of clogging.

Figure 38.2 shows a single vane impeller which is preferable when handling products which are easily damaged, for example, fruit. It is also of value in pumping abrasive liquids. The single vane impeller can be balanced statically but cannot provide balancing

Figure 38.1
Free flow impeller.

Figure 38.2
Single vane impeller.

Figure 38.3
Closed two channel impeller.

Impeller Type	Curve slope H_O/H_{opt} H_r = shut-off head H_{opt} = head at η opt	Efficiency η/η 'a'	NPSH/NPSH 'a' Capacity Q = constant head drop $\triangle H/H$ = 3% gas content < 0.5%	Applications
'a' Closed impeller, standard vane arrangement	1.28	1	1	For final effluent.
'b' 2/3 vane non-clogging impeller	1.56	0.96	1.11	Non-clogging impeller with large free passage; danger of plaiting with textiles and fibrous admixtures. For screened sewage, activated sludge, final effluent; for gentle pumping of product; not suitable for liquids with a high gas content
'c' Closed, single-vane impeller	1.64	0.79	1.37	Single-vane impeller; very large free passage corresponding to pump connection diameter; insensitive to clogging and plaiting when handling liquids containing textile fibres and admixtures; for unscreened sewage and all types of sludge; for gentle pumping of product.
'd' Free-flow impeller	1.64	0.64	2.89	Free-flow impeller; very large free passage corresponding to pump connection diameter; insensitive to clogging; for unscreened sewage and sludges; for liquids and sludges with a high gas content; not suitable where gentle handling of product is required.

Figure 38.4 – Impeller type comparisons.

for the dynamic out of balance of the single jet discharging from the impeller. The increased rigidity of the impeller on the motor shaft construction, however, is able to deal with this problem satisfactorily. The non clogging impeller (Figure 38.3) with a few vanes affords an improvement of efficiency for screened duties but cannot easily handle a high gas content.

Finally, the conventional impeller giving full efficiency affords an opportunity to improve the overall station wire to water power balance.

The advantages and disadvantages of these impellers are shown in the chart of Figure 38.4. A complete station arrangement is given in Figure 38.5 where the package well avoids the need for inlet pipes, valves etc. Pre-cast concrete boxes can be lowered readily into a prepared hole and quickly set to work.

The motor and pump section is shown in Figure 38.6 which indicates the rigidity of the rotor and the grease sealed bearings.

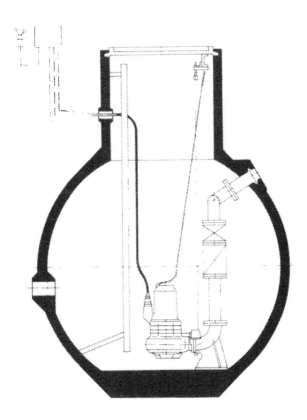

Figure 38.5
Packaged pumping station.

Pumps for Industrial Slurries

For many industrial purposes, it is necessary to pump liquids having a large amount of suspended solids. Suitable pumps may have rubber lined or chilled metal internal parts and be provided with clean water seals to keep the abrasive matter from the fine clearance parts. Typical duties are sand stowage, china clay slurry, coal pumping, coal washing, ash handling, etc.

The metal thicknesses of impeller and casing are very much greater than the hydraulic duty demands in order to provide a margin for wear caused by abrasive particles.

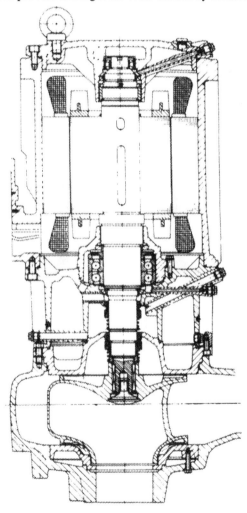

Figure 38.6
Air motor oil barrier submersible
pumpset with single vane impeller.

Material	Loss Rate $(g/m^2 h)$
Cast iron (GG-25)	110
Chrome steel (1.4138)	100
ERN (nickel alloy cast iron)	90
Chrome steel (G-X 190 Cr 16)	18.3
Ni-hard 4	7.5
Norihard	2.7
Abrasive agent:	
silica sand/water	test duration : 2 hours
mixture ratio : 1:1	speed : 3 000 r/min
grain size : 0.9−1.2 mm	material sample : 055 x 5

The laboratory test results have been confirmed by field tests.

Figure 38.7
Loss rates for various materials

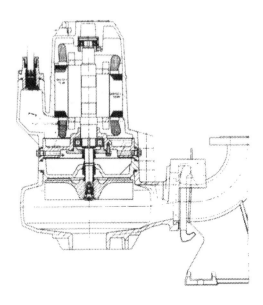

Figure 38.8
Torque flow sewage pump.

In Reference 58, D. H. Hellman discusses the erosion wear on pumps handling aggressive liquids and shows how the wear is affected by solid content, impingement angle and the nature of the pump component surface. The behaviour of various materials is seen in Figure 38.7, Reference 58.

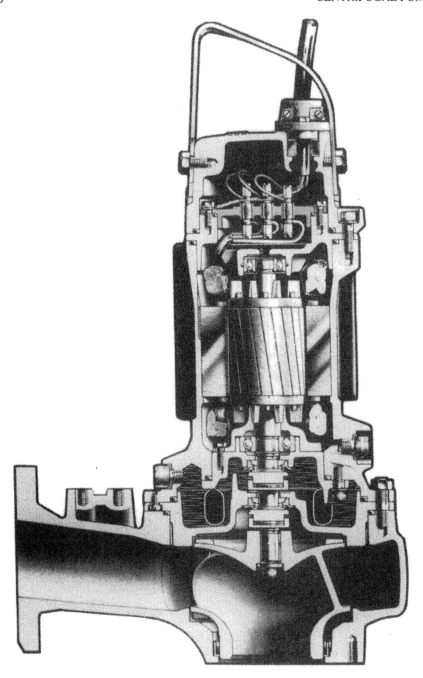

Figure 38.9
Single vane sewage pump.

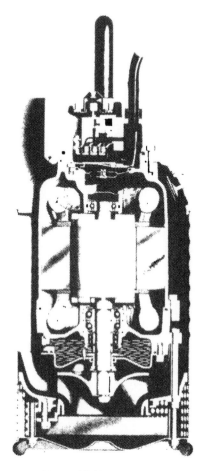

Figure 38.10
Vertical submersible pump.
(Grindex)

References

51. HEWMANN, A., The design and characteristic features of the submersible motor pump, World Pumps, May 1984.

58. HELLMANN, D. H., The influence of the size of submersible pumps on efficiency and erosion wear, World Pumps, September 1984, p 317.

59. RAAB, P., The economics of submersible water pumps, World Pumps, June 1983.

OFFSHORE WATER DUTIES

The primary requirements of water pumps on oil production platforms are the ability to operate reliably tor long periods without dismantling, to occupy the smallest amount of space, especially on the platform itself the installation of as much plant as possible below the platform floor, the greatest rigidity to ensure the correct alignment throughout the working life and the efficient piping, cabling etc to ensure efficient control. These requirements are assured by the submersible units with their unit construction, their high speed and their vertical aspect.

Since the majority of these pumps handle sea water with abrasive content, stainless steel and aluminium bronze are in general use.

Platform Trimming Pumps

A group of trimming pumps is shown in Figure 39.1 embodying a 250 kW submersible pump in each leg of the platform, near its lowest point. The pumps position the platform in the horizontal plane and are used to empty the ballast. Each pump is installed in a canister or pressure jacket and is detailed in Figure 39.2. The canister was initially suspended in the leg thus permitting the installation of the pumpset without the need to send operators below. The water enters the canister from below so as to ensure that it passes over the outer cylinder of the motor to afford adequate cooling. The discharge flow is led either to a point above sea level or to another individual cell.

Since the pumps are adjacent to the sea bed, the full sea pressure operates in the canister and in the motor casing. Two mechanical seals arranged back to back protect the motor from abrasive and corrosive sea water.

The cables are led out of the motor and out of the canister, then via water-filled protection pipes to a water reservoir with a float valve and leakage detector. Motor temperature monitors protect the set from overheating.

Cooling Water Pumps for Gas Compressors and other vital Plant

Submersible pumps for these duties are relatively large, a typical example being rated at 500 kW for a flow of 560 l/s against a head of 61 m (200 ft). Figure 39.3 shows this pumpset

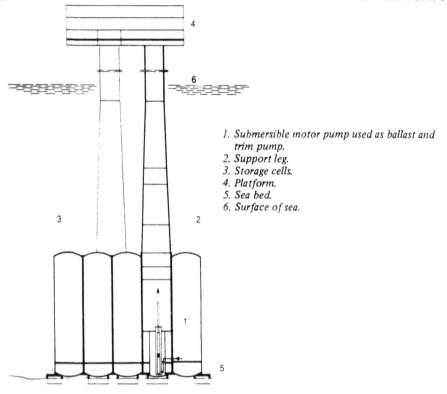

1. Submersible motor pump used as ballast and
   trim pump.
2. Support leg.
3. Storage cells.
4. Platform.
5. Sea bed.
6. Surface of sea.

Figure 39.1
Diagrammatic illustration of a
production platform

which, for simplicity of installation and for motor cooling, is arranged in an outer rising
main.

Submersible pumps offer the following vital advantages for this task:

(i) Exceptionally high reliability and safety of operation ensuring an adequate supply
    of cooling water to the primary installations which the platform depends.

(ii) The location of the pumpset below the platform thus saving precious space.

(iii) The installation of the pumpset with its high tension motor in a flameproof chamber
      remote from the explosion-proof zone thus simplifying the general arrangement.

For a long service life in aggressive sea water all parts of the pumpset were made of
suitably resistant material. The pump impellers, diffusers stage casing and the motor
casing, the rising main and the outer protective casing were made of aluminium multialloy
bronze, which has the strength of steel and the resistance against sea water to afford a long
life in spite of the corrosive and abrasive elements involved

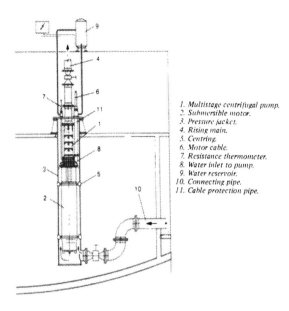

1. Multistage centrifugal pump.
2. Submersible motor.
3. Pressure jacket.
4. Rising main.
5. Centring.
6. Motor cable.
7. Resistance thermometer.
8. Water inlet to pump.
9. Water reservoir.
10. Connecting pipe.
11. Cable protection pipe.

Figure 39.2
Section through an installed
ballast and trim pump.

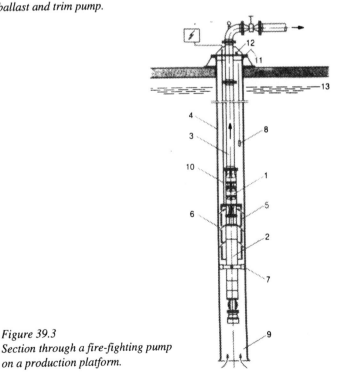

Figure 39.3
Section through a fire-fighting pump
on a production platform.

Fire Fighting Pumpsets

This is a comparable duty with the one described previously for which Figure 39.1 also applies. Monitors to check the water level and the motor temperatures are normally provided.

Water Injection Duties

The extraction of water from a well and its injection into another well to float the oil to the surface can be effected in one operation by the Weir Downhole Pump described in Chapter 40. The water is extracted from one strata and injected into lower strata without ever rising above the first strata.

Hydrostatic Drive for Submersible Fire Pumps (Figures 39.5 and 39.4)

The Weir-lMO Screw Pump and Screw Motor Drive is a versatile system offering a life expectancy of at least 5000 hours between overhauls in spite of operating under sea water and with hydraulic fluid contamination passing I mm (0.04 in) strainer. The screw units have proved their reliability over many decades and can accept non-flammable fluids giving more stable viscosity and harmless spillage The system can withstand frequent stopping, starting and intermittent running and has low seal pressures with leakage indicators

Considerable flexibility is afforded in the positioning of the various components; there is no electrical power cabling or risk of wrong rotation. Since the rising main contains no shaft, complete freedom of angular aspect and location of the main driver is obtained The screw pump can be remote from the top of the rising main and within a safe area.

The screw pump and motor have only three moving parts; they have virtually no pulsation and are self-lubricating. The system complies with the following specifications:

API 676 Positive Displacement Pumps.

API 610 Centrifugal Pumps.

NFPA 20 System Design.

Water Supply on Offshore Platforms (Figure 39.6)

Sea water is pumped up to the platform for:

(i)   Water to be desalinated for drinking, washing, etc.

(ii)   Water for fire extinction and deluge.

(iii)   Water for general deck washing, etc.

(iv)   Cooling water for process, power generation, air compression, etc.

(v)   Drill water for mud mixing and on platforms, for formation injection.

(vi)   On barges and semi-submersibles drilling rigs, water for ballast purposes.

(vii)   Trenching by sea water injection.

(viii) Water jets to render the sea bed semi-liquid for deep installation of structures. Cessation of water jet permits sea bed to reconsolidate

Figure 39.4
Hydraulic drive fire pump, 300 l/s, 62 m head, 265 kW.

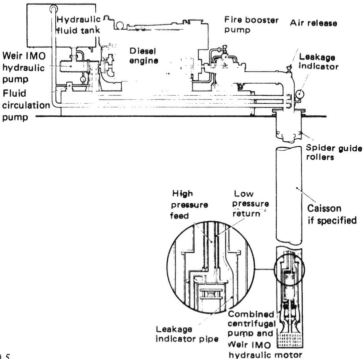

Figure 39.5
Hydraulic drive fire pump system

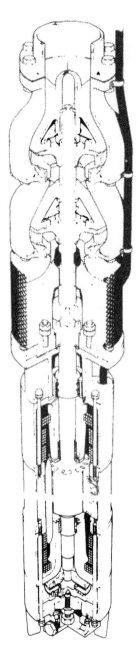

Figure 39.6
Submersible pump and motor

References

50. KUNTZ. G.. Submersible motor pumps for offshore and onshore engineering applications, KSB Technical Report d/2 and Pumps-Pompes—Pumpen 1 979- 1 54.

OFFSHORE OIL DUTIES

This chapter shows Imperial units in addition to SI since the majority of oil activities refer to barrels a day and feet head or depth, 100 000 barrels a day equalling 184 l/s or 2430 gal/min.

Reliable Turbine Driven Downhole Pumpset

Oil recovery is initially effected by underground pressure for example, a gusher or in water terms, an artesian well. Secondary extraction consists of injection of sea water into the well so as to restore the pressure lost by the gusher action. The third stage is by direct pumping from 1520-3050 m (5-10000 ft) below ground. Although extremely reliable within its sphere, the electric submersible set involving a few hundred stages cannot be expected to give reliable service always on these extreme depths.

The oilfield is not as simple as an underground lake. The oil lies within the pores of the rock, usually limestone or sandstone and is free to pass through it in similar manner to a fine grain sponge. The purpose of enhanced recovery is to help the oil out of the rock. Some of the methods are cracking the rock with explosives, applying heat or pumping steam down. Carbon dioxide is also helpful, as are chemicals acting similarly to detergents in washing out the oil. Considerable promise is offered in these advanced methods.

Until a few years ago the conventional method if pumping oil was by electric drive wherein the two pole motor at 60 Hz was inserted in a typical 228 mm (9 in) hole. This imposed a limit of about 25 m (82 ft) of head per stage so that the number of stages could be anything from 100—300 depending on the well pressure and depth. The pump and motor, probably comprising, for example, 200 pump stages and four separate motor units could be as long as 46 m (140 ft), which was entirely unsuitable for insertion down a sharply curving well, necessarily at an angle to the vertical. Differential strain and expansion of rotor and stator due to stress and temperature presented problems. The pumpset would have to be delivered to the rig in parts and assembled thereon, involving risk of damage. The presence of abrasive and corrosive elements in the oil and the curvature of the well would involve a high starting torque which is not available on the very long small diameter submersible motors. Experience on these pumps showed in many cases a life of only a few months before removal for renewal of worn parts. The cost of

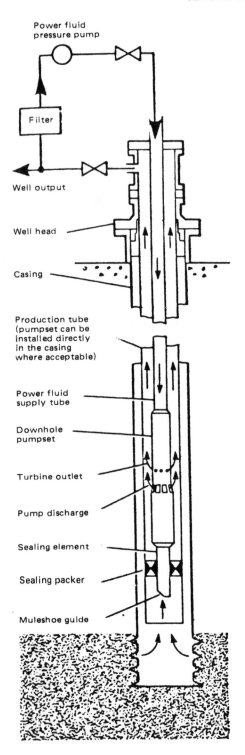

Power fluid
pressure pump

Filter

Well output

Well head

Casing

Production tube
(pumpset can be
installed directly
in the casing
where acceptable)

Power fluid
supply tube

Downhole
pumpset

Turbine outlet

Pump discharge

Sealing element

Sealing packer

Muleshoe guide

Figure 40.1
Hydraulic downhole
pumping system.

removal and reassembly and the consequent loss of production was clearly prohibitive.

A few years ago Weir Pumps plc with sponsorship from British Petroleum and the UK government initiated a development exercise to find a superior method of downhole pumping. Experience with boiler feed pumps for 660 MW turbo generators showed that centrifugal pump stage heads of 1200 m (4000 ft) were entirely practicable compared with the 25 m (82 ft) of a typical electric submersible pump. This meant that the speed of the electric submersible, 3500 r/min could be increased to 11 000 r/min to generate 250 m (820 ft) stage head quite safely.

Figure 40.2
Downhole turbine driven pumps 350 SHP

Increase of Rigidity

It was also shown that a drastic reduction in the number of stages resulted in sufficient rigidity to increase the operational life considerably and to obtain a thirty-fold increase in power from the same volume of pump and driver. A similar improvement of power per unit volume was seen in the use of hydrostatic motors instead of electric motors for aircraft control.

In order to provide high speed power the electric motor was replaced by a turbine, which, like the pump, was easily capable of a three-fold increase in speed for the same diameter, giving at least a thirty-fold increase in power. There was also a three- or four-fold increase in the ratio of starting torque to full load torque which with the long pumpset and curved well, was a vulnerable point: had the electric unit been able to start, the very severe cost of an overhaul could have been postponed. Considerable further advantages arose since the pump and driver length would be only one-tenth of the electric set, thereby being unaffected by the curvature of the well and being able to be delivered to the rig fully assembled.

This turbine was entirely novel. It was equivalent to powering a hydrostation with a 3660 m (12000 ft) waterfall, but it could be done, thereby eliminating most of the problems. There would be no need for the waterproof insulation in the motor, since the fluid powering the turbine could be drawn at rig level from the discharge, filtered and pumped down the hole to drive the turbine and discharged up the rising main with the pumped oil.

Operation

Prototype research for 450 hours showed virtually no wear. Since a full scale test would

*Figure 40.3 – SSP vertical pumps supplied to the
Esmond, Forbes, Gordon and Duncan fields.*

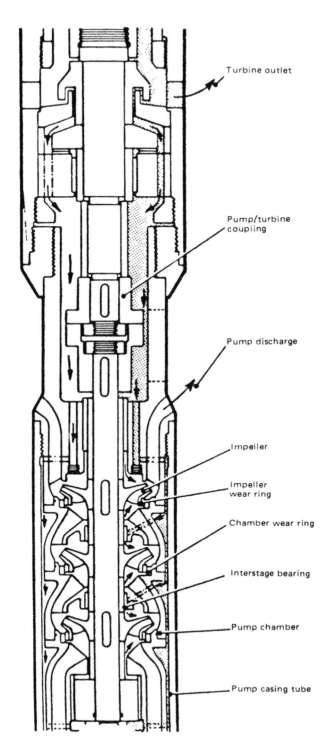

Figure 40.4
Clean turbine power
fluid supply to
all bearings.

necessarily involve some interruption to production it was fortunate that the UK Department of Energy was able to permit trials on a geothermal well 610 m (2 000 ft) below ground level containing brine three times stronger than sea water at 73°C. After 2000 working hours no significant wear or loss of performance occurred. The prototype is now installed raising water from an underground aquifer for reinjection into an oil well in the Middle East.

A most satisfactory result from these tests has been the freedom from starting up problems, the commonest cause of failure in electric submersibles and one of a special importance in fields where the oil is loaded directly from the well into tankers, and stopping and starting is frequent. During the 450 and 2000 hour tests, the pump was placed in various positions from vertical to horizontal and then stopped and started 500 times without failure.

Failures mean lost time and in the North Sea, for example, time is very big money. Even in the case of a relatively small well, stopping for a few days to replace a pump can cost £500 000 in wages, engineering costs and lost revenue. Existing pumps sometimes have a life before overhaul of only a few months. In such cases the new pump with a life measurable in years can save at least £1 million a year on every well. The implications are important, both in the short and long terms. The pump can be used for pumping sea water into the wells or in desert areas raising water from a few thousand metres deep to inject into an oil well or for general water supply. Most Middle East oilfields have a water bearing layer over the oil bearing layer so that one pump in one hole could obtain water and inject it downwards .

Simplification of Design

The long working life and maximum reliability arises from the absence of the electric motor, eliminating cables, sealing joints and temperature and insulation problems. The short rugged pumpset, about a tenth the length of an electric set eliminates handling and installation damage in spite of curved wells from vertical to horizontal axis. The very small size permits the use of the highest quality alloy steel and stellite for maximum resistance to abrasion, corrosion and erosion. All pump and turbine bearings are lubricated by the clean filtered power fluid, there being no mechanical seals because a minor leakage of filtered power fluid via the bearings into the pump discharge can only be beneficial. The pumpset is impervious to high well temperatures and repeated stop-start operations.

Since a fixed electrical frequency is eliminated, the pump speed can be almost infinitely variable for the greater part of the speed range by control at rig level, with economy of power on low rig outputs. The starting torque or the well output can be increased without raising the pumpset, merely by increasing the pressure and flow from the power pump at the rig. The pumpset is easily transported, requiring no site assembly and running into the well on the end of the power fluid supply tube. The set can be withdrawn leaving the production tube and sealing packer in-situ, the setting depth being independent of production tube length. Chemical dosing through the power fluid can protect the whole system against corrosion.

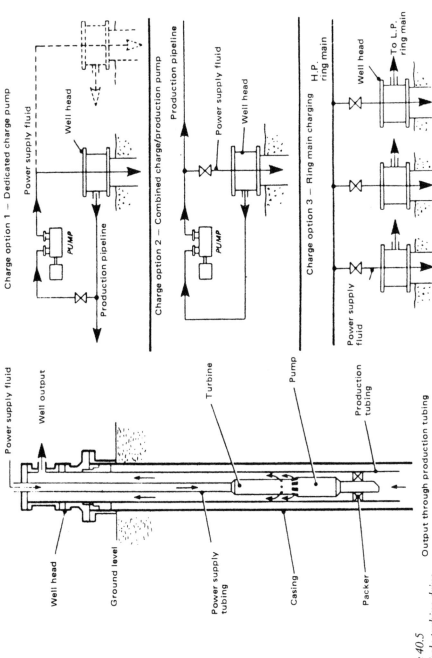

Figure 40.5
Downhole turbine drive
options.

Plate 40.6
Lowering electric motor and seal section
into the well.

High Frequency Drives

Some alleviation of the overall length can be effected by the adoption of a higher electrical frequency, but this involves separate generators, speed control etc and still leaves the electric cable which is liable to be damaged during the insertion of the pumpset into the oil well.

Variable Speed

Present systems convert the total power from a.c. to d.c. then invert the total power to variable frequency a.c. to drive the well pump. The Anderson transmission (Reference 62) converts the slip power only into hydrostatic power which is returned to the input shaft driving the generator feeding the well pump. The reliability and the overall efficiency are about the same as on the inverter drive, but the cost is halved and the volume occupied is only one-tenth of that of the inverter drive, a vital point on an oil rig.

Oil Well Content

Oil wells contain gas and water as well as crude oil. The analysis of oil well data and its interpretation in the selection of the optimum pump size, type speed and power are fundamental problems in oil rig practice. Leo V. Legg in Reference 63 (four papers) presents a comprehensive analysis of all these parameters. Further analyses of these aspects appear in References 64, 65 and 66. Variable speed is discussed in Reference 67.

Monitoring and Analysis of a Large Number of Oil Well Pumpsets

The American Petroleum Institute Document API RP I IS-1982, 'Operation, Maintenance and Troubleshooting of Electrical Submersible Pump Installations' Reference 68, is a vital publication for all oil well engineers since it contains firstly a warning, (which may appear obvious but its very mention shows its importance), to the effect that with the well liquid at the ground surface the full pump head will be imposed on the surface valves etc. Fifteen recording ammeter discs are thoroughly examined to illustrate a variety of faults and to suggest means of rectification. Gas and water content problems, solids difficulties etc are discussed in addition to the electric pumpset behaviour.

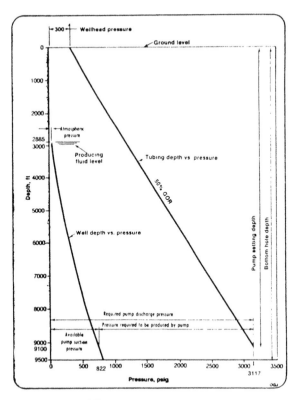

Figure 40.7
Well depth and pressure.

A complete textbook on Submersible Pumps is published by Centri-Lift Hughes (Hughes Tool Co) and gives a comprehensive guide to all aspects of oil well pumping.

References

60. RYALL., M. L and GRANT, A. A., Development of a new high reliability downhole pumping system for large horsepowers, European Petroleum Conference, London, October 1982.
61. Downhole turbine pump aims to cut downtime, The Oilman, February 1983.
62. ANDERSON, H. H. and ANDERSON, C. E. F., Hydrostatic variable speed drive, Patent Application No. 8508654 et al. Variable speed slip recovery drive. Power, May 1985, The Trade & Technical Press Limited.
63. LEGG, LEO V., Electric submersible pumping, Oil and Gas Journal Jl 9 August 27 1979.
64. BEAVERS, J., Gas calculations aid submersible pump selections, Petroleum Engineer, Jl 1981.
65. LEA, J. F., Gas separator performance for submersible pump operation, Society of Petroleum Engineers of AIME, SPE 9219.
66. LEA, J. F., Effect of gaseous fluids on submersible performance, SPE 9218.
67. DIVINE, D. L., A variable speed submersible pumping system, SPE 8241.
68. API RP 11S, Operation, maintenance and troubleshooting of electric submersible pump installations, American Petroleum Institute.
69. Submersible pump handbook, Centri-Lift Hughes (Hughes Tool Co).

DEEP SEA MINING

Considerable mineral wealth exists on the ocean floor, but its recovery presents a most difficult engineering problem. One example is the large tonnage of manganese nodules which are found in all ocean beds. A hundred years ago a manganese nodule was recovered from the sea bed by the British Research Vessel 'Challenger'. The piece of material the size of an egg contained manganese, iron copper and nickel with small traces of cobalt, molybdenum and titanium. It was not until the International Geophysical Year, 1957, that it was found that a vast wealth lay on the ocean bed. What was merely an academic novelty one hundred years ago will become in the future a major source of minerals, provided the technical problems can be overcome.

A very promising start has been made in a pilot exercise carried out by an International Consortium (USA, Canada, Germany and Japan) on the research ship 'Sedco 445' Reference 70 in 1965. The German pump makers KSB were given the task of designing, developing and manufacturing suitable submersible pumps. Deep sea mining is an entirely new field and involves considerable expense and technical research. From the ship a 200 mm (8 in) diameter pipe led vertically to the collector drawn along the sea bed at a depth of 5000 m (16400 ft), the final section of the pipe being flexible to allow for undulations of the sea bed.

Small submersible motors propelled the collector and cleaned the nodules of sea bed deposits. At 1000 m (3280 ft) below the ship three 800 kW, 4kV submersible pumpsets were suspended in the rigid pipe, slightly separated. A spherical clearance of 75 mm (3 in) in all passages of the 550 mm (21 in) diameter pumps was an important parameter in the determination of a vertical system with a flow velocity exceeding the natural sinking velocity of the nodules. The pipe system weighing 1000 tonnes had to remain in a fixed vertical position in order to ensure satisfactory collection. In this pilot exercise, 800 tonnes of nodules were raised.

Preliminary tests carried out twenty years ago confirmed that submersible motors could operate satisfactorily at depths down to 10 000 m (32 800 ft). The PVC and polyethylene windings were tested hydrostatically in autoclaves for mechanical and electrical properties up to 10 kV.

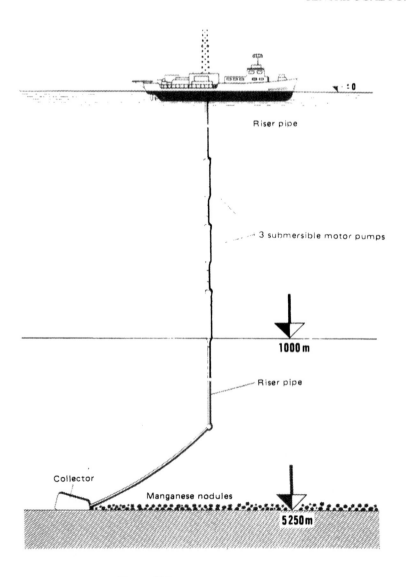

Figure 41.1
Complete diagram of lifting process.

   The submersible pumps were encased in a shroud of adequate area to pass the nodules and most important in tropical areas, to give adequate cooling to the motor. The rated capacity was 500 m cubed per hour (1830 gal/min) against a head of 265 m (870 ft). It is of course obvious that only the friction loss and the velocity head in the riser pipe, which is 5250 m (17 220 ft) long have to be overcome.

It was necessary to develop a special mixed flow hydraulic configuration to pass the 75 mm (3 in) sphere representing the worst flow pattern of the nodules and also to aim at an area ratio of unity which is the largest possible area for the impeller and diffuser passages. This area ratio also gives the same flow velocity to each component .

Cable connection is always a major problem and the final solution was to lead the ends of the winding out through the shroud and connect them to the armoured steel cable in a specially developed junction box which is watertight at a pressure of 100 bar. Submersible pumps have already proved reliable at greater sea depths, for example in a bathysphere at 10 000 m (32 800 ft).

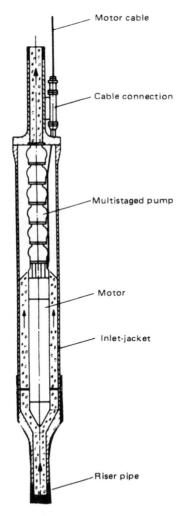

Figure 41.2
Submersible pumpset.

Figure 41.3
Delivery system for manganese nodules
with submersible motor pumps.

The overall picture of the manganese nodule recovery is seen in Figure 41.1 where the three submersible pumps in series occupy the 1000 m (3280 ft) region of the rising pipe. The general disposal of the pump and motor appears in Figure 40.2 with the flow passages shown large enough to pass the nodules but small enough to give a water velocity that will cool the motor and at the same time raise the nodules against gravity. Figure 37.1 illustrates one of the three pumpsets assembled in the factory prior to despatch.

References

70. Undersea mining for manganese nodules. (KSB), Pumps-Pompes-Pumpen. 1979-152.

FLUID STORAGE CAVERNS

Many countries have a target of three months critical reserve storage of hydrocarbons. Caverns, abandoned mines and newly created cavities below the water table serve as the most economical storage for the following reasons:

(i) Total costs are less than those required for conventional surface tanks .

(ii) The space saving is considerable. Land and transport costs are saved by underground storage local to the process.

(iii) The temperature in the cavern is constant at a convenient level.

(iv) The cavern is proofed against terrorists, bombs or vandals.

(v) Maintenance costs are extremely low.

In most cases the products, for example, crude oil or LPG will be floating above a surface of fresh or salt water which, having a higher specific gravity than the product, will stay below; any leakage of ground water into the cavern will also sink below the product and will prevent the product leaking from the cavern. Although shaft driven pumps have been used on caverns, the submersible pump is becoming the only practicable device, used in caverns world wide.

In Scandinavia storage caverns for petrol, diesel fuels, heavy oils. LPG etc are almost universal the technique having been developed in northern countries where the rock quality was good. The rock cavities were located about 30 m (98 ft) below the surface, the cavity being left 'as blasted' without additional treatment.

Pumps can range up to 2 MW for these duties which in the case of high viscosity, for example, 800 centistokes, may require heating to minimize pump power. It is here that the advantage of the cavern as a natural insulator virtually prevents heat dissipation.

Since there is a definite horizontal level separating product from water, submersible pumps, float controlled, can maintain the correct level, generally by means of a weir. The upper part of the cavern is gas tight and extreme care is taken to ensure that any inward loss of oxygen from the atmosphere is minimized to keep the cavern free from any explosion ratio of products and oxygen. Typical applications involve submersible pumps at depths of 500 m (1640 ft) for which 500 kW motors are required. The pumps must be capable of handling both water and light hydrocarbons since the cavity is filled with water during commissioning.

There are three different types of caverns:

(i) Cavern with fixed water bed, the height of which is maintained constant by two submersible pumps and a weir.

(ii) Cavern with fluctuating water bed, the level of the product being maintained at the required minimum height by pumping in or out of the water bed. The product thus rests at sufficient pressure above ground level, preventing ingress of ground water by the back pressure.

(iii) Cavern for storage of liquified petroleum gases. These caverns are equipped with casing pipes inside which the pumps are installed, suspended from their rising main. Installation and removal in LPG caverns are lengthy and involved, so that reliability of operation without removal for many years is essential.

Three different pump duties are involved: the product pumps, the leakage water pumps and the circulating pumps.

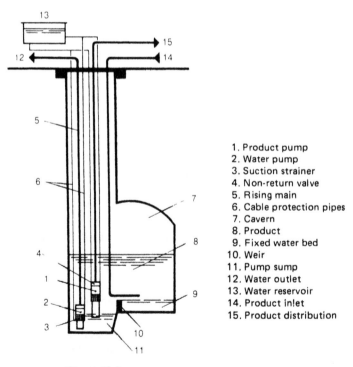

1. Product pump
2. Water pump
3. Suction strainer
4. Non-return valve
5. Rising main
6. Cable protection pipes
7. Cavern
8. Product
9. Fixed water bed
10. Weir
11. Pump sump
12. Water outlet
13. Water reservoir
14. Product inlet
15. Product distribution

Figure 42.1
Storage cavern with fixed water bed and
submersible pumpsets.

In Scandinavia alone there are about one thousand cavern stations in operation, 80% of the plant with submersible pumps. The vertical spindle pumps with the advantage of the motor outside the cavern suffer the disadvantages of the shaft drive and bearing maintenance, particularly on severe depths It is here that the use of submersible pumpset with its motor and bearings always in liquid, compared with the shaft bearings which would start dry, is becoming universal, minimizing foundations and space requirements.

Motor cables are led out of the cavern in water filled protection pipes rigidly attached to the motor, and leading to a water reservoir in an interconnected system always full of water at a greater pressure than the product. This ensures that the product cannot penetrate into the motor, any water leakage from the motor passing harmlessly to the sump below the product.

It is important to note that hydrocarbons have considerably lower heat conductivity than water and that the products are often stored at elevated temperatures, which are vital factors in the power that can be taken from a given motor size. Many plants are operating satisfactorily at a product temperature of 85°C.

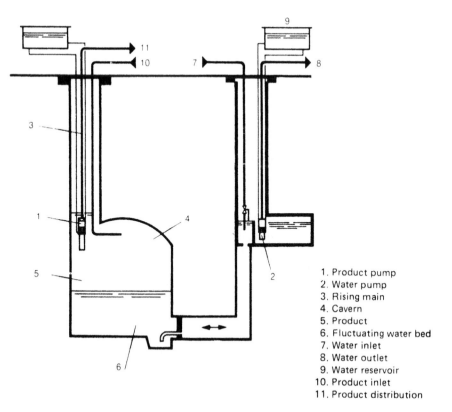

1. Product pump
2. Water pump
3. Rising main
4. Cavern
5. Product
6. Fluctuating water bed
7. Water inlet
8. Water outlet
9. Water reservoir
10. Product inlet
11. Product distribution

Figure 42.2
Storage cavern with fluctuating water bed.

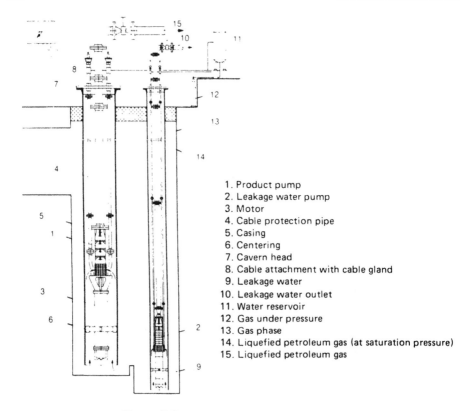

Figure 42.3
Liquified petroleum gas storage cavern
showing product and water pumpsets.

1. Product pump
2. Leakage water pump
3. Motor
4. Cable protection pipe
5. Casing
6. Centering
7. Cavern head
8. Cable attachment with cable gland
9. Leakage water
10. Leakage water outlet
11. Water reservoir
12. Gas under pressure
13. Gas phase
14. Liquefied petroleum gas (at saturation pressure)
15. Liquefied petroleum gas

Tank Storage

A similar duty is the provision of submersible pumps tor emptying and filling tanks of crude oil, LPG and LNG etc. The submersible has the advantage of being installed through the top of the tank, which, by inert flushing, can facilitate an explosion proof situation.

The tanks can be buried or free standing without openings below the liquid level. A parallel duty is that of glandless circulating pumps for transformers.

The usual position of the pumps in an oil storage system is indicated in Figures 42.1 and 42.2, with respect of the product level and the water bed. Alarms and monitors effectively control the storage exercise. The storage system for liquid gas is seen in Figure 42.3, where the pumps are inserted into vertical pipes filled with an inert gas during erection. The whole system is situated below the water table.

References

50. KUNTZ, G., Submersible motor pumps for offshore and onshore engineering applications, KSB Technical Report d/2 and Pumps-Pompes Pumpen, 1979-154

THERMAL POWER STATION PUMPS INCLUDING NUCLEAR

The three basic submersible types. (i) motor in air with oil barrier: (ii) stator and rotor in water, ie wet motor and (iii) stator in air, rotor in water, ie canned motor, render valuable service in thermal power stations offering the advantages of greater rigidity, product lubricated bearings, simplicity of piping arrangements often without inlet pipes and with self sealing main joint minimum maintenance and high reliability.

Figure 43.1 shows the general disposition of power station pumps, the majority involving modest heads, but often with severe abrasion or corrosion problems.

The main circulating pumps generally geared, and the feed pumps of very high powers and pressures are unsuitable duties for submersible pumps, but all the remaining duties are ideally suited to the various advantages offered in the submersible ranges.

Heater drain pumps, normally operating on a fairly high temperature, can be subjected to vacuum inlet pressures during turbo-generator transients. To avoid any risk of air drawing at the glands of a non submersible pump, which would lead to oxygen entering the feed water, a wet motor is preferred as shown in Figure 43.2. Here the inlet is pointing upwards and embodies an inducer or axial impeller to minimize NPSH and a diffuser in the casing to minimize radial thrust. The pump and motor casings are heavy pressure vessels to withstand the heater pressure and are separated by a thermal barrier with watercooling to protect the motor. A cooling water impeller at the bottom of the motor circulates the filtered water.

Boiler circulating pumps handle four times the feed water flow to cope with the very high heat rate of the boiler, but, at a modest pressure, this involves a single impeller. The water temperature is higher than that of the heater drain pumps so that more elaborate cooling and a large separation of pump and motor is required as shown in Figure 43.3. This illustrates the cooling effort being applied primarily to the shaft between the pump and the motor. The pressure vessel of pump and motor is particularly heavy to sustain boiler pressure and is provided with a double volute to reduce radial thrust.

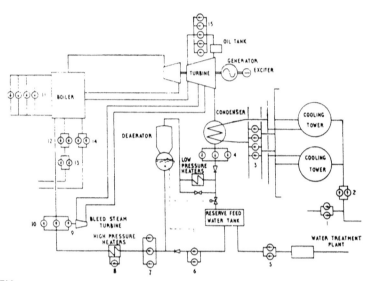

KEY:

1. Screen washing pumps
2. Cooling water make-up pumps
3. Cooling water main pumps
4. Condenser extraction pumps
5. Feedwater make-up pumps
6. Emergency reserve feedwater pumps
7. Booster feed pumps

8. Heater drain pumps
9. Main feed pump
10. Starting and standby feed pumps
11. Boiler circulating pumps
12. Ash booster pumps
13. Ash main pumps
14. Ash sluicing pumps
15. Lubricating oil pumps

Figure 43.1
Diagrammatic arrangement of pumps
in a thermal power station.

Nuclear duties demand complete hermetical sealing of the pumped fluid which may be water or sodium transferring heat energy from the reactor to the steam cycle.

A diffuser type of pump within a forged steel casing embodying an internal thermal shield appears in Figure 43.4. The thrust block acts as a pump to circulate the internal cooling system. The thin nickel alloy can is clearly seen inside a thicker tube to hold the pressure difference outside the stator stampings. There is also an outer lower pressure cooling chamber surrounding the stator.

In contrast, a cast casing primary circulation nuclear heat transfer pump appears in Figure 43.5. The cast casing offers the advantage of a double volute to reduce radial thrust which is not easily arranged in a forging. A separate cooling circuit is provided outside the stator in the form of coils fed by an outside source. The internal bearings are of carbon graphite compacted material lubricated by the pumped water via an internal impeller.

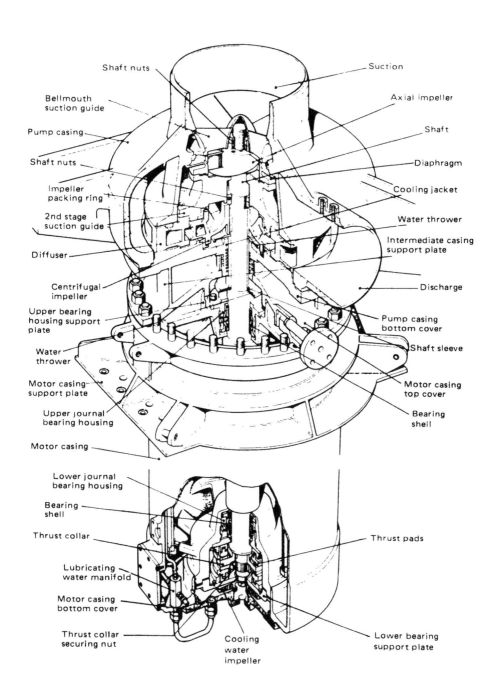

Shaft nuts

Suction

Bellmouth
suction guide

Axial impeller

Pump casing

Shaft

Shaft nuts

Diaphragm

Impeller
packing ring

Cooling jacket

2nd stage
suction guide

Water thrower

Diffuser

Intermediate casing
support plate

Centrifugal
impeller

Discharge

Upper bearing
housing support
plate

Pump casing
bottom cover

Water
thrower

Shaft sleeve

Motor casing
support plate

Motor casing
top cover

Upper journal
bearing housing

Bearing
shell

Motor casing

Lower journal
bearing housing

Bearing
shell

Thrust collar

Thrust pads

Lubricating
water manifold

Motor casing
bottom cover

Thrust collar
securing nut

Cooling
water
impeller

Lower bearing
support plate

Figure 43.2
Wet motor pump for heater drain duties.

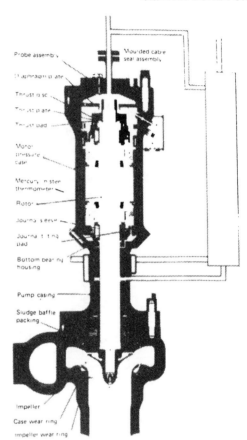

Figure 43.3
Glandless motor pump unit
embodying wet rotor and
stator windings.

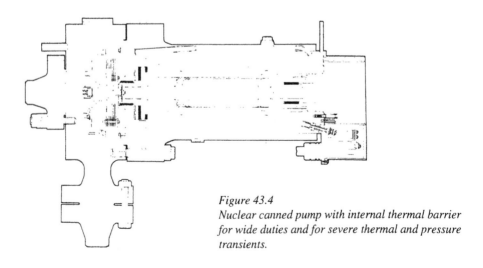

Figure 43.4
Nuclear canned pump with internal thermal barrier
for wide duties and for severe thermal and pressure
transients.

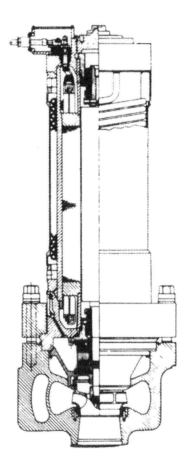

Figure 43.5
Nuclear reactor primary coolant pump,
300 l/sec, 95 m at 3550 r/min.
Wet rotor with can, dry stator windings.

References

71. RODDIS, L. H. and SIMPSON, J.W., Nuclear propulsion plant, *USS Nautilus*, Westinghouse Engineering, March/May 1955.

72. Pumps for nuclear power plants, IME Conference, Bath 1974.

OFFSHORE MARINE AND AEROSPACE GLANDLESS AND SUBMERSIBLE PUMPS

Offshore marine and aerospace duties are more severe than those on land for the following reasons:

- Failure of pumps could have fatal results.
- The pumps must accept changes of gravity direction caused by the rolling, pitching and accelerations of the vessel or aircraft. These changes can severely prejudice NPSH levels.
- During warfare very large inertia forces can be imposed on the pumps from explosives.
- The isolation of the oil rig, vessel or craft in ocean or air seriously limits maintenance and repair facilities.
- Fire possibilities are immeasurably more hazardous than on land.
- Weight and size must be kept to an absolute minimum, especially in the air.
- The normal cooling fluid, instead of being simple water, must be corrosive sea water or rarefied air.

The above considerations have engendered an increase of pump speeds in order to Increase rigidity. For example 60 cycle frequency is normal at sea compared with the 50 cycles of the old world. Steam turbine drive is preferable. Firstly because it lessens fire risk. Secondly because the pump speed can thereby rise to 10000–20 000 r/min compared with the electrical limit of 3600 r/min and thirdly because of the greater availability of steam. Gas turbine drive offers two of these advantages.

Submersible and glandless pumps driven by electric motors at 60 Hz or higher frequencies, or by steam gas or liquid turbines offer the advantages of very high speed and minimum weight ideally suited to the severe environments. The machines are of course totally enclosed.

Glandless Combined Steam Turbine and Pump Lubricated by Pumped Water (Weir Pumps plc)

The steam turbine driven water lubricated feed pump (TWL) is an excellent example of the aforementioned requirements of minimum weight, steam drive and complete simplicity.

Lubrication

In pumping installations where space and weight are critical, oil lubrication creates problems.

Firstly because a constant supply of oil, clean and at the correct temperature must at all times be available to the pump: secondly, as this oil has to be clean and cool, a number of sub-units are required to fulfil these tasks, such as a filter, cooler, pump and drive, etc, each one adding to the weight and size, not to mention cost and maintenance of the installation.

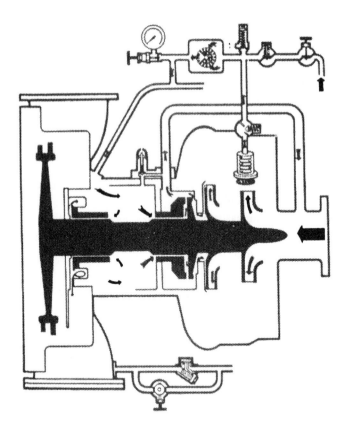

Figure 44.1
Steam turbine driven water lubricated feed pump.
Space rocket pumps have similar configuration.

Today, where continuous service and trouble-free operation are essential, the risk of failure and shutdown must be reduced to a minimum and therefore the fewer the complications the better.

Water lubrication is the obvious solution: it is handled by the pump, therefore it could be employed to do the work instead of oil. Water lubricated bearings had been used successfully on electrically-driven glandless high pressure pumps, and it was recognized that if they could be used on a small boiler feed pump the design would be greatly simplified. The pumped water, at suction pressure, flowing axially through the bearings would provide the necessary lubrication and eliminate the pump and turbine glands, oil pump and drive, cooler, relief valve and sump and would, therefore, reduce the overhang of turbine wheel and pump impellers to a minimum. From this design study emerged the water-lubricated turbo-feed pump now used in many of the world's major marine installations. Figure 44.1 shows water circulation in this combined turbine pump.

This unit is simple, robust, completely reliable and entirely self-sufficient with respect to lubrication, the complete assembly being housed in a single casing with no external openings. Axial movement of the shaft is controlled from within the unit by an automatic hydraulic balancing arrangement, and any residual thrust when starting or stopping is taken by a thrust ring. (See Reference 36).

Bearings and Balance Disc

Two bearings located within the casing between the turbine pump ends support the complete shaft assembly, utilizing split journal bushes surfaced with PTFE. The axial thrust is absorbed by the balance disc and a non metallic thrust ring (Figure 44.2).

An inducer provides the pump with enhanced suction performance which is reflected in reduced suction head requirements and higher running speed, (Figure 44.3).

The Hirth coupling allows relative thermal movement to take place between the wheel and the shaft, during low load conditions, without disturbance of alignment or imposition

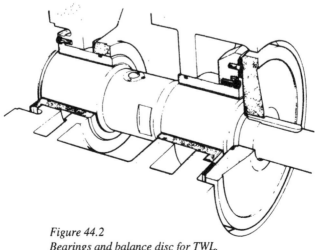

Figure 44.2
Bearings and balance disc for TWL.

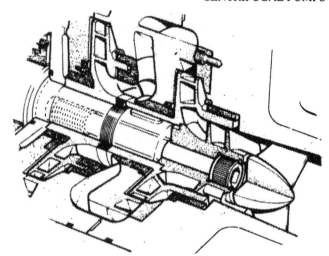

Figure 44.3
Inducer for TWL.

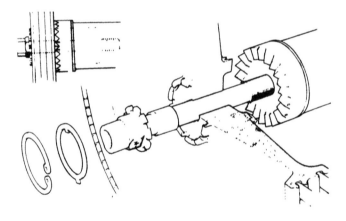

Figure 44.4
Hirth coupling for TWL.

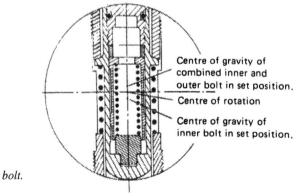

Centre of gravity of
combined inner and
outer bolt in set position.

Centre of rotation

Centre of gravity of
inner bolt in set position.

Figure 44.5
Emergency covernor bolt.

of thermal stresses. The wheel can also, because of this design, be easily separated from the shaft simply by the removal of a single central bolt. (Figure 44.4).

Running speed is normally controlled to suit the boiler demand by a relay steam governor using a signal from pump pressure or flow. An emergency overspeed trip initiated by a diametrical bolt in the middle of the shaft between the bearings. This bolt has a central element which is slightly eccentric, held in radial position by spring. At overspeed the centrifugal force of the eccentric bolt mass overcomes the spring tension and the bolt flies out radially to trip a steam valve. Figure 44.5 shows a relay bolt where the small hermetically sealed inner bolt/spring assembly reasonably free from static friction and corrosion, acts as a trigger to release the larger outer bolt and spring which in turn trips the steam valve.

Figure 44.6
Glandless water lubricated steam turbine driven feed pump.

The water lubricated turbine pump operates on duties up to 100 l/s at 100 bar for marine boiler feeding (Fig 44.6).

Miscellaneous Marine Pumps

There are a host of relatively small pumps on extraction, distillation, transfer, bilge and

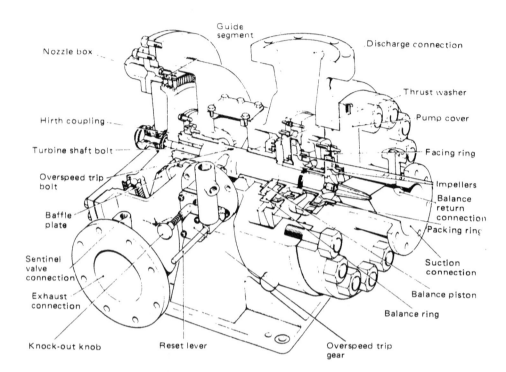

Figure 44.7
Section of water lubricated feed pump (TWL).

fire duties. etc, several of them being self priming and usually having bronze casings in view of sea water and sea atmosphere corrosion. Many of these duties can be met by submersible or glandless pumps.

Bow thrusting duties are a typical field for glandless pumps since they operate necessary under sea level. Reversible motors and variable pitch propeller are involved.

Aerospace Pumps

Here the paramount requirement is minimum weight which necessitates maximum possible pump speed, generally driven by steam or gas turbine. Inducers giving inlet shape numbers in the range of 40 000-60 000 are essential to help solve the problem of weight reduction by speed increase. For space vehicles the operating life of the pump may be less than one hour, permitting a predetermined rate of cavitation attack, but for jet planes the fuel boosting duties demand a longer life, for example, a few minutes several times a day, at take-off, so that operation, reasonably clear of cavitation, is essential. Titanium is preferable to stainless steel for aerospace pumps.

Stage heads range up to 2000 m (6560 ft) in order to minimize weight - the fluid being kerosene or similar hydrocarbon, is less prone to attack metal than water, thereby permitting higher stage heads. The glandless design principles just described apply equally to aerospace pumps.

The future increase of size of aerospace pumps will involve operation up to and beyond the limits of metallurgy and it is here that newer materials, for example the carbon fibre developments and the ceramics described in Chapter 15, will be used. The compressor of the gas turbine, jet plane or space vehicle is really a fluid pump and the liquid pumps for similar duties will tend to follow gas turbine developments.

Numerically controlled generated flow surfaces with high surface finish will be essential.

References

36. ARKLESS. G. F., The development of the water lubricated feed pump, Proc IME 1963, Vol 177, No 691.

73. Pumps and compressors for offshore oil and gas, Conference. IME Aberdeen. 1976.

REFINERY CHEMICAL AND PROCESS PLANT – PROPERTIES OF LIQUIDS

This chapter considers the vast array of submersible and glandless pumps in the general process sense of industries not already discussed. It is also a convenient time to investigate how design and materials of pumps are affected by the liquids handled and, therefore, a study of liquid properties follows.

The special metals and design features required in chemical pumps are reviewed from the aspects of manufacture and operation. Continuous flow production of product has replaced the previous batch production and this has largely arisen through the ability of submersible pumps to give the exact flow required under automatic or computer control. The considerable development of seals, of metals, plastics and ceramics in the last few years has provided the pump maker with the means to supply reliable plant for the most severe chemical duties. In general all Newtonian liquids can be handled by submersible pumps.

Chemical Duties

In general chemical duties include all liquids which present special problems of corrosion, erosion, high or low temperatures, vaporizing, high viscosity and tendency to choking. In this chapter chemical duties are considered in the widest sense. The chief difference between a chemical pump and a water pump is the provision of special materials and designs to resist corrosion. These features must protect not only the flow passages but also the bearings and, most important of all, the personnel who breathe the atmosphere surrounding the pumps.

Nature of Liquids - Acidity and Alkalinity

Liquids can be classified according to the hydrogen ion concentration or pH value. Speaking generally, liquids with pH values below 7 are acidic and require stainless materials, bronze, iron, lead or plastics. The pH value of 7 is neutral or inert. Above pH7 the liquid is alkaline and is liable to attack bronze so that other suitable stainless materials must be used.

Aggressive Nature of liquid

Arranging the liquids in order of severity with respect to pumps we obtain the following divisions:

> Water duties and stable hydrocarbons
>   (aliphatic and aromatic compounds)
> Mild chemicals
> Substituted halogen hydrocarbons
> More severe chemicals
> Acids
> Slurries

Raw Materials

The substances for manufacturing processes are drawn from the atmosphere, the hydrosphere and the lithosphere for example, nitrogen from the atmosphere for producing explosives and fertilizers, magnesium and other metals from the sea, minerals, crude oil and china clay from the earth. These substances may be solid, liquid or gaseous but the easiest way to handle any substance is to pump it in liquid form. It is therefore convenient to liquidize gases to reduce their bulk both for handling and storage, and to convert solids into finely divided powders which can be suspended in liquids to form slurries which can be pumped. Solids can sometimes be melted or dissolved to assist handling. Liquid and gaseous supplies of raw material are therefore preferable, since the whole process can then be handled by pumps which are often the only moving power-driven units in the plant.

Viscosity and Specific Gravity

Submersible and glandless pumps can handle viscous liquids with, however, a reduction of efficiency. Beyond a certain limit of viscosity it may be cheaper to heat the pipes and pump casing with a steam jacket so as to keep the product in a sufficiently liquid state to be economically pumped. Hydrocarbons will be considered in this context since they afford a good example of how viscosities and specific gravities vary with temperature and molecular complexity.

Hydrocarbons

The winning and refining of hydrocarbons and other products of the oil industry include some of the most arduous duties that submersible pumps have to perform. Cracking and other refinery processes introduce problems comparable with those on boiler feed pumps and, generally speaking, their solution requires a pump similar to a boiler feed pump.

   Many other special features of oil, however, call for special consideration in the application and in the design of oil pumps.

   Tetlow (Reference 74) gives a very useful survey of the problems of oil field and refinery duties and of the design of pumps to meet these duties.

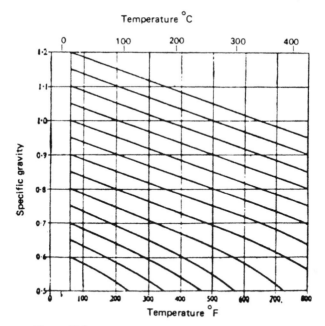

Figure 45.1
Approximate variation in specific gravity of
hydrocarbon products with temperature.

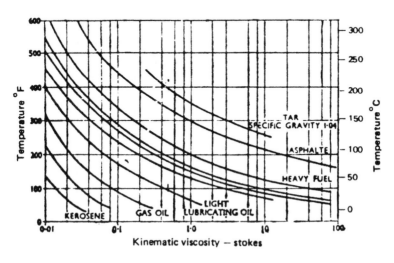

Figure 45.2
Typical viscosity-temperature characteristics
of hydrocarbon distillates.

Properties of Hydrocarbons

Specific gravity, viscosity and saturation pressure of the oil at the pumping temperature must be determined, since these properties vary considerably with temperature and may be specified to the pump manufacturer at temperatures other than the pumping temperature.

Figure 45.1 shows the variation in specific gravity of hydrocarbon products with temperature; Figure 45.2 shows the variation of viscosity with temperature, whilst Figure 45.3 shows the viscosity conversion figures for the various viscometers in service, (from Reference 74).

Variation of Pump Performance with Specific Gravity and Viscosity Effect of Specific Gravity

Since a submersible pump is a dynamic machine, it follows that for a given point on the characteristic and for a given speed the volumetric quantity (l/s or m³/h) and the head in metres (feet) of liquid pumped will remain constant. Reduction of specific gravity will not affect flow or head, but will reduce output if expressed in kilograms of liquid per minute, will reduce pressure in bar, and will reduce power. The reductions are proportional to specific gravity.

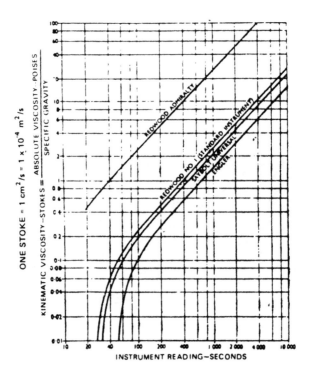

Figure 45.3
Approximate viscosity conversion figures:
Engler degrees = Engler seconds ÷ 51.3.

Effect of Viscosity

The effect of normal water viscosity (one centistoke or 11×10^{-4} m/5m²/s) has already been described in Chapter 7. For greater viscosity, reduction of quantity, head and efficiency occurs, the reduction being more pronounced in smaller pumps.

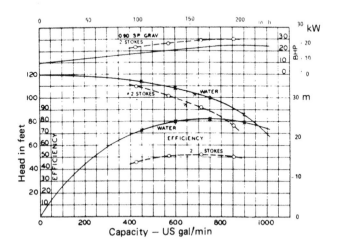

Figure 45.4.

EXAMPLE OF CHART DUTIES

	0.6 x QNW	0.8 x QNW	1.0 x QNW	1.2 x QNW
Water capacity (QW) m3/h	102	136	170	204
Water head in metres (HW)	34	32	30	26
Water efficiency (EW)	72.5	80	82	79.5
Viscosity of liquid	10000 SSU			
CQ – from chart	0.95	0.95	0.95	0.95
CH – from chart	0.96	0.94	0.92	0.89
CE – from chart	0.635	0.635	0.635	0.635
Viscous capacity – QW x CQ	96	130	162	195
Viscous head – HW x CH	33	31	27.5	23
Viscous efficiency – EW x CE	46	50.8	52.1	50.5
Specific gravity of liquid	0.90			
BHP viscous	17.3	19.4	21.4	22

These duties are shown in the dotted line with arrows in Figure 45.4 to explain the procedure.

Correction Charts

Charts for various viscosities appear in Figures 45.4 and 45.5 (Standards of Hydraulic Institute USA).

It is most important to ensure that adequate power for the highest operating viscosity and the highest specific gravity is provided, since viscosity varies greatly with temperature. For certain highly viscous duties, steam jacketing may be required.

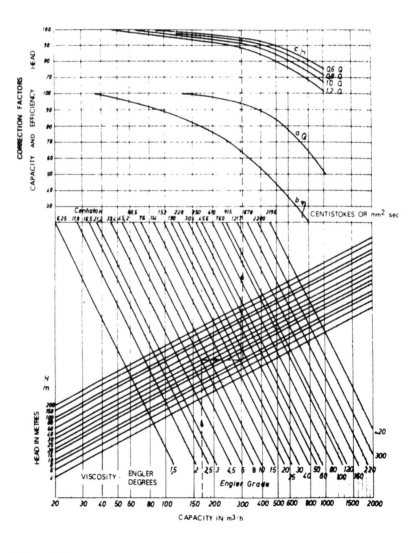

Figure 45.5.

Liquid Properties Affecting the Design of Pumps

The following data are essential to the investigation of any pumping problem:

Quantity
Pressure
Temperature
Specific gravity
Chemical nature
Erosive property
Tendency to coke and/or line the passage surface
Viscosity
Vapour pressure
Inlet conditions
Flammability
Toxic tendencies
Possible lack of lubrication properties
Choice of metals
Nature of drive - flameproof motors
Type of gland seal
Need for thrust bearings
Provision for cooling leak-off
Possibilities of freezing on low specific gravities or liquefied gases
Non-overloading aspects
Quench or smother gland requirement
Need to avoid non-ferrous metals
Radioactive properties
Motor in separate room for safety against fire
Tendency to clog or to change stable nature during pumping eg separation of constituent parts.

Pressure and Temperature

For moderate temperatures, refinery and process engineers will often specify a higher table of flanges than the pressure apparently needs in order to ensure absence or leakage. Studded flanges are avoided since a broken stud is more difficult to replace than a broken bolt.

Many joints may have steam-sealing, that is to say, a steam groove running circumferentially halfway across the joint. For the higher pressure and temperature ranges all-welded equipment is usual.

The design of a pump to suit the combination of pressure and temperature involved, in general, forged steel pressure vessels which are preferable since castings can never be guaranteed entirely free from porosity. Thermal barriers are required on high liquid temperatures.

Corrosion and Erosion

Many crude oils contain acids causing a combination of erosion and corrosion at the impeller periphery and require the use of stainless steels.

Saturation Aspects

Many duties involve handling liquids very close to conditions of saturation, so that the pump is a combination of an extraction pump and a multistage pump. Appropriate impeller design for low NPSH is then needed.

Coking

Certain oils tend to deposit hard carbon on the pump surfaces and for these duties a simple volute casing may be preferable to the use of diffusers. In a similar manner, balance drums for coking duties may be preferable to balance discs. A mechanical thrust disc in addition to the balance disc or drum is often provided.

Flammability, Poisoning, etc.

The slightest leakage or gland loss in an oil pump can be very serious indeed because it may fill the pump house with a toxic vapour. Where oil is pumped above its flash point, the slightest leakage appears as a flame, and any failure of pipe or pump involving copious loss of liquid would be a disaster of the first magnitude since the power house would be filled with flame. This emphasizes the vital nature of the pressure vessel and also illustrates the need for satisfactory seals where the shaft enters the pump, and for smother or quenching chambers where steam or water can carry to a safe destination any small continuous seepage from the seal.

Radioactive liquids require glandless pump driver units since entire freedom from leakage must be guaranteed.

Mechanical seals are almost universal in the oil industry, together with breakdown bushes and adequate disposal arrangements.

Light Fractions and Liquefied Gases

Here the problem is lack of lubrication which, on a multistage pump, calls for a rigid rotor.

Use of Non-Ferrous Metals

On certain high temperature duties (eg 300 degrees C) ferrous metals only are used as they are less likely to melt in case of fire. A non-ferrous portion might melt and permit a further oil leakage which would only feed the fire.

Nature of Drive

Oil pumps are generally driven by a.c. electric motors, or by a steam turbine. In many cases, the motors were housed in a different room from that containing the pumps in order to minimize risk of fire, a flameproof wall box and fire wall separating pump and motor. This is not now required since a submersible pump is generally regarded as flameproof.

For oil pipelines, steam turbine driven pumps have the advantage of flexibility of speed

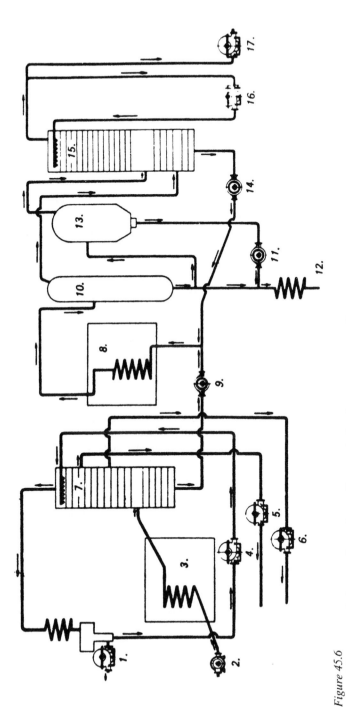

Figure 45.6

Typical diagrammatic layout of a modern refinery unit showing arrangement of pumps.

1. Benzene pump; 2. Crude charge pump; 3. Furnace; 4. Reflux pumps; 5. Kerosene pump; 6. Gas oil pump; 7. Fractionating column; 8. Furnace; 9. Reduced crude high pressure hot charge pump; 10. Reactor chamber; 11. Hot residue pump; 12. Residue; 13. Flash chamber; 14. High pressure recycle oil pump; 15. Fractionating column; 16. Reflux pumps; 17. Pressure distillate pump. Many of these pumps can now be submersible or glandless units.

when the pump characteristic at optimum efficiency will often be found to agree reasonably with the pipeline characteristic. A steam turbine submersible for these duties is not impossible.

Refinery Layout

Figure 45.6 shows a typical diagrammatic layout of a refinery unit with the various pumps involved. Many of these pumps can now be submersible or glandless.

Erosion

The pitting and removal of material from a pump by flow of abrasive and corrosive liquid appears to be minimized as much by stainless properties as by the hardness of the pump materials. This fact accounts for the very considerable use of austenitic stainless steel in pumps since it has high resistance to corrosion and work-hardening properties. For the most abrasive duties, for example, quartz slurries, super-saturated solutions with suspended crystals, etc, the use of manganese steels, rubber lining and ceramics is necessary.

Pumping of Solids

Pumps to handle solids suspended in liquids must obviously have large flow areas in order to permit passage of the solids without risk of choking. Solids may be large, for example, sugar beet, small, for example, various chemical slurries or uniformly suspended, for example, paper pulp. Small solids in general pass through the impeller but in some cases pulps and larger solids are handled by impellers which are out of the main liquid flow and produce a pumping effect by rotating an adjacent vortex of liquid. This has the effect of minimizing any choking risk and permits the handling of very thick pulps and slurries.

High and Low Temperatures

The materials in a pump must obviously have adequate strength and stability at the normal room or site temperature and at the final pumping temperature. Equally vital is the ability of all parts of the pump to withstand thermal shocks and the resulting differential expansions and brittleness risks, without distortion, malalignment or failure. The driving unit must be capable of accepting, often on a short-term basis, the starting and the warming up load associated with the higher viscosity of cold liquid.

Centre line suspension is required on high temperatures to ensure that the shafts of driver and driven unit are always in line irrespective of temperature variations. Spacer shafts and thermal barriers are often incorporated in high or low temperature sets.

Vaporizing Duties

A pump handling a vaporizing liquid must have its centre line at a distance below the free surface of the boiling liquid equivalent to the NPSH required by the impeller. Net Positive Suction Head (NPSH) is the head difference required to pass the specified quantity of liquid into the impeller without vaporizing in the inlet passages to the prejudice of performance. Figure 45.7.

Freezing Problems

At the other end of the liquid range, consideration must be given to the risks of pumped

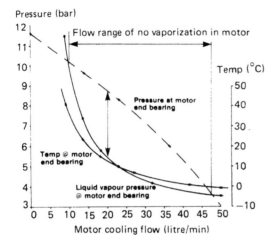

Figure 45.7
Fluid conditions inside a 55 kW
motor – chlorine @ 10°C pumping
temperature.

liquid freezing at any points of contact with the atmosphere (for example, at glands), and also of local freezing of water vapour in the atmosphere to the possible prejudice of performance. Freezing can often be of value as in the case of pumping liquid sodium, where liquid metal escaping at a shaft bushing solidifies to produce a close clearance minimum leakage seal, local heating providing sufficient liquid to lubricate, but avoiding excess leakage.

Properties of Water

Natural water from rainfall, boreholes and rivers contains dissolved and suspended matter especially compounds of calcium, carbon, chlorine and sulphur; the proportions and states of these impurities determine the nature of attack or of deposition of protective or of obstructive matter. The following useful evaluation is given by courtesy of EMU Unterwasserpumpen of HOF Germany. Further authoritative data appears in H.S. Campbell's paper to the British Non Ferrous Metals Research Association, Reference 76.

Natural Water-Evaluation of Water Analysis

An approximate evaluation of a chemical water analysis is possible by the following procedure:

Eliminate successively the criteria which do not apply until you get evaluation applying to the remaining data. See Figure 45.8.

Evaluation for Corrosion of Iron

Carbonate hardness* < 6° dH
Free oxygen < 4 mg/l
pH ≯GW
harmless

pH < GW
iron is increasingly attacked as pH value decreases

*The carbonate hardness is determined by the calcium bicarbonate contents $Ca(HCO_3)2$ and is part of the total hardness. The total hardness itself is not decisive for corrosion.

°dH	°F	°UK	°USA
1.000	1.79	1.25	17.85
0.560	1.00	0.70	10.00
0.800	1.43	1.00	14.28
0.056	0.10	0.07	1.00

Carbonate hardness < 6° dH
Free oxygen > 4 mg/l
   Iron is increasingly attacked as O_2 content increases

Carbonate hardness ⩾ 6° dH
Free oxygen ≈ 0
   pH ⩾ GW (pH)
   harmless
   pH < GW (pH)
   iron is increasingly attacked as pH value decreases

Carbonate hardness > 6° dH
Free oxygen ≈ O

Free CO_2 ⩽ GW (free CO_2)
or pH > GW (pH)

Free CO_2 > GW (free CO_2)
or pH < GW (pH)
   iron is increasingly attacked as O_2 content increases.
   In case of blisters: corrosion holes
   GW is the saturation index

Dry evaporation residue (salts) ⩽ 500 mg/l
   harmless

Dry evaporation residue > 500 mg/l
   avoid certain metals as electro-chemical corrosion will increase with increasing content of dissolved salts (higher electric conductivity).
   Particular attention has to be paid to choice of materials, avoiding metals of differing potential.

Chlorides: < 150 mg/l Cl
   harmless

Chlorides > 150 mg/l Cl
   Avoid differing metals. Hole corrosion is possible.
Free acids: Humic acids, eg organic acids
Hydrogen sulphide, attacks iron

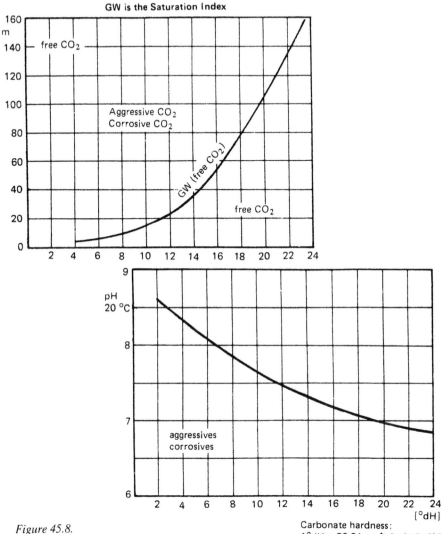

Figure 45.8.

Carbonate hardness:
$1°dH = 29.91$ mg/l Ca $(HCo3)2$
1 mg/l Ca$(HCo3)2 = 0.033°dH$

Evaluation of Danger of Deposits

Lime
Free $CO_2 > GW$ (free CO_2)
 no deposits
Free $CO_2 < GW$ (free CO_2)
 Deposits increasing CO_2 admixture. If free $O_2 \approx O$, deposits will be like mud, it free
 $O_2 = 0$, there will be scales.

Iron < O.2 mg/l
no precipitation: > 0.2 mg/l
  ochre mud deposits increasing with increasing Fe and O_2 contents.

Manganese: < 0.1 mg/l
  no precipitations
  >0.1 mg/l

Manganese deposits increasing with increased Mn and O_2 contents.

Iron and manganese precipitations cannot be avoided by any means. A mechanical cleaning of pumps and motors is indispensable from time to time.

Evaluation of expected wear

Sand contents in g/m^3
  Sandy water will cause erosion of rotating parts and casings even in low concentration.
  This danger is influenced by consistence of sand, grain size, type of grains and minerals and quantity.
  Sand contents up to 25 g/m^3:
  Protect pumping unit against sand.
Sand contents > 25 g/m^3: harmful

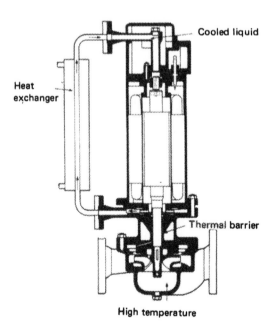

Figure 45.9
Canned unit for hot liquids.

Thermal Controls

Liquids may require temperature control to avoid vaporizing, solidifying, depositing choking layers on the passage surfaces, or to avoid changing the chemical nature of the liquid or the passage surfaces.

Such temperature change could be up or down, provided by steam or electric heating or by water or other liquid cooling. Figure 45.9.

In particular, cooling may be necessary on canned and wet motors to carry away the heat due to electrical resistance and any possible heat conducted from the pump when hot liquids are being handled. In the case of cryogenic fluids it may be necessary to heat the motor to prevent deterioration of the winding insulation.

Hydraulic Configuration

For the majority of process duties the flow quantity is low enough to permit single entry impellers at 2-pole motor speeds, thus permitting the simplest design of combined motor pump with right angle branches for horizontal shaft and I- or U-branches for vertical in-line arrangement.

Configuration of Pumps

In general the majority of chemical pumps are single stage single entry, driven by 2-pole motors at 2900 r/min or, in the USA at 3500 r/min. The pump arrangements to integrate most effectively into a process plant are frame mounted with separate motor, in-line U-type and in-line I-type. (Figure 45.10).

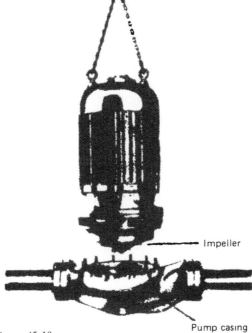

Impeller

Pump casing

Figure 45.10
Canned motor pump.

Figure 45.11
In-line process pump.

The frame mounted pumps generally have a horizontal shaft carried in two bearings with axial inlet branch and radial upward discharge branch. The drive may be via a spacer coupling thus permitting the withdrawal of the impeller shaft, bearings and seal out of the pump towards the motor without disturbing motor or pipe joints. This is referred to as back pull out and can also be arranged on vertical sets.

In-Line Pumps

In-line pumps with I-configuration are bolted into a straight run of pipe line which often takes the weight of the set. The U-configuration is similarly arranged but in this case, the inlet and outlet pipes are at the same side of the pump hence the U-designation. In-line pumps generally are combined with the motor to a compact set, the pump impeller being on the motor shaft with back pull out, but in this case the motor and the cabling must be disturbed.

Shaft Material

Where an impeller for chemical duties is fitted to a motor shaft it is usual to make the pump portion of stainless steel but the motor portion, for magnetic reasons, must be of mild steel. This problem is generally solved by the use of a friction weld between the two portions of the shaft or by some form of mechanical attachment.

Canned and Wet Motors

For entire freedom from gland leakage a totally enclosed motor pump unit with the rotor in the pumped liquid (canned) or the whole motor in the pumped liquid (wet) may be used, particularly on radioactive liquids.

An *Artificial Heart Power Source* appears in Figure 45.12.

Detail Design of Pump for Non-Abrasive Liquids

This pump would adopt a conventional closed impeller of stainless steel, with the usual neckrings, and a volute casing of cast iron, bronze or stainless steel. The stainless steel shaft would be carried in ball roller bearings.

For more severely corrosive duties titanium is used, particularly on small sizes. Titanium is half the weight of stainless steel and about four times the cost per kilogramme but only twice the cost per cubic centimetre. This can often prove economic as the material cost is a small proportion of the total cost of the pump.

An interesting process is to cast iron around a thin shell of titanium sheet to minimize the amount of the more costly material. The thin sheet metal can readily withstand the internal pump pressure but has insufficient rigidity to prevent gross distortion in the case of poorly aligned pipes being tightened to pump branches. Ideally, such malalignment of piping should not occur, but it often does and, in any case, the in-line I-arrangement involves springing pipes in order to provide space for insertion of gaskets.

Pumps for Abrasive Duties (Reference 75)

It is usual to provide open type impellers of simple construction for abrasive liquids since the fine cylindrical clearances of the conventional neckring would soon be enlarged by wear to an uneconomic size. Clean liquid injection at the glands or mechanical seals is advantageous but must be controlled to avoid undue dilution of product. In many cases an iron casting is provided with a lining of rubber, chilled manganese iron, plastic or ceramic material. Critical parts are the impeller eye where cavitation is most likely, and the casing throat where flow velocities have their greatest value. Smooth surfaces are helpful in avoiding local turbulence which can erode the passage surfaces.

An extreme case can be cited of a ground flat side plate which had studs for attachment to the casing wall. The stud ends were not visible on the ground surface but the difference of metal grain flow between the studs and the plate caused local turbulence which eroded through the plate in four circumferential holes following the stud holes. The stud ends themselves were only slightly affected. The problem was cured by arranging for the studs to be blind so that the wearing plate surface adjacent to the impeller was entirely smooth and had the single grain flow of a flat rolled plate. Glass plates give good service on these duties since they are hard and perfectly smooth.

Tangent Pumps

Low quantity high pressure duties can be handled by small single stage pumps running at 10000-25000 r/min via gears from 2-pole motors. These pumps have the advantages over multistage pumps of smaller size, less expensive where stainless materials are concerned, and greater clearances which permit dry running without risk of seizure. They are also

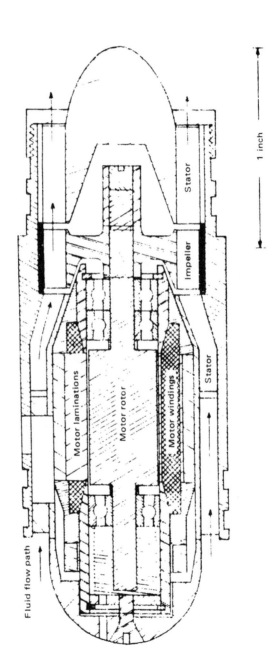

Figure 45.12
New Jarvik-7 heart power source.

easier to align than a horizontal multistage pump and motor, particularly where higher temperature is involved.

There appears to be no objection to the use of geared high speed pumps in a submersible version.

Pipe Loadings

The branch configurations of chemical pumps tend towards the minimizing of casing distortions, since the in-line I- and U-branches are the strongest position to resist pipe loads.

Combined motor pumps where the motor aspect is dictated by the pump position minimize risk of malalignment since the pump and motor portions can move together in case of pipe movements.

Power Recovery

Many processes require high pressure liquid which is afterwards rejected whilst still representing considerable energy. This rejected liquid can be discharged through a water turbine to utilize the energy of flow times pressure.

For example, on nitrogen fixation, a pump of 2000 kW may raise water to 300 m (1000 ft) head for scrubbing CO_2 out of a hydrogen, nitrogen and CO_2 mixture. Afterwards the water and CO_2 at 250 m (820 ft) head can be led to a water turbine which thereby develops 1000 kW, so that the net power required is only 1000 kW from the electric motor. The motor would be rated at 2000 kW for the short term starting load.

Over the last fifty years, some 50 such units have been supplied of various powers comprising pump, motor and turbine. No clutch is necessary between turbine and motor and the sets are entirely stable in operation.

In several cases the 'turbine' is actually a normal submersible pump operating in reverse with the same efficiency turbining that it would develop as a pump.

A pump giving 100% flow, 100% head at 100% speed will require minus 120% flow at minus 125% head when running as a turbine at minus 100% speed.

In similar manner, power recovery sets are used on circulating duties where the static lift exceeds syphon recovery level.

CODE OF PRACTICE

For safe and reliable design:

American Petroleum Institute Standard 610 (API 610)

References

74. TETLOW, N., A survey of modern centrifugal pump practice for the oil field and oil refinery services Proc IMechE. 1943.

75. TRUSCOTT, G.F., Literature survey on abrasive wear in hydraulic machinery, TN 1076, British Hydromechanics Research Association, November 1970.

76. CAMPBELL, H.S., Corrosion of metals in the water supply industry The British Non Ferrous Metals Research Association, 1968.

VARIABLE FLOW PUMPS

Variable Flow performance is essential for the following reasons:

1. Variation of the required flow duty
2. Margin on flow duty required to give greater output
3. To avoid any damaging starting or transient conditions
4. To avoid loss of efficiency or energy on lower duties.

Provision of variable flow by variable pump geometry is simpler and more efficient than by the usual variable speed, and is in the pump's control; it is as effective mechanically and as efficient as the water turbine flow variation by its gates. The pump impeller is surrounded by a close fitting, non rotating, axially moving sleeve which isolates it from the volute or diffuser, avoiding any damaging turbulence and taking only four percent of power, instead of the usual 50 percent of power at zero flow. When moved entirely clear of the impeller, normal operation at full efficiency is obtained. Between these limits, variable flow, always at the best efficiency according to the area ratio occurs, since the device acts by varying the area ratio. Since the pump is never operating off peak, the Fraser recirculation is avoided.

About 1920, a UK firm fitted two rings to the periphery of an overloading impeller to reduce the outlet width and thereby make the power curve turn down at the best efficiency. The author told the Managing Director of his firm, in the early 1930s, that, according to the area ratio theory, a normal impeller of one third width and a narrower casing would be much cheaper and slightly more efficient. He agreed, but said that the efficiency was acceptable and that many orders were obtained thereby. In the late 1930s, similar mid flow large area plenum chamber impellers with small outlets were giving falling head characteristics smoothly and reliably on boiler feed pumps for 25 MW turbo-generators. This shows that impeller outlet areas can be gradually reduced in axial width, and still give turbulence free operation and acceptable efficiency according to the general formula, earlier in this book, of flow and type number.

On Advanced Class Feed Lumps axial diffusers with large impeller periphery gap reduced inlet attack by an order of magnitude at all flows, showing that inlet operation is improved by the avoidance of outlet turbulation. Palgrave (ref) helps to assess the sleeve valve concept. On zero flow, the complete isolation from the diffusers from the impeller eliminates the destructive turbulence, reduces the closed valve power to only four percent and gives full low flow protection with a modest heat dissipation which can be led to a acceptable destination. Radial thrust on the impeller is reduced to one fifth, and hydraulic pulsations are largely eliminated. The line of flow and head reduction, controlled by the area ratio values, follows the head/flow characteristic of 40 percent of the site head as friction, a closer approach to a typical operation duty than the 100 percent of total head friction line of a variable speed drive. The efficiency, first cost and bulk will be lower than that on the inverter drive. Some twenty years ago, a four inch impeller embodied a sliding sleeve, but the operating piston was of relatively large diameter and narrow width.

In order to avoid any tendency to lock, it is suggested that a preferable operating device would be three fixed parallel screws, around the axis of the pump. The screws engage sprockets with screwed bores, actuated by an enveloping chain, to maintain a correct sleeve plane normal to the axis at all times. The complete elimination of off peak operation very much minimises cavitational attack, and permits higher heads. Drax boiler feed pumps have operated at almost 4000 feet per stage for 27 years. Space rocket liquid oxygen pumps operate at 10 000 ft per stage, a six inch impeller taking 28 000 HP.

Impellers have disintegrated at 100 000 ft head. A single stage feed pump at 4000 feet head for a one gigawatt turbo generator is entirely feasible, giving considerable simplification. Energy losses due to over specification will be eliminated since the duty is readily matched to the demand by the sliding sleeve. Diffusers are only required where it is necessary to return the flow to the next impeller for more than one stage. For a single stage, a double volute would be a further simplification. Inducers to reduce HPSH are valuable at best efficiency, but on off peak duties, can suffer hydraulic pulsations which the sleeve would eliminate. The sliding sleeve will be of value on water turbines, giving flow variation, in addition to that provided by the gate regulation. This offers the very full shape efficiency characteristic of Kaplan and Deriaz turbines, especially valuable on the matching of pump to turbine operation on energy storage pump turbines.

Starting a pump turbine as a pump, full of water, will only take 4 percent of power, compared to the 7 percent power pony motor and the 14 percent shock power demand from the grid when admitting the water as on present schemes. The sliding sleeve will eliminate the need for the pony motor, its switchgear, bulk, etc.

ELECTRIC MOTORS

The large majority of pumps are driven by alternating current electric motors, generally squirrel-cage with fixed speed.

Where a considerable variation of duty occurs and the cost of power justifies it, variable speeds, (for example, wound rotor or commutator a.c. motors) may be used. An other means of obtaining variable speed is by the use of rectified direct current, but this imposes a limit on speed and power.

Next in popularity to motor drives is steam turbine drive where flexibility of speed is readily arranged with a negligible sacrifice of efficiency. Engine drive, either direct or by belt and gearbox is important, particularly in remote areas, where it is necessary to make general provision in the coupling for the minimizing of cyclic irregularity.

Although drives other than squirrel-cage motor are generally regarded as giving flexibility of speed, it does not follow that such flexibility necessarily permits full power to be utilized. It should be remembered that most of these variable speed devices are constant torque machines, so that a reduction of speed involves a greater reduction of pump power than of available power.

Coupling of Driver to Pump

For modest duties — up to a few hundred kilowatts — pin type couplings with rubber bushes are used. Although these couplings are referred to as flexible, it is not important that a dial indicator gauge should be employed so as to line up the pump and driver as accurately as possible, preferably within two hundredths of a millimetre, as measure at the coupling peripheries, both by straight edge and by measurement of the axial gap.

For larger powers, gear type couplings are used, and here again the greatest accuracy of lining up is essential. Spring drive couplings, using either short rectangular springs parallel to the pump shaft or a continuous wound spring are helpful in reducing cyclic irregularity on engine drives. A recent development is the membrane coupling, wherein the drive is transmitted through flexible plates of steel of the order of a quarter of a millimetre thick, thus giving a certain amount of axial flexibility.

Where the pump has its own mechanical thrust bearing, for example ball or Michell bearing for axial location, the axial float provided in a gear, a pin or a membrane coupling, will permit the breathing, under load and temperature changes, of the pump and its driver. In the case of multistage pumps, however, the axial alignment of the shaft is determined by the balance disc and, in general, a clearance of the balance disc away from its face of the order of one millimetre and provision for wear of the order of two millimetres are provided. In this case, it is possible for a thrust to be transmitted to the driver. For example, on a gear coupling it is possible to transmit the tooth load multiplied by the coefficient of friction between the gears. The balance thrust may be anything up to about 100 tons, but of this very large force, only the above tooth friction can be transmitted. (See below).

If, however, we consider the case of a pump stopping with its shaft remote from the balance face, this shaft will move towards the balance face immediately on starting, thereby transmitting the above tooth friction thrust in the direction of the driver. For motor drives, in order to prevent this happening, a limited float coupling is often provided. Alternatively, a membrane coupling can carry the motor a few millimetres either way without damage, the motor bearings in these cases having a generous provision for axial float.

Determination of Pump Efficiency

Figs 7.2 and 7.3 represent tests of several thousand pumps on which it can be assumed that test error is eliminated, since each particular point represents the average of a large number of pumps tested. The scatter of Fig 7.3 is therefore due to the variation of friction of cast surfaces and indeed, efficiency variation is seen in the chart to be proportional to variable losses. (See Chapters 7 and 8).

Pump, Motor and Overall Efficiencies

Determination of pump performance is described in detail by Young and Nixon of the National Engineering Laboratory in the paper 'Power, Flow and Pressure Measurements in Pump Testing'. Ref 77 and the following extracts from the discussion of that paper illustrate present day procedure.

H.H. Anderson and W.G. Crawford wrote — 'The fact that the capitalized value of one per cent efficiency is very roughly equal to the price of the pump, illustrates the importance of this excellent paper.

The production test bed with which we are associated measures flows of 18l/s to 2 m³/s by eight instruments, of which the 460 mm venturi, covering medium flows, has been calibrated at the National Engineering Laboratory and serves as a standard for the other instruments which are similarly check by series tests on the same flow.

The efficiencies resulting from production tests are subject to random errors of test observation, to variation due to the waywardness of water and to variation due to the differing cast surfaces of the various pump internals, these variations covering, in general, plus or minus five per cent efficiency. The amount by which the efficiency attained on the test bed exceeds the efficiency guaranteed in the contract is statistically analyzed, from which it appears that the average value of this excess is remarkably consistent from year to year when considered over the years covering several thousand pumps. Because of the total scatter of efficiency mentioned above, the quoting level is appreciably lower than the average test level in order to keep rejects to an absolute minimum.

Centrifugal pumps are normally driven by electric motors. From the user's point of view, the overall or wire to water efficiency is what really matters as this determines the power bill. Accurate electrical measurements are, therefore, as important as accurate hydraulic measurements. Fortunately, high precision wattmeters which are sufficiently robust for commercial testing are available and if these are used with the most accurate grades of current and potential transformers, high accuracy can be attained when reasonable care is taken. The errors in the electrical measurements are, therefore, small compared with the random variations in pump efficiencies.

When a pump is driven by an electrical motor and input/output readings taken, the pump efficiency cannot be assessed until the motor efficiency has been determined. The conventional procedures for determining the efficiencies of electric motors in this country are laid down quite clearly in BS269. It is well known that these procedures are conventional and do not necessarily give the true efficiency of the motor in every case; nevertheless, there are a number of arguments in favour of the conventional procedure.

First, the measurements are simple and can be carried out on a complete machine in any manufacturer's works and in most pumping stations. This makes it easy to verify the conventional efficiency at any time.

Second, at any particular time, the electric motors made by different manufacturers are designed with the same general background of theoretical knowledge and practical experience. Consequently if two motors of similar rating have equal conventional efficiencies, then their true efficiencies will be very similar also. This is fully confirmed by many years of experience in testing one manufacturer's range of pumps with their own range of motors and with motors made by almost every manufacturer in the country. Consistency of this kind is as useful as accuracy to the pump designer, because he is still able to assess the improvement in pump efficiency resulting from any change in design as soon as sufficient results are available for statistical analysis.

A large number of pumpsets are supplied to major Waterworks Authorities whose independent site testing approaches the accuracy of the National Engineering Laboratory described by the authors. Almost 100 recent pumpsets averaging 300 kW each have shown on these independent site tests an appreciably higher efficiency than was obtained on the Production Test Bed. Experience has thereby been built up over the years of wire to water efficiencies of pumping plant rendering the true value of the motor efficiency of little importance to the purchaser of a pump and motor set'.

K.F. King wrote — 'Whilst there appeared to be a case for revision of the allowances in BS269 to make them more realistic, it was doubtful whether manufacturers would willingly agree to actual measurements of stray losses on each and every machine, but there was no reason why standard test motors should not have their losses accurately measure. (See below).

If that were done, however, and if pump efficiencies were quoted on the basis of such a measurement, pump manufacturers would have to bear in mind that when they were called upon to give a guaranteed consumption of, say, kilowatts input to liquid power, they would still have to work on motor efficiencies in accordance with BS269, and it would therefore be up to them to ensure that they used well designed motors which would have their stray losses kept down to a minimum value'.

Pump makers who do not make motors sometimes use dynamometers, but a set of dynamometers giving consistency of readings as accurately as electric motors would be far too expensive for commercial testing and, in any case, these dynamometers would be driven by electric motors or embody electric motors. (Note: accurate torque meters are now available).

Conventional electric motor tests to BS269 are, therefore, convenient for the vast majority of applications and are slightly conservative on efficiency for pumps which will finally be driven by other means. BS269 tests avoid the expensive dismantling of the motor to determine exact stray losses.

Starting Conditions, etc

Fig 37.4 shows starting torques on full and reduced voltages for low and high static lifts. See also motor data in Chapter 33 Boiler Feed Pumps, and Chapter 37 Mine Pumps.

Variable speed with a.c. supply is obtained by hydraulic couplings or by slip ring would rotor motors, x% speed drop resulting in x + 2% efficiency drop. Commutator a.c. motors of various types give variable speed at higher efficiency but with greater cost and complication. See Figs 43, 78 and 108.

Flexible couplings permit axial freedom and some measure of alignment error, although it is desirable to have alignment correct to 0.05 mm at all times, particularly when operating temperature is reached.

Pin and rubber bush couplings are used for low power duties.

Gear and steel spring element couplings for larger powers permit axial movement but may transmit some thrust during direct on-line start up, when torque is at a maximum.

Membrane and contoured disc couplings anchor the driver axially to the pump but may allow some relative axial movement with a predetermined transmitted thrust which is proportional to the axial deflection.

Space couplings halve alignment errors and facilitate gland and impeller access. Care is necessary to avoid excessive mass which could cause vibration.

Detail Practice

The motors are initially filled with clean water and a corrosion inhibitor. After some years of operation it is obvious that the clean water mixes with the pumped water, but the fact remains that the motors still give reliable service. Distilled water was used as an initial filter, but at present clean water appears to give satisfaction and may indeed be less corrosive than the original distilled water. The mechanical seals to separate the clean water from the pumped water certainly are helpful in excluding abrasive matter.

The pump thrust is carried by the water lubricated motor bearing which in some cases involves the use of double neck rings or double entry impellers to limit the thrust to an acceptable value.

Reverse rotation capability of the thrust bearing is necessary if no foot valve is fitted and if the rising main is allowed to empty on shut down. Double thrust bearings are advisable where the pumpsets have to operate across a range from zero flow to maximum flow.

The thrust bearing in some cases was fitted at the top of the motor which kept it less liable to the effects of abrasion, but present practice is to put the thrust bearing below the motor where there is no risk of it being starved of water.

On wet winding motors the stator is entirely dependent on the integrity of the insulation: polyvinyl chloride has been established as the preferred insulation for thirty years, sometimes strengthened by a spray coating of polyamide. So as to provide a smooth bore for minimum power loss, tunnel slots are used in the stator involving pull through windings and thin plastic liners. The consequent looser assembly means that the end windings must be fully braced with a suitable plastic tape not likely to stretch under the transient loads of starting. Pull through windings minimize the number of joints; eg on mining applications where there is no size limitation on the junction box, only one joint inside the motor is required at the star point at or about star potential.

The motor shaft is generally of stainless steel, as are the rotor and stator stampings for a short distance at each end. In severely aggressive waters a clean water supply to the motor casing may be required. The windings in these cases can be further protected by the epoxy resin encapsulation.

Carbon or carbon loaded PTFE journal bearing bushes have the advantage of minimizing stray leakage of currents and indeed bearing materials of the unusual lead bronze, plastics impregnated with metals or strong fibres have proved of value.

Although the submersible motor shaft is long and of relatively small diameter in order to give overall minimum diameter of the unit, the presence of water damps vibrations in similar manner to the case of neck rings (see Mechanical Design of Centrifugal Pumps) to a limited, but sufficient extent to permit adequate operation (see Critical Speeds, Vibration and Noise).

The cooling time constant of a submersible motor is generally sufficiently long to permit the filling of the rising main shown in Figure 24.8 when the cooling time constant can be used to give some idea of the heating time. Normal practice shows that the slight overloading whilst filling an empty rising main does not cause undue overheating.

Here it is essential to be reminded of the fact that a mining type motor in a large shaft perhaps 2m diameter may need a shroud to increase the water velocity entering the pump and to localize the flow to cool the motor adequately.

The starting torque of the motor is adequate on mining duties, but on oil well pumps of perhaps a few hundred stages, the electric motor starting torque is often inadequate and the cause of failure.

The relationship between motor rotor diameter and length is a compromise of the rigidity of the short rotor and the better heat dissipation of a long rotor. The small diameter of the long rotor minimizes powers loss, which improves efficiency, largely affected by liquid drag.

Motor winding standards may be Class Y (90°C), Class E (120°C), Class B (130°C), Class F (155°C), or Class H (180°C), the higher qualities giving a better factor of safety to the machine. It is essential that the motor should not run for long periods on very low water flows since the motor carcase will not be properly cooled. Where the water quality demands a bronze pump, a plastic disc between pump and motor is advisable to prevent electrolytic action.

Variable Speed

The large majority of pumps are driven by alternating current electric motors, generally squirrel cage with fixed speed. Where a considerable variation of duty occurs, variable speed devices such as wound rotor or hydraulic coupling are unsuitable for submersible pumps. The inverter system of converting the a.c. current to d.c. then back to variable frequency a.c. is used for the slave pumps driving the downhole pumps and for driving the alternators powering the electric oil well submersibles. An alternative is the high efficiency hydrostatic drive, which has about the same reliability and efficiency as the inverter drive, but is about half the cost and occupies one-tenth the space, a vital aspect on an oil platform.

Reference

77. YOUNG, L., and NIXON, R.A., 'Power, Flow and Pressure Measurement in Pump Testing' Proc IME. 1960. Vol 174 No. 15.

VARIABLE SPEED FOR CENTRIFUGAL PUMPS

In order to match the flow of the pump to the demands of the system in which it operates, variable pump rotational speed is often adopted. The reliable, economic, universally used alternating current cage motor has an almost fixed speed. A wound rotor motor can insert rotor resistance to vary the speed and to increase the torque. This, however, reduces the motor efficiency by 2 percent plus the percentage speed reduction, ie, by 12 percent for a 10 percent speed reduction.

The provision of a hydrodynamic coupling between a cage motor and the pump involves a similar efficiency reduction according to the speed reduction. Time is taken in filling and emptying the coupling, and on seizure of a driven machine (which can, perhaps, even if rarely, happen), twelve times full load torque can be applied to the driven machine, which may cause severe damage. On speed reduction, the resulting heat dissipation is confined to the coupling casing. Feed pump drives can, for some installations, involve considerable additional machinery in extra converters and couplings. The inverter drive has high cost and bulk and involves a derating of the motor power. A paper in the BPMA 13th International Conference, describes the noise and the complexity of the inverter drive wave forms.

The efficiency of the inverter drive is quoted as 91 percent at full speed and load, and 80 percent at 70 percent speed. A novel proposed transmission uses three element screw pumps running also as screw "turbines" (to avoid confusion with AC motors). A cage motor drives a screw pump with its oil flow feeding a screw turbine, which in turn drives a centrifugal pump. By passed oil leakage, to reduce the driven unit speed, will reduce the drive efficiency in similar manner to the above slip drives. The screw units are virtually free from pulsations, and can accept high speeds and pressures. On fully open control valves, there will be zero oil pressure and no torque can be transmitted. On closing the bypass valve, instant stock free torque will be transmitted.

Unlike the hydrodynamic coupling, the heat dissipation on speed reduction can be led to a suitable sink. The hydrostatic screw pump will be about half the diameter of the hydrodynamic coupling, has instant valve response and on seizure of the driven load, can only apply the extra 20 percent torque corresponding to the system relief valve pressure setting. Pole changing motors are a reliable means of changing the driven unit speed. They are combined with an epicyclic gear box, (its cost covered by the efficiency saving) and screw units to provide a reliable, highly efficient, economic variable speed transmission.

A main motor, 80 percent of the total power, rotates the driven unit via the epicyclic gear. The slip element of the epicyclic engages a 20 percent power pair of screw units, which in turn, engage a two and four pole changing 20 percent power pony motor/generator. The two pole pony motor draws power from the mains, drives a screw pump feeding oil to the screw turbine engaging and processing the slip element, to raise the driven unit speed to 100 percent. At 80 percent speed, the pony unit is inoperative, and the screw unit, on closed valve, holds the slip element stationary, apart from a very small creep, due to internal clearance, which avoids gear tooth fretting. At 60 percent speed, the slip element drives the screw pump feeding oil to the screw turbine driving the pony generator which thereby returns the power from the slip element to the mains. At four pole pony speeds, 10 percent speed changes will be obtained at 10 percent of power. A valve manifold ensures that, as the screw pumps and turbines exchange functions, each high pressure region handles the high pressure. This gives five speeds, but five further speeds are obtained by leaking 5 percent of oil flow to give ten modes of instant shock free infinitely variable speed down to 60 percent speed at an average of 96 percent efficiency. This gives about 10 percent higher efficiency than the inverter drive, with very low cost and bulk, high reliability, no derating of main motor, no excessive noise or complex electrical wave patterns. Any malfunction of controls cannot exceed the usual four times full load starting current, or the torque corresponding to 20 percent relief valve pressure setting. The very high efficiency is due to the ten modes of only 4 percent speed change, the average efficiency loss across the 40 percent speed reduction being 2 percent.
Rail Drives

It is apposite here to make a very brief mention of rail drives since the screw pumps and turbines are so basic to the concept. Many electric and diesel trains now adopt all electric drives or three full power torque converters and two hydrodynamic couplings for braking. The pole change, epicyclic gear and screw unit drive can offer advantages of simplicity, reliability and high efficiency.

An 80 percent power two and four pole main motor and a 2,4,6 and 12 pole double winding pony motor can go down to 20 percent speed at five times full speed torque with an average efficiency of 95 percent in 24 modes of infinitely variable speed. The main motor is pony motor started, with very high low speed torque, at Euston, and is never switched until reversal at Inverness, idling at 50 hz at stations. This gives a very great increase in life and in catenary traffic capacity., with no starting loads. Regenerative braking and the use of motor inertia for acceleration and climbing make large energy savings. No extra installed power is involved, except on diesel drives, where a 25 percent of power extra pony unit is required. Pole changing motors and screw units can have an important part to play in future large power installations.

Frequency Converters

Electronic systems, which can give very high frequencies and speeds will apparently have similar losses to the inverter variable speed drive. Epicyclic gears have very high reliability and efficiency. In addition, combined with pole changing motor/generators, they also facilitate the provision of variable speed with little extra cost as described above.

OIL, WATER, GAS AND STEAM TURBINES ON GLANDLESS DUTIES

From submersible pumps there developed the great advantage of glandless units wherein the inlet pipes, pump and driver, and outlet pipes made a closed box entirely free from leakage. It was this glandless pump and this alone that permitted the development of nuclear power. The electrical drive is described in the preceding chapter, but here we consider the role of turbines.

These machines have the advantage, especially when powered by variable pressure surface supply, of giving variable speed and, most important of all, of providing rotational speeds of several times the two pole electrical speeds. This can give a thirty fold increase of power per unit volume and a ten fold length reduction, which is a vital element in the rigidity and resulting reliability of pumpsets on curved wells several thousand metres (feet) deep.

Displacement motors of the helical screw type have high powers per unit volume, but are mainly of value on liquids free from abrasive matter.

Oil and water turbines operating on the equivalent of 4000 m (13 000 ft) head of filtered liquid offer the attraction of the highest power to unit volume ratio and are described in Chapter 40. Typical flow, head, torque, power and efficiency characteristics are given in Chapter 24.

Steam turbines are used on the marine glandless pumps thereby minimizing volume and weight, eliminating the complexities of glands, oil pumps and drives, filters etc, reducing overhang and increasing rigidity. They are described in Chapter 44.

Vast increases in power per unit volume were seen one hundred years ago in the replacement of a 300 ton reciprocating steam pump by a one ton centrifugal pump, by the Parsons steam turbine, by recent boiler feed pumps and by the jet engine. Basically, all these changes occurred because of the increase of speed, bringing also the reliability advantages of rigidity, of small size permitting the highest quality expensive materials, and facilitating transport and erection. Glandless units can operate as a closed box as far as the operating staff are concerned, since the fluids involved can escape freely from pump to driver or driver to pump through the separating wall without distress. For example, filtered turbine outlet oil or water escaping via the pump bearings into the pump mainflow is valuable in ensuring reliability; in the case of steam turbine driven pumps the leakage of balance water from the pump into the steam exhaust evaporates quickly and harmlessly. Gas turbine driven glandless rocket pumps similarly permit pump shaft leakage to escape into the main jet.

Liquid boost storage space rocket pumps involve powers comparable to boiler feed pumps with the highest power concentrations and minimum weights so that the vehicle can succeed in its purpose to escape the earth's gravity. Propellants may be liquid oxygen, liquid hydrogen, methane etc. Figure 49.1.

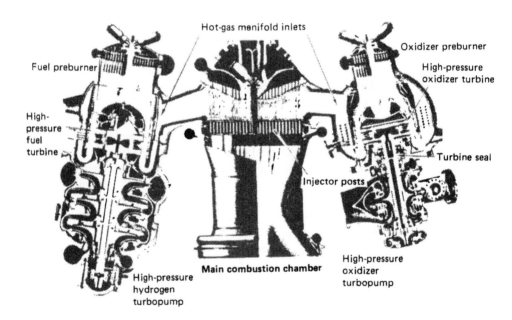

Figure 49.1
Space rocket pumps.

References

80   Reliability - the user-maker partnership, British Pump Manufacturers Association 9th Conference Warwick University, April 1985. Several submersible pump aspects recorded.

Level Control

The need for a dependable control to automatically switch submersible pumps has led to the development of mercury activated level alarm systems. A particular switching device consists of a heavy duty mercury to mercury type capsule with omni-directional on/off tumblers so that the float is not sensitive to rotation.

This together with a novel wiring circuit prevents relay contact arcing when starting and stopping the pump. Figure 1 shows the principle of operation for a double float pump switch. Installation of these devices is usually quite simple with cord, plug and adjustable mounting straps and clamps as shown in Figure 2. Switches of this type are capable of controlling pumps up to 1 hp in single switch versions and 2 hp in double switch versions.

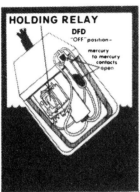

Off position – float is tipped down and mercury remains in non-contact position.

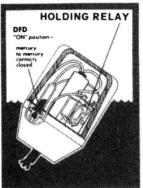

On position – float is tipped up and mercury moves to contact position.

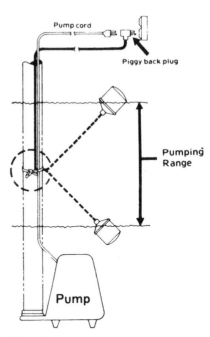

Figure 2
Typical installation showing pumping range.

Figure 49.2

PUMPS FOR VARIOUS SPECIFIC DUTIES

This chapter covers various special pumps not already described in previous chapters. Space does not permit a chapter on each type and therefore references are given for further study.

Paper Mill Pumps

Paper stuff (pulp suspended in water) contains gases and is akin to a viscous liquid in resisting shear. Paper pumps have single entry open impellers and an entire absence of air pockets, particularly in the suction passage and in the casing volute. Bronze parts are essential for sulphite liquids; cast iron for certain chemical liquids. For back water and very low consistencies of stuff, split casing pumps with shroudless, open type impellers or adjusted neckring clearances are used, particularly for large quantities.

Quantity, head and efficiency are reduced as the consistency of stuff increases, the reduction being more marked on smaller machines. The mature of the reduction of quantity, head and efficiency is rather similar to the reductions shown in Chapter 45 on chemical pumps — 1% of paper pulp being similar in effect on performance to one stoke of kinematic viscosity.

Clean water seals to the glands of paper pumps are usually provided.

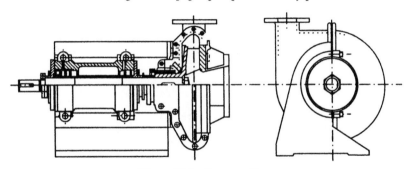

Figure 50.1 – Paper stuff pump.

A section of a paper stuff pump is shown in Fig 50.1, which illustrates the single entry impeller which is open at one side where it runs against a wearing plate, often adjustable, and the back blades at the other side of the driving shroud in order to minimize pressure on the gland and keep the gland packing free from paper pulp.

Paper pumps are generally liable to be hosed down and the bearings are, therefore, fitted in a cartridge, permitting removal of the rotating element for servicing.

C.G. Evans (Ref 69) describes the design of pumps for paper mill duties.

Steelworks Pumps

Steel works require general water supplies for which the single stage and two stage pumps already described are provided. The presence of mill scale may, however, call for special materials, or for clean water seals to the pump glands.

The descaling of steel bars and plate is an essential part of the process in order to give a perfect surface (for example, on the tin plate used for canning food), and for this duty, high pressure centrifugal pumps are used. Descaling pumps are rather similar to the multistage boiler feed pumps, but without any of the temperature features. The presence

Figure 50.2 – Sewage pump.

of mill scale and the frequent and sudden opening and closing of the nozzles demand the provision of an externally oil-lubricated Michell thrust bearing, in addition to the balance disc — the thrust bearing being rated for the full hydraulic thrust.

Evans and Fairfield describe the design of steelworks pumps in Ref 70.

Descaling pumps are described by P.L. Fairfield in Ref 71.

Sewage Pumps and Storm Water Pumps

Unchokable features with heavy construction are essential for sewage pumps. To this end, impellers are of the single entry type and generally have two or three blades (see Fig 50.2). Provision must be made for clean water seals to the glands and for the escape of gases from the casing. Vertical spindle pumps, usually of cast iron construction, are favoured. Storm water pumps must handle large quantities with occasional solids or grit and are generally of the single entry of axial flow type with partially unchokable features. (See Chapter 31).

Sewage pumps are described by Kershaw in Ref 72.

Pumps for Industrial Slurries

For many industrial purposes, it is necessary to pump liquids having a large amount of suspended solids. Suitable pumps may have rubber lined or chilled metal internal parts and be provided with clean water seals to keep the abrasive matter from the fine clearance parts. Typical duties are sand stowage, china clay slurry, coal pumping, coal washing, ash handling, etc. (See Chapter 47).

The metal thicknesses or impeller and casing are very much greater than the hydraulic duty demands in order to provide a margin for wear caused by abrasive particles.

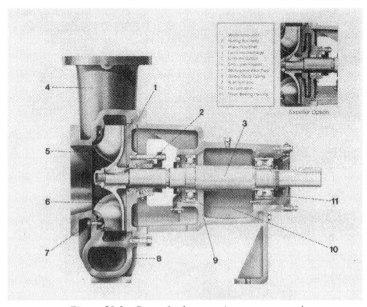

Figure 50.3 – Pump for low consistency paper pulp.

Plate 50.4
Tehachapi pump.

The handling of solids in water is described by Bonnington in Ref 73 and by Julian Nardi in Ref 74.

Molten Metal Pumps

Either vertical glandless centrifugal pumps are used or electro-magnetic pumps in which the flow is induced by electric currents in the liquid metal. The centrifugal pumps are submerged in the liquid metal and driven by a long vertical shaft to the motor with suitable heat baffles. They must be self-draining for maintenance and allowance provided in their stressing for the specific gravity of the metal; a rugged simple design is essential.

Hammitt describes liquid metal pumps in Ref 75.

Food Industries: Milk and Beer Pumps, etc

For handling food liquids, pumps must be suitable for quick dismantling and daily scalding. On some duties it is essential for all internal surfaces to be highly polished so as to avoid the lodging of harmful bacteria. These pumps are generally constructed of bronze with chromium plating, or stainless steel, of glass or other plastics or of ceramic material. Various foods, eg whole sugar beet, apples, potatoes, coffee beans, hops, fish, etc, are transported in the raw state by pumping, with water. It is important to ensure entire freedom from choking, and to provide against possible abrasion.

Drainage and Irrigation Pumps

Drainage and irrigation pumps are generally of the low lift type as shown in Chapter 31. Applications of these pumps are described by Fawcett (Ref 76).

Self Priming Pumps

There are several ways in which a centrifugal pump can be primed. A vessel on the suction side of the pump, for example, may contain a supply of water sufficient to prime the pump, after which circulation can take place, gradually drawing air from the suction pipe and discharge it to atmosphere. When the pump and suction chamber are full of water, a float valve cuts off the discharge to atmosphere, thus permitting the pump to operate normally. The suction tank remains filled for the next start up. In certain designs, internal by-passes permit sufficient circulation to get rid of the suction pipe air; other designs have independent priming rotors. Priming is more difficult as the static head increases since the impeller head depends upon the amount of water in the passages, any air present reducing the head. On higher static lifts the provision of by-passes to get rid of this air and permit generation of full head is essential. Here various devices are available reprime the pump periodically from the delivery pipe until the suction pipe is filled. Self-priming pumps run on 'snore' at the coal face for the first stage of mine drainage, delivering the water as and when it flows in to the pump.

References

69. EVANS, C.G. 'Pumps for Paper Mills'. Proc. Technical Section, British Paper and Board Makers Assoc 1952, Vol 33, Part 1.

70. EVANS, C.G. and FAIRFIELD, P.L. 'Pump Design and Application in Iron and Steel Works'. Journal of Iron and Steel Institute, Vol 170, April, 1952.

71. FAIRFIELD, P.L. 'Hydraulic Descaling of Hot Steel', Iron and Coal Trades Review, July, 17-24, 1953.

72. KERSHAW, M.A. 'Sewage Pumping'. Pumping June, 1960 et seq.

73. BONNINGTON, S.T. 'Transportation of Solids'. Pumping, September 1960.

74. NARDI, Julian. 'The Past and Future of Pipeline Transportation of Solids'. Pumping, February, 1961.

75. HAMMITT, F.G. 'Considerations for Selection of Liquid Metal Pumps', Chemical Engineering Progress (USA), May 1957, Vol 53, No 5.

76. FAWCETT, J.R. 'Land Drainage', Pumping, February, 1960.

MISCELLANEOUS ASPECTS

This chapter gives various data which have proved of value in the design and application of centrifugal pumps.

Noise

Fig 51.1 (Ref 78) shows typical noise values of pumps on tests in an anechoic chamber and on site.

Frame Selection Chart

Fig 51.2 shows a logarithmic chart of flow and head on which is drawn a speed line at 2:1 gradient corresponding to flow increasing as the first power and head increasing as the second power of speed. On this speed line are marked appropriate speeds from 300 to 4000 rev/min.

A chart may be drawn on a very transparent medium so that it can be overlaid on Fig 51.2.

Lines at a gradient of two to one are drawn for various numbers of stages with two vertical lines, one to be set on the operating speed and the other to be set on the operating speed divided by ten. The appropriate stage line is placed above the diagonal line on Fig 51.2 with the point where the vertical speed line crosses the appropriate stage line positioned above the running speed of the pump. On the transparent chart a complete range of pumps can be plotted against the flow and head readings of the underneath chart, Fig 51.2.

The transparent chart shows typical type number lines on which a range of pumps could be plotted.

This charge gives a quick and convenient way of laying out a new range and of showing the performance of any particular pump at a given speed and number of stages.

Plotting and Analysis of Pump Tests

The test results on each size and type of pump are plotted as follows:

All tests are corrected to a basis speed, for example 1470 rev/min depending upon the

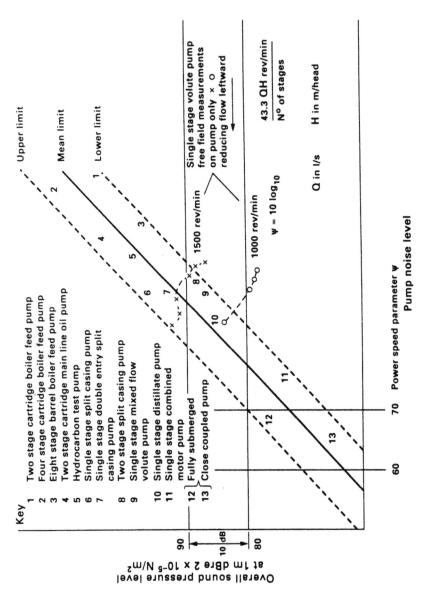

Figure 51.1 – Typical noise values of pumps.

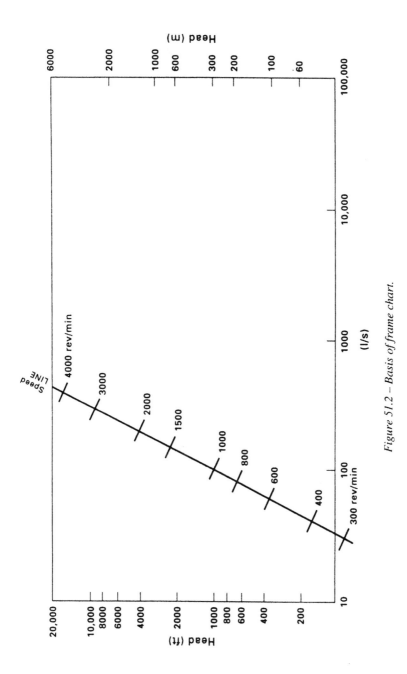

Figure 51.2 – Basis of frame chart.

size of the pump, and a family of curves obtained for the head quantity characteristics of the pump at various diameters of impeller. These curves will be similar to the variable speed curves shown in Fig 22.1, but will show diameter changes instead of speed changes. It is convenient to group the tests into the relatively small number of diameters making a diameter correction in a similar manner to the speed correction and keeping different colours for different diameters. The efficiency curves are then transposed by increasing the flow of each test reading proportional to the impeller diameter so that all efficiency curves, irrespective of diameter, have the same value of flow at their peak, thus making it much easier to view the performance with the colour range than would be the case if the efficiency curves crossed as in Fig 22.1.

From these charts isoefficiency lines can be drawn on the head quantity curves for the various diameters.

Cavitation tests are plotted in the same manner although, of course, they are little affected by diameter change.

The pump frames are designed to cover a wide range of heads and quantities, so that it is economic to vary the impeller diameter on each frame to avoid gaps in the field to be covered.

Pipe Friction

A chart of friction loss in piping appears in Fig 51.3.

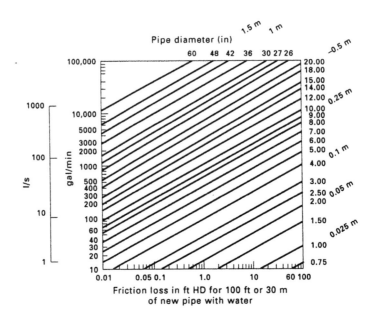

Figure 51.3
Characteristics of pipe system.

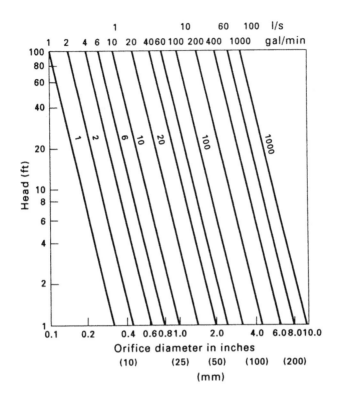

Figure 51.4
Discharge from orifices. gal/min = D2 √H x 9.83.
Coefficient = 0.6. Quantity scale can be increased by 10 while head increases by 100.

Flow Through Orifices and Nozzles

Orifice flow is important in pump work, for example, on the breakdown orifices for boiler feed leak-offs, etc.

Fig 51.4 shows orifice flows against various pressures based upon a sharp edged orifice with a coefficient of discharge of 0.6. For nozzles, increase the flow as the coefficient of discharge increases.

Economical Assessment of Efficiency

As mentioned in Chapter 7, the capitalized value of 1% efficiency can very often be comparable to the price of the pump. On the other hand, on certain applications where operating hours are short, where fuel is cheap and capital is more important, the efficiency assessment carries less weight.

In order to assess financially the value to the pump user of 1% higher efficiency (or the disadvantage of 1% lower efficiency) an exercise in discounted cash flow is involved. Where the rate of interest is assumed to be fixed for the anticipated or assessed life of the pump (and complexities such as tax relief, investment grants etc are not involved) the capital required to produce an annual payment equal to the reduction or increase in fuel cost of the assessed life of the pump may be determined.

On certain contracts where efficiency is vital, no tolerance, penalty or bonus payments, based on the above calculations, have been imposed. Whilst this procedure has been followed in the past on duties up to about 10 MW, it is hardly practicable on duties in the region of 100–200 MW, since 1% efficiency might then be worth 4 million, an impossibly large sum compared to the value of the contract.

It is possible to calculate the relationship between capital payment and a revenue payment such as would arise in assessing the capital equivalent of an improved efficiency, which brings, in its turn, a reduction of electric power bills. The basic formulas

$$r^n \quad \text{and} \quad \frac{r^n - 1}{r - 1}$$

apply also to many other transactions. For 5% interest $r = 1.05$.

Discounted cash flow calculation for pump fuel cost, pension or annuity.

Capital x r^n =

$$\text{Annual cost} \ \times \ \left[\frac{r^n - 1}{r - 1} \right]$$

Endowment   Total amount in n years =

$$\text{Annual premium} \times \ \left[\frac{r^n - 1}{r - 1} \right]$$

Mortgage   Borrow £X repay X x r^n at the end of n years.

Annual premium for n years =

$$X \times r^n \ \left[\frac{r^n - 1}{r - 1} \right]$$

Quality Assurance

Variation of cast surfaces demands considerable vigilance in ensuring that quality is maintained. It is desirable to record the scatter of performance in the form of a histogram, so that the variation is continually under review and the rejects can be kept to a minimum. (See also comment on electric motor testing and variation of pump efficiencies in Chapters 7, 8 and 47).

The scatter of pump efficiencies and errors of observation in tests must be taken into account in assessing performance. The incidence of scatter with respect to efficiency and to the tolerances of the pump test code ISO 3555 is discussed in Chapter 7.

Scatter of head quantity characteristic is comparable to the scatter of efficiency shown in Fig 7.3 but it can, of course, be adjusted by altering the impeller diameter. For this reason it is usual to design the pump to be slightly high in head; designing for the exact head may scrap the impeller if the performance is on the low limit of the scatter. On the high limit, of course, the impeller can be easily turned down. See Appendix D.

Calculation of Combined Stresses

Determination of the principal stresses on ductile materials resulting from a combination of torque stress on a shaft (shear) and end stress on a shaft: for example, bending stress due to radial loading of a volute pump (unbalanced hydraulic load) or from axial end loading (balance disc thrust) is discussed below.

From strength of materials (CASE, Ref 79).

Principal stress =

$$\frac{\text{Longitudinal stress}}{2} \pm \sqrt{\frac{\text{Long stress}}{4} + \text{Shear stress}^2} \quad \dots\dots\dots (1)$$

The plus and minus signs give two values, referred to as F_1 and F_2 in the two dimension formula below:-

Mise Hencky Ductile Material Formula

Stress in reference to uni-axial loaded bar yield strength = F_y

Two Dimensional Stresses

$$F_y = \pm \sqrt{F_1^2 - F_1 F_2 + F_2^2} \quad \dots\dots\dots (2)$$

These values F_1 and F_2, are true stresses, but since failure is due to the combined effect of two (or three) stresses at right angles to each other, the Mise Hencky formula derives from these stresses $F_1 F_2$ an equivalent stress F_y which can be referred to the simple tensile yield stress of the ductile material.

Application to Shafts

The plus and minus of F_y in formula (2) gives two values to be used as follows:

In the case of a shaft with torque due to power transmitted and end thrust due to balance disc, no variation in stress occurs and a plus value is taken. This stress F_y is then referred to the yield stress in tension to determine the safety factor.

In the case of a shaft subject to bending stress: for example, the unbalanced hydraulic pull of a volute pump, the bending portion of the stress reverses at each revolution. The plus and minus values are, therefore, taken as representing the extreme range of stress and are referred to the fatigue strength of the material to determine the safety factor.

Final Formula for Shear and Longitudinal Tensile Stress on a Shaft

Formulas (1) and (2) are combined in the Mise Hencky formula as follows.

Maximum stress F_y to be referred to tensile yield stress of the material

$$F_y = \sqrt{\text{Longitudinal stress}^2 + 3 \, (\text{Torque shear stress})^2} \quad \dots\dots\dots\dots\dots\dots\dots\dots \quad (3)$$

Three Dimensional Stresses:

This covers the case of a pump barrel which is subject to hoop stress due to internal pressure, longitudinal stress due to internal pressure, and radial stress due to the differential thermal expansion consequent upon different of inner/outer wall temperatures.

Differential thermal stress for the same reason is, of course, added to the hoop and longitudinal stress.

Here, the maximum total stress to be referred to yield stress is given by the following

$$F_y = \sqrt{\frac{(F_1 - F_2)^2 + (F_2 - F_3)^2 + (F_1 - F_3)^2}{2}} \quad \dots\dots\dots\dots\dots\dots\dots\dots \quad (4)$$

$F_1 \, F_2 \, F_3$ represent the three principal stresses at right angles.

Hoop and Longitudinal Stresses in Brittle Materials

For a single pipe or barrel under internal stress, maximum stress with respect to ultimate tensile strength can be based on maximum principal stress, ie hoop stress. The end compression of cast iron multistage chamber rings does not weaken their hoop stress capacity, but end compression stress must also be referred to ultimate stress of material in compression.

Design of Single Stage Pump Castings

For medium pumps, casings having circular volute cross-section open to the impeller are used in cast iron up to about 12 bar and in cast steel up to about 20 bar depending on size.

Fabricated (ie welded) steel casings for combined motor pumps have trapezium cross-section of volute as this facilitates construction by avoiding double curvature surfaces. Double volute or external ribbing is necessary above about 12 bar to prevent the volute section from opening under pressure.

Very large pumps follow water turbine construction, namely a stay ring of welded plate steel or cast steel to which is fixed (by welding) a scroll or volute of plate steel, each section of this volute being a C-shaped rolled plate. The diameters of the C-plates increase progressively towards the throat of the casing after which the discharge taper leads the water to the delivery branch.

The stay ring vanes complete the circle of the C-plates so that the metal thickness is calculated on the basis of a pipe, the circular cross-section giving minimum thickness, whereas the open volute sections, (circular and trapezium) mentioned above, must have sufficient rigidity for the C-portion to hold against internal pressure without undue deflection.

The reason for the stay vanes is readily seen when we consider that for a large pump or turbine of 100 MW at 300 m head, the completed circle of the volute section, C-plate and stay ring would require a thickness of at least 5 cm in high tensile steel plate. To hold the C-section rigid without the link effect of the stay ring would be entirely impracticable on units of this size and power.

Transport to site involves splitting the casings (and sometimes the impellers) of these large units with consequent site welding and stress relieving by peening, by flame or by electrical heating.

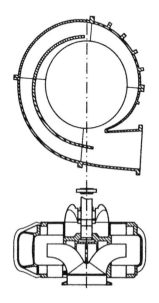

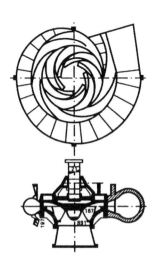

Figure 51.5 – Worthington twin-volute vertical-shaft high-head pump.

Figure 51.6 – Byron-Jackson, Pelton twin-volute vertical shaft pumps with multi-vane diffuser for the Grand Coulee plant.

Figs 27.3, 34.1, 35.14, 35.15, 51.5 and 51.6 (Ref 80) show that the shape of the volute and the axial width of the volute opening, with respect to the impeller width, are far from critical. The four pumps in Fig 51.7 show how size is reduced with increase of area ratio for same duty.

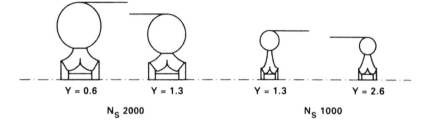

Figure 51.7 – Effect of area ratio on size of pump for the same duty. The smaller pump in each case has a lower pressure rise from duty to zero flow.

Heating Effects

When a pump is operated on closed valve, nearly all the power input is converted into heat, and the temperature of the pump rises. This phenomenon has been used in the prototype testing of high temperature feeders. (See Chapter 33 Boiler Feed Pumps).

An approximate rule for preliminary determination of leak-off flow is as follows: The quantity in l/s to be leaked away is equal to the kW of the motor divided by 53 which results in a temperature rise of approximately 10 deg C to the balance water. This 10 deg C rise must then be checked against the available NPSH in order to avoid vaporization and consequent risk of pitting of leak-off orifice or balance disc.

Specific Gravity and kinematic viscosity of water at various temperatures are shown in Fig 51.8.

Pressure Test

Pump casings are hydraulically tested to 150% of maximum working pressure as normal standard. Higher pressure tests are sometimes specified for severe duties.
Packed Glands are described by Thomson in Ref 81.

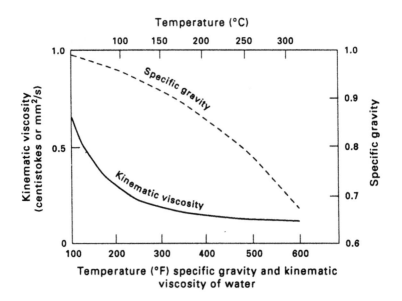

Figure 51.8

References

78. FRANCE, D., 'Noise, A Philosophy of Pump Manufacturers' Fourth Tech. Conference, British Pump Manufacturers Association, Durham 1975.

79. CASE, J., 'Strength of Materials', (Arnold).

Figure 51.9 – Geared oil pipeline pump.

80. FRASER, W.H., 'Model Tests of the 84'. Tracy Pumps, ASME 54A 225.
81. THOMSON, J.L., 'Packed Glands for High Pressure Pumps', Pumping, May 1959 et seq.

THE PAST AND THE FUTURE

This chapter, from the second and third editions, was originally a paper to the British Hydromechanics Research Assn. Conference in 1967.

The author started apprenticeship on pumps, turbines and gear generation in 1921, hence the generated impeller flow passages. About that time, an overloading impeller was fitted with peripheral rings to make the power characteristic turn down at best efficiency flow, foreseeing the variable flow pumps in chapter 46.

Pump power then ranged up to one megawatt. An irrigation pump was pictured with a Morris car in its delivery branch. The typical turbo-generator in the 1930s was 30 MW, involving a 700 kW boiler feed pump.

The mid 1930s 107 MW turbo-generator was a very large and successful forward step. The 2.2 MW feed pump and the similar oil pipeline pump are probably still in operation.

Vertical pumps in a borehole ran satisfactorily, sight unseen. In a deep well, the vertical pipes and bearings could be monitored, and were found to deflect markedly in a sine wave nodal pattern causing anxiety. The stresses were, however, quite low, and the pumps ran for many decades.

The author used ejectors to raise the pump pressure tenfold at zero flow for press duties, which avoided the need for a reciprocating pump to give the last squeeze after the centrifugal pump had lifted the ram. A four pass pump to give four times the head from a single impeller was also tried out, and showed that, with a 45 degree barrier between the stages, full head was generated.

In 1947, the author suggested for the first high speed geared feed pump, 2.5 MW at 6000 rev/min, a 250 kW pony motor on grid power engaging the gear box annulus, a brake and the pony motor as generator returning power to the grid. This gave 6600 rev/min motoring, 6000 r/m braked and 5400 r/m generating, with improved efficiency, using direct on line switching.

The Twenty Mode High Efficiency Variable Speed of chapter 48 is a further development.

Area Ratio performance determination originated in the early 1930s from the need to obtain varying characteristics at the same shape number, and is now the universal design system.

Two aspects of the possible changes in fluid machines over the next twenty years are the increase of power, speed, pressure and temperature of existing designs, and the emergence of entirely new designs, metallurgy and plastics. Possible changes in the pump field are surveyed, first with respect to increased consumer demand on present functions, and secondly with respect to new untried processes.

Introduction

The year 2001, the third millennium, is nearer to us than the mid 30s when the usual turbo-alternator was 30 MW — now 660 MW is normal. The increase in the next 20 years to 2000 MW will involve 50 MW turbine driven feed pumps at 6000–9000 rev/min. Refineries, chemical plant, steel works, desalination plant, paper mills and waterworks will similarly increase their liquid pump demands, but present designs and metallurgy can cope with such anticipated increases. New and untried processes, and nuclear, MHD, space travel, gas turbine and deep sea developments are certain to require pumps — but the problems to be met are more difficult to foresee.

Escalation of Present Activities

Increase of throughput of oil refineries, chemical plants, steel works, paper mills, etc will not present difficulties in the pump field since, in general, these duties now involve relatively low shape number pumps — the increase of flow will raise the shape number and demand larger machines of more favourable proportions.

Waterworks pumps are already in operation at 200 MW 600 metres head, so that they now are to be considered more in the field of water turbines and reversible pump turbine energy storage machines. These last machines are already in operation at 700 MW and may be expected to reach 1 000 MW by the third millennium. No insuperable problem arises here — the basic limitation is the problem of transporting larger pieces, but even this problem is easier than the transport of the associated electrical components. A 7 metre impeller is in operation for a 200 MW 260 metre reversible machine — the impeller being split diametrically for transport.

Any anticipation of materially larger rotors must consider site welding and machining, particularly as the British Rail loading gauge is only 3.7 metres compared to the American 5 metres. For example, a 20 metre diameter impeller and casing could be built up by welding on site. A boring mill would then apply an orbiting cutter to machine the casing bores in situ, and a non-rotating cutter assembly to machine the rotating impeller. The generator/motor would be similarly constructed. A portable boring mill of these dimensions is entirely practicable.

Major global projects to change continental and oceanic configurations, humidities and temperatures are probable in the future: for example, the provision of pump operated tidal flaps across the Bering Straits to provide more temperate conditions all over the world by circulating the North Atlantic and Pacific, the surplus water due to partial ice cap melting being offset by pumping to the Sahara, Australian and Gobi deserts to avoid flooding of maritime cities. A portion of the Nile could be diverted to the Sahara. (See Lunaheat Ref 82).

Water demand is increasing so fast that ultimately all rivers will be pumped into water supply mains. This state of affairs is almost here in the case of the River Thames. Since, for example, the Mediterranean Sea is an evaporating basin its isolation by these means will provide large amounts of land and also of minerals to be pumped away as heavy brine. Such projects will be greatly needed in the future and will be no more costly than space travel.

Boiler Feed Pumps

Here the anticipated developments in metallurgy and bearing practice suggest that a two stage 50 MW feed pump at 6000 rev/min will be a practicable proposition for 2000 MW supercritical steam turbo-generators. Adequate booster systems and inducers will of course be required. The velocities in the feed pump will then be well below sonic velocity.

Space Rocket Pumps

Similar conditions to boiler feed pumps will arise here, but on smaller sized units at speeds up to 100 000 rev/min, single stage still predominating.

Production and Performance

The enormous demand of developing countries over the next 20 years will involve mass production with batch runs of 10 000 components of a very wide range of pumps compared to the present relatively small runs of a few hundreds. For stage heads up to 1000 m sophisticated moulding techniques will produce impellers, diffusers and casings of adequate surface finish, but in the range of stage heads from 1 000 to 4 000 m the high flow velocities up to 300 metres/sec will demand fully machined surfaces on solid forgings. Pumps will then be produced in numbers comparable to present car manufacture.

Statistical Control of Performance

These very large runs permit elaborate and accurate moulding procedure so that the cheapest components become the best as far as high performance and consistent performance are concerned. Testing would then involve perhaps one pump in a batch of 100 to ensure that quality is maintained. These trends are already noticeable in such long runs as are now obtained in small and popular sizes.

Cost of Mass Produced Pumps

At present, pump prices are very roughly 3000 per ton. Cars, made of very much more expensive materials and more intricate geometry, are roughly the same price per ton showing the effect of hundred-fold production runs. Over the next 20 years, with increasing demand, pump costs should, by larger runs, ultimately reflect the fact that their basic materials, cast iron and bronze, are so much cheaper than the basic materials of a car and that their geometry is very simple.

Production of Flow Passage Surfaces

Since impeller passages represent a combination of impeller rotation about its axis and toroidal rotation of liquid through a quadrant, it follows that the impeller is a gear wheel and should be produced as a gear wheel. This method has been proved over several years on a very large aggregate of power, to give improved performance.

For stage heads up to about 1000 metres generation of core box would appear adequate, but on anticipated heads of 3000 to 4000 metres, complete machine from forgings of the impeller and diffuser would seem to be necessary.

Limits

The foregoing has dealt with the largest size (20 metre impeller), the highest stage head (4000 metres), and the highest speed (100 000 rev/min). All smaller and easier duties are thereby covered.

Metallurgy and Plastics

We visualize the ultimate development of steel to give a proof stress of at least 225 hectobar combined with stainless properties, weldability and acceptable thermal expansion. This material would accept the most severe duties mentioned above, surface finish and geometry being of a very high order; eg 0.3 micron for the highest stage heads. These super steels would only be justified on the more sever duties.

Pump components for the more normal duties, (eg, up to 300 m head per stage) would in the future probably be made of plastic since this material is expected to replace mild steel and bronze to a major degree.

Recent tests to destruction of steel impellers running in air, show that they burst at speeds corresponding generally to stage heads of 25 000 to 30 000 metres. This suggests that the centrifugal stresses at 4000 metres head are reasonable and that the predominant stress problems at these high heads will be concerned with pressure fluctuations and torque loadings.

Future Development of Pumps

The Author's dream of a rotating tube to produce pressure without the need for liquid to enter and leave a vane system, has not yet been fulfilled. The aim of this device would be to eliminate losses and cavitation so that efficiency would be almost 100% and the generation of head would be almost unlimited. It would therefore appear that the vaned impeller will always be with us, and that progress will depend upon higher speeds and improved metallurgy, the basic principles of the pumps that we now know remaining unchanged. Volute pumps replaced diffuser pumps for modest duties some 60 years ago and a further development to vaneless diffusers may occur in pumps as in turbo-blowers.

Mechanical Seals

At present in almost universal use on process work, these seals can show considerable development to meet the severe duties now demanded. In parallel, devices will be developed without any rubbing contact for nontoxic duties.

Submersible Motors and Pumps

These are rapidly replacing shaft driven borehole pumps and could in future be in universal use for such duties.

Canned Motors

For toxic radio-active liquids this type is essential, but as in the case of submersible units, the design problems arise in the motors and not in the pumps.

Pressure Vessel Design

For the highest pressures of the future (say, for example, six kilobar) bolted joints will give way to some form of pressure assisted joint.

Gearing

It would appear that frequencies of 50 cycles in the old world and 60 cycles in the new world and at sea will remain with us perhaps for a few hundred years. It is fortunate that gearing has reached the stage of high reliability. In general the reduction of capital cost of a smaller pump pays for the gearbox, and improved shape number pays for the gearbox power losses.

Liquid Metal Induction Pumps

The do not show any immediate promise of competing with centrifugal pumps for really large duties.

Stage Numbers

The only significant trend in the rotodynamic world appears to be the reduction of a number of stages for any given duty, both on multistage pumps and on water turbines where Francis machines are performing duties previously done by Pelton wheels (the equivalent turbine to the multistage pump).

In a similar trend axial flow pumps, Deriaz units and Kaplan turbines are now performing duties previously in the field of centrifugal pumps and Francis turbines.

Inducers

Axial inducers to give the inlet conditions that alone will permit the higher speeds involved in the above changes, are bound to develop from their present infrequent use to more general application.

Ecology

The vital problem of the future is the avoidance of pollution of the rivers, the lakes and the atmosphere. To this end power stations of 20 gigawatts in ten units will probably be constructed on artificial islands 40 kilometres from the coast with high tension d.c. links.

Hydro-power causes negligible pollution but is insufficient in magnitude, even fully developed, to provide more than a small part of world power needs. Reversible pump turbines will probably develop up to 2 gigawatts on high heads and one gigawatt on low head bulb units in tidal schemes for peak load duties.

New and Untried Processes

It is comforting to the pump engineer to remember that, whatever changes occur in the future development of power, pumps will always be indispensable, and the future challenge can probably be estimated within the forseeable ranges of speed, power, stage head, corrosion tendencies, temperature etc, as suggested above.

Ultimately, coal and oil supplies will be exhausted. Even then the beaming of the sun's heat to earth via satellite and microwave, to provide power without air and river pollution, would still require cooling pumps. (See also Lunaheat Ref 82).

Nuclear duties

No insuperable pumping problem appears on nuclear duties other than the possibility, raised some years ago, of the need to pump abrasive radio-active slurries. The difficulties of internal bearings in close contact with such pumped liquids can readily be imagined.

Space and Deep Sea Duties

The primary considerations here are that volume and weight of pumps and drivers must be reduced to the absolute minimum, which involves the ultra high speeds already described, using steam or gas turbines or gearing.

MHD

In whatever direction the generation of electricity from fuel develops, it can be safely assumed that heat transfer pumps will be required, only their form and duties would be the subject of speculation.

Conclusion

The foregoing notes represent the thoughts of one who has been concerned with pumps and turbines for 60 years, and having seen so many major changes in that time has been rash enough to visualize the changes of the next 20 years in this field. It is hoped that, at least, this chapter will give rise to some useful and interesting thought.

The Author acknowledges and appreciates the advantages he has enjoyed in working in an industry that is changing so rapidly.

He is grateful for the help and encouragement that he has enjoyed from the manufacturers, the research associations, the Institution of Mechanical Engineers, The American Society of Mechanical Engineers, many Universities and the Publishers, still continuing. All his life he has never met other than helpful colleagues.

Reference

82. ANDERSON, H.H. 'Lunaheat: Warming the Northern Regions in the Third Millennium', Chartered Mechanical Engineer, December 1976.
83. ANDERSON, H.H. 'Prediction of Head, Quantity and Efficiency in Pumps — the Area Ratio Principle'. ASME 22nd Fluids Conf New Orleans 1980 Book No 100129.

Appendices

APPENDIX A

Effect of Pump Geometry on Pump Performance

FIRST EXPLANATION OF AREA RATIO

(from 'Mine Pumps' Ref 2 1938)

Inlet Conditions

The inlet portion of a pump comprises the path of the liquid from suction branch to the point at which it is flowing through the passages of the impeller. Between these two points a pressure drop occurs due to friction and increase of velocity of flow. This pressure drop increases as the quantity pumped increases.

The liquid entering the suction branch will have a pressure which is greater by a certain margin than its saturation pressure for the particular temperature. If the pressure drop between inlet branch and impeller passages becomes as great as the margin above saturation, then saturation conditions will obtain in the impeller and the liquid will tend to vaporize with consequent loss of pressure and risk of erosion of the impeller. The reason for this loss of pressure is as follows:

The head generated by a centrifugal pump is expressed as the height to which a liquid of given specific gravity would be raised. Consequently, if the specific gravity of the fluid is reduced by passing from liquid to vapour, the pressure generated will be reduced similarly, although the head expressed as feet of fluid will remain unaltered.

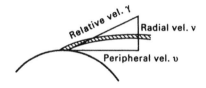

Figure A.1 – Typical inlet velocity diagram.

It is therefore the function of the inlet of the pump to feed the impeller with an adequate supply of liquid without risk of vaporization. To this end, the inlet vanes should be designed to give minimum shock at the required duty. A typical inlet diagram for radial entry is shown in Fig A.1. The velocities concerned are the peripheral velocity of the inlet tips, the relative velocity through the impeller passages and the radial entry velocity. It will be seen that the diagram is a right angled triangle.

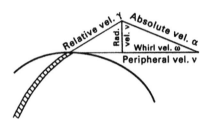

Figure A.2 – Typical outlet velocity diagram.

Outlet Conditions

An outlet velocity diagram is shown in Fig A.2. The triangles of velocities comprise the peripheral velocity of the impeller, the relative velocity of flow through the outlet passages of the impeller and the absolute velocity of the liquid when leaving the impeller. The diagram also indicates the radial and the whirl, or tangential, velocity of the liquid. These two velocities are combined in the absolute leaving velocity.

Assuming that in the theoretical pump, the absolute leaving velocity of the liquid coincides with the velocity and angle of flow in the diffuser, or volute collecting chamber, this would, in theory, indicate that no shock occurs at exit from the impeller. The quantity of liquid passing through the pump at this velocity is the designed quantity of the pump and at this point in the ideal pump gives the maximum hydraulic efficiency.

Symbols

Let the peripheral velocity of the impeller be u m/sec

Radial velocity	v	m/sec
Relative velocity	r	m/sec
Absolute velocity	a	m/sec
Diffuser or volute velocity	t	m/sec
Whirl velocity	w	m/sec
Acceleration due to gravity	g	m/s²

Quantity of liquid at maximum hydraulic efficiency by Q m³/sec

For maximum hydraulic efficiency a = t.

Let the ratio

$$\frac{\text{Outlet area between vanes of impeller}}{\text{Diffuser area}} = Y$$

Then since impeller and diffuser or volute, are running full the velocities in these parts will be inversely as the areas – ie

$$\frac{t}{r} = Y \qquad (1)$$

Torque Equation

The generated head, based upon the relation between torque and angular momentum, is given by the equation

$$\text{Head} = \frac{u\,w}{g} \qquad (Ref\ 24)$$

Head Characteristics

$$\text{Head generated by impeller} = \frac{u\,w}{g} \text{ metres}$$

From the two right-angled triangles in Fig A.2 when a = t

$$v^2 = r^2 - (u - w)^2$$

$$w^2 = t^2 - v^2 = t^2 - r^2 + (u - w)^2$$

$$w^2 = t^2 - r^2 + u^2 - 2\,u\,w + w^2$$

$$2\,u\,w = t^2 - r^2 + u^2 \quad \text{thus} \qquad \frac{u\,w}{g} = \frac{t^2 - r^2 + u^2}{2\,g}$$

$$r^2 = \frac{t^2}{Y^2} \qquad \text{from (1)}$$

$$\therefore \quad \frac{u\,w}{g} = \frac{t^2\left(1 - \dfrac{1}{y^2}\right) + u^2}{2g} = \text{Head generated}$$

If Y = 1 ie, Impeller area = diffuser area

$$\text{then head generated} = \frac{u^2}{2\,g} \text{ m.}$$

If Y is less than 1 eg ½

$$\text{Head} = \frac{u^2}{2\,g} - \frac{3t^2}{2\,g}$$

If Y is greater than 1 eg 2

$$\text{Head} = \frac{u^2}{2\,g} + \frac{^3/_4\,t^2}{2\,g}$$

Thus we find that if the area of the impeller is greater than the diffuser area the head generated at maximum hydraulic efficiency will exceed $\dfrac{u^2}{2g}$.

If the impeller area is less than the diffuser area the head generated will be less than $\dfrac{u^2}{2g}$.

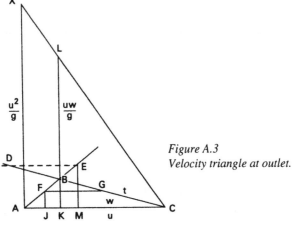

Figure A.3
Velocity triangle at outlet.

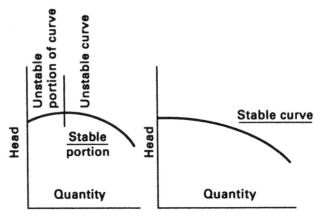

Figure A.4
Stable and unstable head curves.

In Fig A.3, A, B, C represent the triangle of velocities at maximum hydraulic efficiency quantity Q.

At quantities less than this, a typical diagram will be AFC where relative velocity is AF, the whirl is CJ and the absolute velocity CF, the peripheral velocity u remaining constant.

A line FG parallel to AC will cut off CG

$$= \frac{\text{Diffuser velocity CG}}{\text{Relative impeller velocity AF}}$$

$$= \frac{CB}{AB} = \frac{t}{r} = \frac{\text{Impeller}}{\text{Diffuser area}}$$

Similarly for quantities greater than Q

$$= \frac{CD}{AE} = \frac{CB}{AB} = \frac{t}{r} = \frac{\text{Impeller area}}{\text{Diffuser area}}$$

The absolute velocity CF is reduced by shock to the diffuser velocity CG. There are several ways of interpreting this shock loss. If the pressure remains unaltered the loss in energy will be given by the algebraic difference of the squares of the velocities. There is, however, a definite changing of the angle of the flow, and therefore it is proposed to assume, for the purpose of these pages, that the energy loss is given by the line FG which is the closing line in the velocity diagram AFGC. This completing line FG connects the peripheral velocity AC, the relative velocity AF and the final diffuser velocity GC.

Fig A.3 also indicates the values of $\dfrac{uw}{g}$ in the triangle AXC.

Since $AX = u^2/g$ and $CK = $ whirl, then a perpendicular KL represents

$$\frac{\text{whirl velocity}}{\text{peripheral velocity}} \times \frac{u^2}{g} \quad \text{from similar triangles} \quad = \frac{wu}{g}$$

We now find it convenient to regard the maximum efficiency quantity Q as zero on our chart of characteristics, quantities less than Q being indicated by negative values and quantities greater than Q being indicated by positive values.

For example, in Fig A.6 the origin represents a quantity Q. The point marked -1 represents zero quantity and $+1$ represents 2Q quantity. Basing the curve in terms of Q and $\dfrac{u^2}{g}$ it can be stated that $Q = 1$ and $\dfrac{u^2}{g} = 1$ for the sake of simplicity.

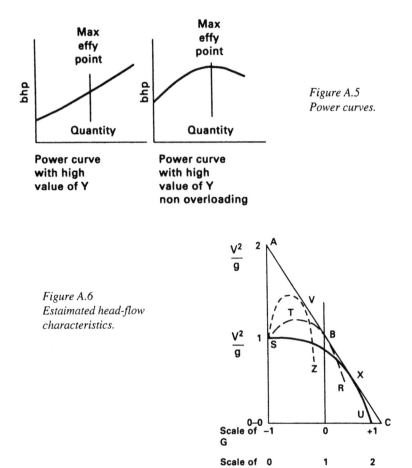

Figure A.5
Power curves.

Figure A.6
Estaimated head-flow
characteristics.

A distance G on the curve will represent G + 1 quantity where G may be positive or negative. When G = 1 quantity G + 1 = 0, when G = 0, quantity is 1.

Consider the special case where Y = 1 and Head $= \dfrac{u^2}{2g}$

The curve Head $= \dfrac{uw}{g}$ is indicated on the graph as the straight line ABC, since head at zero quantity is $\dfrac{u^2}{g}$ when w = u and head at quantity Q is $\dfrac{u^2}{2g}$.

The equation to ABC is Head $= (1 - G) \dfrac{u^2}{2g}$ where (1 + G)Q is quantity flowing.

The velocity FG $= \dfrac{FB}{AB} \times AC$ (Fig A.3)

$= \dfrac{AB - AF}{AB} \; AC$

$= \dfrac{\text{Rel vel at Q} - \text{Rel vel at quantity } (G + 1)Q}{\text{Relative velocity at quantity Q}} \times u$

$= \left[\dfrac{r - (G + 1)r}{r} \right] u = \left[1 - G - 1 \right] u = -Gu$

Energy lost $= \dfrac{G^2 u^2}{2g}$

Similarly the velocity DE can be equated to Gu. If we subtract this loss from the expression head $= (1 - G) \dfrac{u^2}{2g}$ we obtain head $= (1-G-G^2) \dfrac{u^2}{2g}$, and the curve is indicated by STBR, this loss occurring for quantities greater and less than Q.

Where a quantity $\dfrac{u^2}{2g}$ is subtracted from ABC note that whether G is negative or positive, G^2 is positive, and therefore subtracted in each case.

The theoretical curve from a centrifugal pump for given values of Y can now be considered. For other values of Y the head at Q is

$$\dfrac{t^2 \left[1 - \dfrac{1}{Y^2} \right] + u^2}{2g} = \dfrac{u \, w}{g}$$

Now quantity flowing as a ratio of Q is $(1 = G)$, and actual diffuser velocity as a ratio of diffuser velocity at Q is $1 + g$.

Then Head $= \dfrac{(1 - G - G^2) \; u^2 + (1 + G) \, (1 - \dfrac{1}{Y^2})t^2}{2g}$

If quantity is zero $1 + G$ is zero and head is $\dfrac{u^2}{2g}$ irrespective of value of Y.

Hence all pumps whatever the area ratio Y will give a head $= \dfrac{u^2}{2g}$ at zero quantity.

If Y is greater than 1 the curve will rise, then fall as shown in SVZ Fig A.6 (this may be unstable).

If Y is equal to 1 the curve is STBR is already shown.

If Y is less than 1 the curve will fall more steeply and will be in the stable region as shown SXU.

In order to draw the curves to illustrate the stable and unstable characteristics it is necessary to move the origin 0 to points corresponding to X and V respectively, the zero quantity remaining at −1 as before and the scale being altered to suit.

Power and Efficiency Characteristics

Since the relation $\dfrac{uw}{g}$ = head is based upon torque on shaft it follows that the torque taken by an ideal pump with no losses will be proportional to uw x quantity.

The work done per kilogram of liquid, based on torque on shaft is $\dfrac{uw}{g}$.

Hence work down /kg x quantity/second x constant will represent the power taken by the pump assuming no losses.

The quantity will be proportional to u − w since this represents the projection AJ of the relative velocity AF on the line AC (Fig A.3). It can thus be said that the power is (u − w) $\dfrac{(uw)}{g}$ x constant where w varies between zero and u. Differentiating with respect to $\dfrac{w}{u}$ and equating to zero we obtain the value $w = \dfrac{u}{2}$ for maximum power. The power curve is therefore a parabola whose value at zero quantity is zero. A quantity to give head equal to $\dfrac{u^2}{2g}$ the power is maximum and at double this quantity the power is zero. Hence for all pumps the maximum power occurs when the head generated is $\dfrac{u^2}{2g}$.

If, due to the varying value of Y, the quantity Q for maximum hydraulic efficiency is greater or less than quantity to give $\dfrac{uw}{g} = \dfrac{u^2}{2g}$ then the power curve will fall or rise after maximum efficiency power. This is the determining factor of non-overloading power characteristics. (Fig A.7).

The value of Y determines the position of maximum hydraulic efficiency, but the maximum power will always occur at head = $\dfrac{u^2}{2g}$.

When Y = 1, maximum power occurs at maximum hydraulic efficiency quantity.

Efficiency

In order to determine overall efficiency including losses, values must be assigned to these losses. When differentiating to obtain maximum efficiency it is not always possible to give

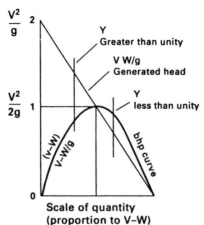

Figure A.7
Power characteristics.

a general form to the efficiency, including losses, of a pump where Y is not equal to 1, since the actual diffuser velocity t is required; therefore only cases where Y = 1 will be considered.

Efficiency Taking into Account Shock Losses Only

The following relations have already been obtained:

$$\text{Head} = (1 - G - G^2) \ \frac{u^2}{2g}$$

$$\text{Power} = (u - w) \ \frac{uw}{g} \ \times \ \text{constant}$$

(neglecting mechanical losses and friction losses in the pump)

$$= (1 + G) (1 - G) \times \text{constant}$$
$$= (1 - G^2) \times \text{constant}$$
$$\text{Quantity} \quad = (1 + G) \times \text{constant}$$

$$\text{Efficiency} = \frac{\text{Liquid Power}}{\text{Input Power}} = \frac{\text{Head} \ \times \ \text{Quantity}}{\text{Input Power}}$$

$$= \frac{(1 - G - G^2) \ \times \ (1 + G)}{(1 - G^2)}$$

$$= \frac{1 - G - G^2}{1 - G} = 1 - \frac{G^2}{1 - G}$$

The hydraulic efficiency has its maximum at the designed quantity, is 50% at no flow and zero at 1.62 x the designed quantity.

Other Losses

A further hydraulic loss is the friction loss through the passages of the pump. This loss generally varies as the square of the quantity passing through the pump.

Mechanical losses comprise disc friction, bearing and windage losses, gland friction. Leakage losses through glands and neckrings, etc, also must be taken into account. Mechanical and leakage losses are roughly constant.

In order to obtain some idea of the shape of the efficiency curve it is necessary to assess these losses. For any given pump it is possible to estimate the losses from data based on actual tests. It will be assumed that the friction loss, varying as the square of the quantity, is equal to 5% of the generated head at theoretical maximum efficiency quantity and also that the contact losses represent an increase of input power from 100% to 110% of the power taken at the theoretical maximum efficiency quantity. This power curve will be given by a constant x $(1 - G^2) + \dfrac{1}{10}$.

One can now plot the head, power and efficiency curves of an ideal pump and subtract the head losses, and add the power losses to obtain the corresponding three curves for an actual pump. It will be noted that the friction loss through the pump makes the head curve fall more steeply. The addition of power losses to the input power curve reduces the pump efficiency at zero quantity from 50% to zero. At the same time the portion of the pump

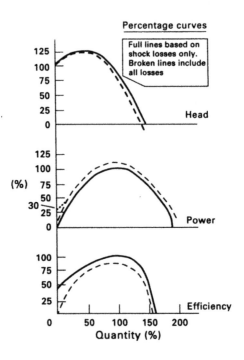

Figure A.8 – Percentage curves.

efficiency curve between zero quantity and maximum efficiency becomes 'fuller' in shape, giving the contours corresponding to actual tests.

Once particular discrepancy between the calculated curves on Fig A.8 and the test curves of a typical pump is the shape of the power curve for very small quantities. The calculated power at zero quantity is $\dfrac{10}{110}$ of maximum power, that is, approximately 9%. In an actual pump this ratio is nearer 30%. It would appear that in addition to the losses already mentioned, a further loss occurs on very small quantities. It is probable that this further loss is due to wasteful circulation and eddies.

This additional loss apparently disappears beyond about 50% of maximum efficiency quantity, so that it will affect the portions of the curve which are of less consequence in average pumping problems.

In drawing the curves an indication is made of additional power loss but it is not possible to calculate what this loss would be, the curves in full lines and dotted lines indicating only 5% head loss and the 10% power loss.

It is therefore found that the maximum overall efficiency of the pump with $Y = 1$ and losses as selected, would occur approximately at 90% of the quantity Q. The value of this efficiency would be 86% and the maximum input power would occur at $\dfrac{100\%}{90\%}$ x maximum efficiency quantity. This maximum power would exceed the maximum efficiency power by 2%. The characteristics of such a pump are shown on a percentage basis in Fig A.8.

Shape Number

It is evident that the shape of characteristic is dependent primarily on area ratios of impeller and casing. Therefore, one should expect to obtain any desired shape characteristic, whatever the shape number of the pump, by adopting a suitable area ratio.
This fact is borne out in practice by the following examples;-

1. 8% and 30% pressure rise from maximum efficiency to zero quantity obtained from two different pumps at 700 shape number.

2. 10% and 30% pressure rise obtained from two different pumps at 1200 shape number.

    (Note on Shape Number).

    $$N_s = \text{Shape number} = \text{rev/min} \; \frac{\sqrt{\text{flow}}}{(\text{Head})^{3/4}}$$

The shape number of a pump indicates the shape of the pump and is sometimes defined as the speed at which a pump would run if reduced geometrically to give 1 l/s at 1 m head. As this is rather difficult to visualize the following may be considered:

1. The term 'shape number' could be discarded and the shape number called instead 'type quantity'. This ratio would then be regarded as the quantity handled by a pump when reduced geometrically to give 1 m head at 1 rev/min. The impellers shown

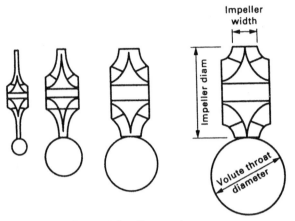

Figure A.9 – Change of type.

in Fig A.9 have the same diameter and are all designed to give the same head at the same rev/min. The alteration in shape to give increased quantity is evident, the width and eye diameter of the impeller and the diameter of the volute throat progressively increasing as the shape number increases, whilst the impeller outlet diameter remains unaltered. It should be noted that if the term 'type quantity' is adopted the present values of shape numbers should be squared to give the true quantity ratios, since shape number is a function of the square root of the quantity.

2. For a certain design of pumps; eg a range of pumps where $Y = 1$, the ratio diameter of volute throat/diameter of impeller is obtained as follows:

Head = u^2 x constant = Dia^2 x (rev/min)^2 x constant

Diameter = $\sqrt{H}$/rev/min x constant

Absolute velocity leaving impeller = volute throat velocity = constant x u, ie, is proportional to $\sqrt{H}$.

Quantity pumped = throat velocity x area of throat, therefore is proportional to H x area of throat.

Hence, area of throat is proportional to $\dfrac{Q}{\sqrt{H}}$ and diameter of throat is proportional to $\dfrac{\sqrt{Q}}{H^{1/4}}$.

Therefore, diameter of throat/diameter of impeller is proportional to

$$\frac{\sqrt{Q}}{H^{1/4}} \bigg/ \frac{\sqrt{H}}{\text{rev/min}} = \frac{\text{rev/min} \sqrt{Q} \text{ x constant}}{H^{3/4}}$$

It is therefore possible to indicate the change of type of pumps shown in Fig A.9 by the varying ratio-throat diameter/impeller diameter. From the above it is seen that this ratio is identified with the shape number. In practice, it is found that the shorter length of vane of the larger eye impellers modifies the relation between shape number and diameter ratio on larger values of shape number.

Shape number based on the best efficiency duty of the pump is a relationship which defines the pump shape at constant $N_s y$, eg the 45° lines on Fig 13.3.

It should, however, be noted that a range of pumps at constant shape number can have various shapes and characteristics according to the area ratio, ie a vertical line on Fig 13.3.

Assumptions

It has been necessary to make the following assumptions to prepare this analysis. Although these are only assumptions and cannot be proved mathematically, there is no doubt that they explain the particular contours of centrifugal pump characteristics and indicate the variation of these contours for various designs of pumps.

1. The loss of energy due to shock at impeller outlet in a centrifugal pump is represented as a function of the side completing the velocity polygon whose other sides are relative impeller outlet velocity, peripheral impeller velocity and absolute diffuser or volute throat velocity.

2. It is necessary to assume that the loss of head due to friction through the pump varies as the square of the quantity, and that the power losses (other than hydraulic) are approximately constant.

3. It is necessary to assume some value for the volumetric efficiency of the pump.

The two latter assumptions are based upon actual test results.

Conclusions

Head, Power and Efficiency

(a) In an ideal pump, ie, taking into account shock losses only, the head generated is represented by:

$$H = \frac{u^2 (1 - G - G^2) + (1 + G) \left[1 - \frac{1}{Y^2}\right] t}{2g}$$

where t is a constant, being the velocity in the diffusers at the maximum efficiency quantity. In the particular case where $Y = 1$

$$H = (1 - G - G^2) \frac{u^2}{2g}$$

The power absorbed by the pump is given by

$$\text{Input power} = \text{constant} \times (1 - G^2)$$

$$\text{Efficiency} = 1 - \frac{G^2}{1 - G}$$

If Y is greater than 1 the head curve will tend to be unstable and the power curve will continue to rise as the quantities exceed maximum efficiency quantity. If Y is less than 1 the head curve will be in the stable region, and the power curve will fall beyond maximum efficiency quantity, ie, the pump will be non-overloading.

Shape number over the whole characteristic from zero flow to zero head varies from zero to infinity, but shape number is always determined from flow. head and speed at best efficiency point and is a useful arbiter of efficiency and of guidance in design.

Actual Tests of Pumps

Comparing the actual tests with the calculated curves it is found that it is necessary to make one further correction, namely, that the impeller and passages must be assumed to run partly empty. The value of the friction losses through the pump cannot be calculated nor can the volumetric efficiency or degree of fulness of the passages. When comparing the test results with calculations, values must be assumed for these unknown factors, hence it is not claimed that this analysis gives absolute values. Nevertheless, it will be seen from Figs A.10 and A.11 that the shape of the characteristic given by test agrees very closely with the shape given by calculation according to the preceding analysis.

For all pumps the maximum input power will occur at a quantity such that the theoretical generated head equals $u^2/2g$.

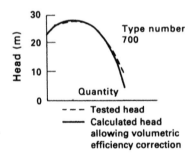

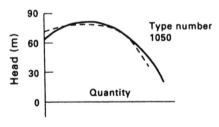

Figures A.10 and A.11
Comparison of test and calculated
head curves to demonstrate shape
of characteristics.

Figure A.12
Erosion in half an hour due to impeller-diffuser
proximity and zero flow operation.

(b) In an actual pump

In the analysis one example based on assumed losses is indicated.

The test characteristics of two actual pumps are shown in comparison with estimated characteristics in Figs A.10 and A.11.

References

2.   ANDERSON, H.H., 'Mine Pumps, Jl Min Soc, Durham University. July 1938.
24. GIBSON, A.H., 'Hydraulics and Its Applications', Constable 1920.

APPENDIX B

Materials for Various Liquids
from Standards of HYDRAULIC INSTITUTE USA

The standard material classification of pumps follows. Their listing in the material column is abbreviated through the use of the associated initials:

(1) Standard and Bronze Fitted SF

(2) All Iron AI

(3) All Bronze AB

(4) Types 4,5,6 and 7 AA

To simplify recording, the symbol AA is used in those cases where types 4, 5, 6 and 7 would normally be listed. This does not necessarily mean, however, that all are equally effective in all environments. It merely means that each type has been satisfactorily applied in handling that liquid under some, possibly all, conditions.

Other materials, including corrosion resisting steels, are listed by number in accordance with the tabulation opposite.

Type 6 covers a series of alloys developed by different manufacturers to augment the standard 4 and 5 types and to provide superior corrosion resisting properties. Type 7, special alloys, covers a wide range of non-ferrous materials of less than 20% iron, containing nickel and chromium or molybdenum, or both, in major amounts, and copper, tungsten, silicon and manganese in smaller percentages. As these are special materials for special services, their effective application requires close cooperation between the producer, pump manufacturer and user, each giving due consideration to the local factors and economic aspects involved. It should be pointed out, however, that a pump of the type 6 material will not necessarily cost more than one in, say, type 4, as a manufacturer having standardized one the more expensive material may be able to sell it as a standard product at a lower price than if he brought through a single pump in a less costly material.

Paragraph E-18, 'Data Required by Pump Manufacturers for Proper Selection of Material', carefully filled out, will assist the pump manufacturer in recommending a suitable material for the particular application. Similar sheets may be obtained from the pump manufacturer concerned.

TABLE I

Type No.	Commercial Designation	Carbon	Chromium	Nickel	Molybdenum	Remarks
1	AISI 410	0.15 max	11.5/13.5	—	—	Free machining type is 416
	ACI CA14	0.14 max	11.0/14.0	1.0 max	—	
2	AISI 442	0.35 max	18.0/23.0	—	—	
	ACI CB30	0.30 max	18.0/23.0	2.0 max	—	
3	AISI 446	0.35 max	23.0/27.0	—	—	
	ACI CC35	0.35 max	27.0/30.0	3.0 max	—	
4	AISI 304	0.08 max	18.0/20.0	8.0/10.0	—	Free machining type is 303
	ACI CF 7	0.07 max	18.0/20.0	8.0/10.0	—	Free machining type is CF7Se
5	AISI 316	0.10 max	16.0/18.0	10.0/14.0	1.75/2.5	
	ACI CF7M	0.07 max	18.0/20.0	8.0/10.0	1.5/3.5	
6		0.07 max	15.0/28.0	22.0/36.0	1.5/4.0	Optional elements Cu, W, Si, Mn, Ti, Cb
7	A series of non-ferrous alloys, of less than 20% iron, containing nickel and chromium or molybdenum, or both, in major amounts, and copper, tungsten, silicon and manganese in lesser percentages.					
8	High silicon iron. 14.25% silicon minimum					
9	Austenitic cast iron, total nickel, chromium, copper contents, 22% minimum.					
10	Monel metal					
11	Lead					
12	Non-metallic					
13	Nickel					
14	Steel					

AISI = American Iron and Steel Institute
ACI = Alloy Casting Institute.

DATA REQUIRED BY PUMP MANUFACTURERS FOR PROPER SELECTION OF MATERIAL

1. **SOLUTION TO BE PUMPED** (Give common name, where possible, such as 'spinning bath', 'black liquor', 'spent pickle', etc) ....
....

2. **PRINCIPAL CORROSIVES (H_2SO_4, HCl, etc)** ....
.... % by weight.
(In the case of mixtures, state definite percentages by weight. for example: mixture contains 2% acid, in terms of 96.5% H_2SO_4).

3. **pH** (if aqueous solution) .... at ....°C

4. **IMPURITIES OR OTHER CONSTITUENTS NOT GIVEN IN '2'** (List amounts of any metallic salts, such as chlorides, sulphates, chromates, and any organic materials which may be present, even though in percentages as low as 0.01%. Indicate, where practical, whether they act as accelerators or inhibitors on the pump material). ....

....

5. **SPECIFIC GRAVITY** (solution pumped) .... at ....°C

6. **TEMPERATURE OF SOLUTION**   Maximum.................................. °C
   Minimum .... °C   Normal .... °C

7. **VAPOUR PRESSURE AT ABOVE TEMPERATURES**
   Maximum..................................... °C   Minimum .... °C
   Normal .... °C
   (Indicate units used, such as bar, pounds gauge, inches water, millimetres mercury).

8. **VISCOSITY** .... SSU; or .... centistokes; at .... °C

9. **AERATION**   Air-Free .... Partial............................
   Saturated ....
   Does liquid have tendency to foam?...

10. **OTHER GASES IN SOLUTION** .... ppm,
    or .... cm^3/l

    ....

11. **SOLIDS IN SUSPENSION** (state types) ....

    ....

    Specific gravity of solids ....
    Quantity of solids... % by weight
    Particle size .... mesh.. % by weight
                   .... mesh.. % by weight
                   .... mesh.. % by weight
    Character of solids   Pulpy .... Gritty ....
                          Hard .... Soft ....

12. CONTINUOUS OR INTERMITTENT SERVICE ....

Will pump be used for circulation in closed system or for transfer? ....

....

Will pump be operated at times against closed discharge? ....

....

If intermittent, how often is pump started? .... times per ....
Will pump be flushed and drained when not in service? ....

....

13. TYPE OF MATERIAL IN PIPE LINES TO BE CONNECTED TO PUMP ....

....

If desirable, are insulated joints practical? ....

14. IS METAL CONTAMINATION UNDESIRABLE? ....

If so, what percentage of element (Fe,Ni,Cu,etc) is objectionable? ....

....

....

15. PREVIOUS EXPERIENCE   Have you pumped this solution previously? ....

....

If so, of what material or materials was pump made? ....

....

Service life in months? ....
In case of trouble, what parts were affected? ....

....

Was trouble primarily due to corrosion? .... Erosion? ....
galvanic action? .... stray current? ....
Was attack uniform? ....
If localized, what parts were involved? ....

....

If galvanic action, name materials involved ....

....

If pitted, describe size, shape and location (A sketch will be helpful in an analysis of
problem) ....

....

16. WHAT IS CONSIDERED AN ECONOMIC LIFE? ....
(Sometimes a pump in a relatively inexpensive material, if the replacements are not
too frequent, may work out to be the most economical).

MATERIALS FOR PUMPING VARIOUS LIQUIDS

Liquid	Condition	Chemical Symbol	Specific Gravity	Materials Commonly Used
Acetaldehyde		CH,CHO	0 78	AI
Acetate Solvents				AI SF AB AA
Acetone		CH,COCH,	0 79	AI SF
Acetic Anhydride		(CH,CO),O	1 08	AA 8
Acid, Acetic	Conc Cold	CH,COOH	1 055	AA 8
Acid, Acetic	Dil Cold			AB AA 8
Acid, Acetic	Conc Boiling			5 6 7 8
Acid, Acetic	Dil Boiling			5 6 7 8
Acid, Arsenic		As05		AA 8
Acid, Benzoic		C,H,COOH		AA
Acid, Boric	Aqueous Sol	H,BO,		AB AA 8
Acid, Butyric	Conc	CH,(CH,),CO,H	0.96	AA
Acid, Carbolic	Conc (M P 106 F)	C,H,OH	1 071	AI AA
Acid, Carbolic	Aqueous Sol			SF AA
Acid, Carbonic	Aqueous Sol	CO,+H,O		AB
Acid, Chromic	Aqueous Sol	CrO,+H,O		AA 8
Acid, Citric	Aqueous Sol	C,H,O-H,O		AB AA 8
Acids, Fatty (Oleic, Palmitic, Stearic, etc)				AB AA
Acid, Formic		HCOOH	1 2	5 6 7
Acid, Fruit				AB AA 10
Acid, Hydrochloric	Coml Conc	HCl	1.16 (20°Be)	7 8 12
Acid, Hydrochloric	Dil Cold			6 7 8 10 12 13
Acid, Hydrochloric	Dil Hot			7 8 12
Acid, Hydrocyanic		HCN	0 70	AI AA
Acid, Hydrofluoric	Anhydrous, with Hydro Carbon	HF + HC		10 14
Acid, Hydrofluoric	Aqueous Sol	HF + H,O		AB 10
Acid, Hydrofluosilicic		H,SiF,		AB 10
Acid, Lactic		CH,CHOHCOOH	1 249	AB AA 8
Acid, Mine Water				AB AA
Acid, Mixed				AI AA 8 14
Acid, Muriatic	(See Acid, Hydrochloric)			
Acid, Naphthenic				AI AA 1
Acid, Nitric	Conc Boiling	HNO,	1.41	2 3 6 8
Acid, Nitric	Dilute			1 2 3 4 5 6 8
Acid, Oxalic	Cold	C,H,O,2H,O		AA 8
Acid, Oxalic	Hot	C,H,O,2H,O		6 7 8
Acid, Ortho Phosphoric		H,PO,	1.36 – 1.4	5 6 7
Acid, Picric		(NO,),C,H,OH		AA 8
Acid, Pyrogallic		C,H,(OH),		AA
Acid, Pyroligneous		H,C,H,O,	1.018/1.03	AB AA
Acid, Sulphuric	77% Cold	H,SO,	1.69/1.835	AI 6 7 8
Acid, Sulphuric	65/93% 175 F			7 8
Acid, Sulphuric	65/93% 175 F			6 7 8
Acid, Sulphuric	10/65%			6 7 8 11
Acid, Sulphuric	10%			AB 6 7 8 10
Acid, Sulphuric (Oleum)	Fuming	H,SO,+SO,		6 7 14
Acid, Sulphurous		H,SO,		AB AA 11
Acid, Tannic		C,,H,,O,,		AB AA 10
Acid, Tartaric	Aqueous Sol	C,H,O,		AB AA 10
Alcohols				AB SF
Alum	(See Aluminium Sulphate and Potash Alum)			
Aluminium Sulphate	Aqueous Sol.	Al,(SO,),		6 7 8 10 11
Ammonia, Aqua		NH,OH		AI
Ammonium Bicarbonate	Aqueous Sol	NH,HCO,		AI
Ammonium Chloride	Aqueous Sol.	NH,Cl		5 6 7 8 10
Ammonium Nitrate	Aqueous Sol.	NH,NO,		AI AA 10
Ammonium Phosphate	Aqueous Sol.	(NH,),HPO,		AI AA 10
Ammonium Sulphate	Aqueous Sol.	(NH,),SO,		AIAA
Ammonium Sulphate	With H,SO,			AB 5 6 7 8 11
Aniline		C,H,NH,	1.022	AI SF
Aniline Hydrochloride	Aqueous Sol.	C,H,NH,HCl		7 8 12
Asphalt	Hot		0.98/1.4	AI 1
Barium Chloride	Aqueous Sol.	BaCl,		AI AA
Barium Nitrate	Aqueous Sol	Ba(NO,),		AI AA
Beer				AB 4
Beer Wort				AB 4
Beet Juice				AB 4
Beet Pulp				AB SF AA
Benzene	(See Benzol)			
Benzine	(See Petroleum ether)			

Liquid	Condition	Chemical Symbol	Specific Gravity	Materials Commonly Used
Benzol		C H	0 88	Al SF
Bichloride of Mercury	(See Mercuric Chloride			
Black Liquor	(See Liquors, pulp mill)			
Bleach Solutions	(See type)			
Blood				SF AB
Boiler Feedwater	(See Water, boiler feed)			
Brine, Calcium Chloride	pH 8	CaCl		Al
Brine, Calcium Chloride	pH 8			AB 6 7 9 10
Brine, Calcium & Magnesium Chlorides	Aqueous Sol			AB 6 7 9 10
Brine, Calcium & Sodium Chlorides	Aqueous Sol			AB 6 7 9 10
Brine, Sodium Chloride	Under 3% Salt, Cold	Na Cl	1 02	Al AB 9
Brine, Sodium Chloride	Over 3% Salt Cold		1 02 1 20	AB AA 9 10
Brine, Sodium Chloride	Over 3% Salt, Hot			5 6 7 8 10
Brine, Sea Water			1 03	Al SF AB
Butane		CH (CH) CH	0 60 @ 32 F	Al SF 14
Calcium Bisulphite	Paper Mill	Ca (HSO)	1 06	5 6 7 11
Calcium Chlorate	Aqueous Sol	Ca (C10) 2H O		6 7 8 12
Calcium Hypochlorite		Ca (OCl)		A1 6 7 8
Calcium/Magnesium Chloride	(See Brines)			
Cane Juice				SF AB 9
Carbon Bisulphide		CS		Al
Carbonate of Soda	(See Soda Ash)			
Carbon Tetrachloride	Anhydrous	CCl.	1 58	Al SF
Carbon Tetrachloride	Plus Water			
Caustic Potash	(See Potassium Hydroxide)			AB 4
Caustic Soda	(See Sodium Hydroxide)			
Cellulose Acetate				5 6 7
Chlorate of Lime	(See Calcium Chlorate)			
Chloride of Lime	(See Calcium Hypochlorite)			
Chloride Water	(Depending on conc)		1 1	5 6 7 8 11 12
Chlorobenzene		C. H Cl		SF AB 4
Chloroform		CHCl.	1 5	AB AA 10
Chrome Alum	Aqueous Sol	CrK(SO) 12H O		6 7 8
Condensate	(See Water, Distilled)			
Copperas, Green	(See Ferrous Sulphate)			
Copper Ammonium Acetate	Aqueous Sol			Al AA
Copper Chloride (Cupric)	Aqueous Sol	CuCl		7 8 12
Copper Nitrate		Cu(NO)		AA
Copper Sulphate, Blue Vitriol	Aqueous Sol	CuSO.		AA 8 11
Creosote	(See Oil, Creosote)			
Cresol, Meta		CH.C H OH	1 04	Al 6 7
Cyanide	(See Sodium Cyanide and Potassium Cyanide)			
Cyanogen	In Water	C N (gas)		Al
Diphenyl		C H C H		Al 14
Enamel				Al
Ethanol	(See Alcohols)			
Ethylene Chloride (dichloride)	Cold	CH CICH Cl	1 28	AB AA 10
Ferric Chloride	Aqueous Sol.	FeCl.		7 8 12
Ferric Sulphate	Aqueous Sol	Fe (SO).		AA 8
Ferrous Chloride	Cold, Aqueous	FeCl.		7 8 12
Ferrous Sulphate (Green Copperas)	Aqueous Sol	FeSO.		5 6 7 8 10 11
Formaldehyde		HCHO	1.075/1 081	AB AA
Fruit Juices				AB AA 10
Furfural		C H OCHO	1 16	Al AB AA
Gasoline			0.68/0.75	Al SF
Glaubers Salt	(See Sodium Sulphate)			
Glucose				SF AB
Glue	Hot			Al SF
Glue Sizing				AB
Glycerol (Glycerin)		C H (OH)	1,262	Al SF AB
Green Liquor	(See Liquors, Pulp Mill)			
Heptane		C H	0.69	Al SF
Hydrogen Peroxide	Aqueous Sol.	H O		AA
Hydrogen Sulphide	Aqueous Sol.	H S		AA
Hydrosulphite of Soda	(See Sodium Hydrosulphite)			
Hyposulphite of Soda	(See Sodium Thiosulphate)			
Kaolin Slip	Suspension in Water			Al 14
Kaolin Slip	Suspension in Acid			6 7 8
Kerosene	(See Oil, Kerosene)			
Ketchup				AB AA
Lard	Hot			Al SF
Lead Acetate (Sugar of Lead)	Aqueous Sol.	Pb(C H O) 3H O		5 6 7 10
Lead	Molten			Al 14
Lime Water (Milk of Lime)		Ca(OH)		Al

Liquid	Condition	Chemical Symbol	Specific Gravity	Materials Commonly Used
Liquors Pulp Mill				
Black				Al 5 6 7 9 10 14
Green				Al 5 6 7 9 10 14
White				Al 5 6 7 9 10 14
Pink				Al 5 6 7 9 10 14
Sulphite				5 6 7 11
Lithium Chloride	Aqueous Sol	LiCl		Al
Lye, Caustic	(See Potassium and Sodium Hydroxide)			
Magnesium Chloride	Aqueous Sol	Mg Cl		6 7 8 12
Magnesium Sulphate				
(Epsom Salts)	Aqueous Sol	$MgSO_4$		Al AA
Manganese Chloride	Aqueous Sol	$MnCl_2$		AB 9A 8
Manganous Sulphate	Aqueous Sol	$MnSO_4.4H_2O$		Al AB AA
Mash				SF AB 4
Mercuric Chloride	Very Dilute Aqueous Sol	$HgCl_2$		5 6 7 8
Mercuric Chloride	Coml Conc aqueous Sol	$HgCl_2$		7 8 12
Mercuric Sulphate	In H_2SO_4			6 7 8 12
Mercurous Sulphate	In H_2SO_4	Hg_2SO_4		6 7 8 12
Methyl Chloride		CH_3Cl	0 92	Al
Methylene Chloride		CH_2Cl	1 26	Al 4
Milk			1 028/1 035	4
Milk of Lime	(See Lime Water)			
Mine Water	(See Acid, Mine Water)			
Miscella	(20% Soyabean Oil and Solvent)		0 75	Al
Molasses				SF AB
Mustard				AB AA 8
Naphtha			0 78/0 88	Al SF
Naphtha, Crude		$(C_{10}H_{11}N.)^4.H_2SO_4$	0 92/0 95	Al SF
Nicotine Sulphate				6 7 8 10
Nitre	(See Potassium Nitrate)			
Nitre Cake	(See Sodium Bisulphate)			
Nitro Ethane		$CH_3CH_2NO_2$	1 041	Al SF
Nitro Methane		CH_3NO_2	1 139	Al SF
Oil, Coal Tar			1 04/1 10	Al SF
Oil, Coconut			0 905	Al SF AB AA 10
Oil, Creosote	Cold		1 04/1 10	Al SF
Oil, Crude	Cold			Al SF
Oil, Crude	Hot			14
Oil, Differential				Al SF AB
Oil, Fuel				Al SF
Oil, Kerosene				Al SF
Oil, Linseed			0 94	Al SF AB AA 10
Oil, Lubricating				Al SF
Oil, Mineral				Al SF
Oil, Olive			0 90	Al SF
Oil, Palm			0 895	Al SF AB AA 10
Oil, Quenching			0 912	Al SF
Oil, Rapeseed			0 92	AB AA 10
Oil, Soya Bean				Al SF AB AA 10
Oil, Turpentine			0 87	Al SF
Paraffin	Hot			Al SF
Perhydrol	(See Hydrogen Peroxide)			
Peroxide of Hydrogen	(See Hydrogen Peroxide)			
Petrol			0 68/0 75	Al SF
Petroleum Ether		—		Al SF
Phenol	(See Acid, Carbolic)			
Pink Liquor	(See Liquor, Pulp Mill)			
Photographic Developers				AA 12
Plating Solutions	(Varied and complicated, consult pump mfgrs.)			AB AA 9 10
Potash	Plant Liquor			AB AA 9 10
Potash Alum	Aqueous Sol	$Al_2(SO_4)_3.K_2SO_4.24H_2O$		AB 5 6 7 8 9 10
Potassium Bichromate	Aqueous Sol	$K_2Cr_2O_7$		Al
Potassium Carbonate	Aqueous Sol	K_2CO_3		Al
Potassium Chlorate	Aqueous Sol	$KClO_3$		AA 8
Potassium Chloride	Aqueous Sol	KCl		AB AA 10
Potassium Cyanide	Aqueous Sol	KCN		Al
Potassium Hydroxide	Aqueous Sol	KOH		Al AA 1 9 10 13
Potassium Nitrate	Aqueous Sol	KNO_3		Al AA 1
Potassium Sulphate	Aqueous Sol	K_2SO_4		Al AA
Propane		$CH_3CH_2CH_3$	0 585 @ 48 F	Al SF 14
Pyridine		$CH(CHCH)_2N$	0 975	Al
Pyridine Sulphate				6 8 11
Rhidolene				SF
Rosin (Colophony)	Paper Mill			Al

Liquid	Condition	Chemical Symbol	Specific Gravity	Materials Commonly Used
Sal Ammoniac	(See Ammonium Chloride)			
Salt Cake	Aqueous Sol.	Na_2SO_4 + impurities		AB AA 8
Salt Water	(See Brines)			
Sea Water	(See Brines)			
Sewage				AI SF AB
Shellac				AB
Silver Nitrate	Aqueous Sol.	$AgNO_3$		AA 8
Slop, Brewery				AI SF AB
Slop, Distillery			1.05	AB AA
Soap Liquor				AI
Soda Ash	Cold	Na_2CO_3		AI
Sodium Bicarbonate	Aqueous Sol.	$NaHCO_3$		AI AA 9
Sodium Bisulphte	Aqueous Sol.	$NaHSO_4$		6 7 8 11
Sodium Carbonate	(See Soda Ash)			
Sodium Chlorate	Aqueous Sol.	$NaC10_3$		AA 8
Sodium Chloride	(See Brines)			
Sodium Cyanide	Aqueous Sol.	$NaCN$		AI
Sodium Hydroxide	Aqueous Sol.	$NaOH$		AI AA 1 9 10 13
Sodium Hydrosulphite	Aqueous Sol.	$Na_2S_2O_4 2H_2O$		AA 11
Sodium Hyprochlorite		$NaOCl$		6 7 8 12
Sodium Hyposulphite	(See Sodium Thiosulphate)			
Sodium Meta Silicate				AI
Sodium Nitrate	Aqueous Sol.	$NaNO_3$		AI AA 1
Sodium Phosphate:				
Monobasic	Aqueous Sol.	NaH_2PO_4		AB AA
Dibasic	Aqueous Sol	Na_2HPO_4		AI AB AA
Tribasic	Aqueous Sol.	Na_3PO_4		AI
Meta	Aqueous Sol	$NaPO_3$		AB AA
Hexameta	Aqueous Sol.	$(NaPO_3)_n$		AA
Sodium Plumbite	Aqueous Sol.			AI
Sodium Sulphate	Aqueous Sol.	Na_2SO_4		AB AA
Sodium Sulphide	Aqueous Sol	Na_2S		AI AA 11
Sodium Sulphite	Aqueous Sol.	Na_2SO_3		AB AA 11
Sodium Thiosulphate	Aqueous Sol.	$Na_2S_2O_3 5H_2O$		AA 12
Stannic Chloride	Aqueous Sol.	$Sn Cl_4$		7 8 11 12
Stannous Chloride	Aqueous Sol.	$Sn Cl_2$		7 8 11 12
Starch		$(C_6H_{10}O_5)x$		SF AB
Strontium Nitrate	Aqueous Sol.	$Sr(NO_3)_2$		AI 4
Sugar	Aqueous Sol.			AB AA 9
Sulphite Liquors	(See Liquors, Pulp Mill)			
Sulphur	In Water	S		AI AB 9
Sulphur	Molten	S		AI
Sulphur Chloride	Cold	S_2Cl_2		AI 11
Syrup	(See Sugar)			
Tallow	Hot		0.895	AI
Tanning Liquors				AB AA 8 10
Tar	Hot			AI 14
Tar and Ammonia	In Water			AI
Tetrachloride of Tin	(See Stannic Chloride)			
Tetraethyl Lead		$Pb(C_2H_5)_4$	1.65	AI SF
Toluene (Tuluol)		$CH_3C_6H_5$	0.86	AI SF
Trichlorethylene		C_2HCl_3	1.47	AI SF AB 4
Urine				AB AA
Varnish				AI SF AB 4 10
Vegetable Juices				AB AA 10
Vinegar				AB AA 8
Vitriol, Blue	(See Copper Sulphate)			
Vitriol, Green	(See Ferrous Sulphate)			
Vitriol, Oil of	(See Acid, Sulphuric)			
Vitriol, White	(See Zinc Sulphate)			
Water, Boiler Feed High Makeup	Not evaporated pH>8.5		1.00	AI See Chapter 26
	pH<8.5			SF
Water, Boiler Feed Low Makeup	Evaporated, any pH		1.00	5% Cr 1 4 10
Water, Distilled	High Purity		1.00	AB 4
	Condensate			SF AB
Water, Fresh			1.00	SF
Water, Mine	(See Acid, Mine Water			
Water, Salt and Sea	(See Brines			
Whiskey				AB 4
White Liquor	(See Liquors, Pulp Mill)			
White Water	Paper Mill			AI SF AB
Wine				AB 4

MATERIALS FOR PUMPING VARIOUS LIQUIDS (cont)

Liquid	Condition	Chemical Symbol	Specific Gravity	Materials Commonly Used
Wood Pulp (Stock) Wood Vinegar Wort Xylol (Xylene) Yeast Zinc Chloride Zinc Sulphate	(See Acid Pyroligneous) (See Beer Wort) Aqueous Sol. Aqueous Sol.	$C_6H_4(CH_3)_2$ $ZnCl_2$ $ZnSO_4$	0.87	Al SF AB Al SF AA SF AB 5 6 7 8 AB 5 6 7

ABRASIVE ASPECTS* (Ref 68)

Owing to the large number of factors affecting abrasive wear, it does not appear possible to make a few overall hard-and-fast rules as to the best way of reducing it; each case will still have to be treated on its merits, not least of which must be economic. However, it is worth noting some general trends derived from the literature for the designer's consideration:

1. Wear increases rapidly when the particle hardness exceeds that of the metal surface being abraded.

2. Wear increases generally with grain size, sharpness and solids concentration. Rubber lining is particularly vulnerable to large, sharp particles.

3. Metal hardness is not an absolute criterion of wear, although for ferrous metals, the expected trend for wear resistance to increase with hardness applies very generally. A reasonable resistance appears to be achieved above about 300 Brinell. The very hard alloys (eg tungsten carbide) and surface treatments are extremely resistant.

4. Chemical composition, microstructure and work-hardening ability all play an important part in wear resistance of metals. Austenitic Cr-Ni (18-8% cr) and Mn alloy steels are good, as in 'Ni-hard' (Ni-Cr) cast iron. 18/8 stainless steel (though resistant to cavitation) and most non-ferrous metals, (except cupro-aluminimum), have rather poor abrasion resistance.

5. Soft rubber appears generally more resistant than hard.

6. Plastics coatings do not appear very promising so far, except possibly in particular applications; bonding can also be a problem. Ceramics are very wear-resistant, but their use to date has been limited by brittleness and susceptibility to thermal shock. New developments in small pump applications may show improvements.

7. Wear increase rapidly with flow velocity, and is often reported as being approximately α (velocity)^3, or α (pump head)$^{3/2}$, from both theoretical considerations and test results. The actual value of the index, for any given conditions, probably depends on at least some, if not all, of the other factors involved in the overall wear process.
    Head limits quoted are up to about 100 m/stage for all-metal pumps, and 50 m/stage for rubber-lined.

8. Impact angle has a marked effect on wear; metals and rubbers behave in opposite ways.

9. Good hydraulic design, particularly by avoiding rapid changes in flow direction, decreases wear, and should be compromised as little as possible by solids-handling considerations. Shrouded impellers are generally favoured.

10. Rubber lining can give a much-increased life compared to that for metal, provided that the solids are not large or sharp, bonding is good, and heads and temperatures relatively low.

11. Soft-packed shaft glands require a grease or clean water supply; scraper-vanes on the impeller, or separate centrifugal seals, are also used to protect the glands. Mechanical seals with special materials, and usually with a flushing supply, are sometimes fitted.

PUMP MATERIALS (from Ref 29)

Cast Irons

Cast iron is a standard material choice for general-purpose pumps and also for alkaline services (pH 9-14). Grey cast irons are specified in BS1452:1977 and form by far the bulk of the iron castings produced in this country, although spheroidal graphite (SG) irons are generally to be preferred for higher duty applications on account of their higher mechanical properties (Ref BS2789:1973).

More limited use is made of alloy irons (austenitic cast irons) of which the following are particularly worth of note:-

NI-Resist AUS 101 - with good resistance to corrosion and especially weak acids, extending the application of cast iron down to about pH = 3. This alloy iron does however, have a fairly high coefficient of thermal expansion.

Ni-Resist AUS 102 - with good resistance to alkalis.

Ni-Resist AUS 105 - with excellent resistance to thermal shock (see BS 3468).

Ni-Resist D-2M - with good resistance to mechanical shock down to very low temperatures, making it particularly suitable for use in contact with liquid gases.

Particular mention must also be made of silicon cast iron, which is a low cost material with good resistance to corrosion. It is widely used for wetted parts in petrochemical applications with a silicon content of 5-7%. Increasing the silicon content to 14-16% makes this iron suitable for use with a wide range of mineral acids. The chief disadvantage of silicon cast iron is that it is a hard, brittle material. When used for pump casings, therefore, it is commonly used as a liner set in a housing of ordinary cast iron.

High chromium white iron is similar in application to silicon cast irons in that it is suitable for acid services. Chromium content may range up to 35% and is usually associated with a high carbon content (up to 3%), so the material is brittle, although it has excellent resistance to wear and abrasion. High chrome white irons with reduced carbon content (under 2%) are more suitable for higher temperature applications and also have good resistance to sulphurous fumes.

Malleable cast irons are ductile and offer a wide choice of mechanical properties, with a tensile strength range of from about 30 to 70 h bar. A particular virtue is that they are

TABLE 2 – Typical chemical composition (castings)

| | Ferralium | Langalloy | | | Hastelloy Alloy B* | Hastelloy Alloy C* |
		3V	20V	7R		
Nickel %	5	10	29	BAL	BAL	BAL
Chromium %	25	18	20	23.0	–	16.5
Copper %	3	–	3.5	6.0	–	–
Iron %	BAL	BAL	BAL	5.0	5.0	5.0
Molybdenum %	2	3.0	3.0	6.0	29.0	17.0
Tungsten %	–	–	–	2.0	–	4.5
Silicon %	1.0	1.0†	–	3.0	0.75	0.75
Manganese %	1.0	1.0	–	0.75	0.75	0.75
Columbium %	–	–	10xC min	–	–	–
Carbon %	0.06	0.08	0.05	0.08	0.08	0.08
Vanadium %	–	–	–	–	0.3	0.35†

† maximum

*Hastelloy is the registered trade mark of the Cabot Corporation

readily surface-hardened after machining and can thus provide low cost, long-life replaceable wear rings, etc, for process pumps where the liquid concerned is not too acidic.

Cast Steels

Carbon steel castings are generally chosen for higher strength and provide greater design flexibility than cast irons. They are, however, particularly limited as regards corrosion resistance, even the low alloy types. Low alloying is employed solely to improve the mechanical or working properties of the steel, although a series of steels, of this type is produced specifically for the petrochemical industry (BS 1504) and they are thus generally suitable for process pump components in such applications

High alloy steels cover a wide range of compositions, selected to enhance particular properties. Only the corrosion resistant alloys are of real interest to chemical and process pump manufacturers and normally are base don high alloy contents of chromium and nickel, with a smaller addition of molybdenum and possibly traces of other metallic elements. They are often described as acid-resisting steels. There are three main types of such alloys:-

(i) **Ferritic steels** alloyed with chromium, becoming more ferritic (and thus less amenable to heat treatment) with increasing chromium content. Annealed castings, however, are generally resistant to most mineral acids and oxidizing acids.

(ii) **Austenitic chromium-nickel steels.** These are not hardenable. The addition of nickel increases resistance to mineral and organic acids. 20/30 chromium nickel content produces an alloy that will withstand concentrated sulphuric acid and most hot, concentrated, weakly oxidizing solutions. Chromium-nickel alloy steels are

TABLE 3 – Service properties of selected plastics

Material	Resistance To*					Max Service Temp°C	Limitations
	Weak Acids	Strong Acids	Weak Alkalis	Strong Alkalis	Solvents		
Acetal Polymer	S	X	S	L	S	120	High permeability to water
Acetal Copolymer	S to L	X	E	E	L to S	110	Limited dimensional stability and creep resistance
Nylon 66	S	X	S	L to S	S	100–130	Prone to stress corrosion cracking with certain chemicals
Polycarbonate	S	L	L to S	X	X to S		
Polyethylene LD	S	L to X	S to X	S	S	85	Permeable, prone to stress cracking
Polyethylene HD	S	S	S to L	S	S	120–130	Prone to stress cracking
Polypropylene	S	S to L	S	S	S	100–120	More permeable than HD polythene
PTFE	E	E	E	E	E	260	High cost
FEP	E	E	E	E	E	200	High cost
PVC Rigid	S	L to S	S	L to S	varies	60–100	Subject to backing action, low softening temperature
PVC Flexible	S	L	S	L	L to X	70	
PVC Chlorinated	E	S	E	S	S to E	–	
DAP	S	L	S	X	S	up to 100	High cost, abrasive
Epoxy	S	S	S	L	S		High cost, careful processing required
Phenolic	S	S	L	X	varies	190	Brittle
Polyester	S	S	L to X	L to X	S	150–220	Exothermic
Alkyd	S	L	S	L	varies	up to 200	Very high cost
ABS	E	S	S to L	L	L	85–90	Brittle
UF	S to L	X	S to L	X	S	75	Brittle
MF (Amino Melamine)	S	X	E	X	S	100	Low impact strength
Chlorinated Polyether	E	E	E	E	E to S	45–60	
EVA (Copolymer)	S	S	S	S	S	130	
PETP	S	L	S	–	S	130	Poor resistance to steam and some acids
PPO	E			S	–		Stress cracking possible
Polysulphones	E	S	E	S	X	150–165	Subject to stress cracking

*Key. E = excellent   S - satisfactory   L = limited   X = unsuitable (attacked)

TABLE 4 – Selected plastics physical aspects

Material	SG	Tensile Strength h bar	Remarks	Application(s)
Acetal	1 425	7	rigid, excellent dimensional stability, high resistance to abrasion	small pump components
Nylon 66	1 09- 1 14	up to 9	low cost, easily moulded	small impellers, casings, etc
Polycarbonate (filled)	1 2 / 1 4	6	good dimensional stability glass filled	small pump casings, impellers, etc
Polyethylene LD / HD	0 918–0 94 / 0 95–0 965	1 / 2 6	low cost, flexible / improved chemical resistance and permeability	linings and coatings / linings and coatings
Polypropylene	0 90	3 4	very light, outstanding chemical stability	linings and coatings
PTFE filled	2 17	1 7 / 3 4	poor mechanical properties / improved strength and rigidity	moulded and fabricated components
FEP		1 8 / 2 0	very high cost	—
PVC Rigid / Flexible	1 34 1 4 / 1 2 1 5	4 0 5 5 / 0 27 1 8	low cost, easily moulded / low cost, easily moulded	moulded pump components, pipes linings, coatings, flexible lines
ABS	1 02 1 06	3 4 5 2	suitable for rigid mouldings (glass filled)	linings and pipes for higher temp corrosive fluids
Alkyd	1 7 2 2	up to 17 (filled)	speciality high cost material	
DAP (Fabric)		3 4		
Epoxy glass filled	1 18 1 75	6 0 8 5 / 14 0	good adhesion to metals care needed in processing	coatings and linings moulded casings and impellers
UF	1 4 1 55	5 0 7 0	low cost thermoset	
MF	1 5 1 55	5 0 8 5	slightly better properties than UF	
Chlorinated Polyether	1 4		speciality material with outstanding chemical resistance	linings and coatings (Penton)
EVA	0 93 0 95	1 4	good low temperature characteristics	
PTCFE			lowest water vapour transmission of all plastics	possibility for non-toxic linings and coatings
Phenolics	1 36	4 5 6 0	cotton or mineral fillers for maximum chemical resistance	gears and small pump components
Polyester	1 12		high dimensional stability low water absorption	coatings, glass fibre mouldings
PETP	1 37	under 0 7 to 7	glass filled and unfilled good hydrolytic stabilities	gears (gear pumps)
PPO	1 06		outstanding thermal stability for a thermoplastic	
Polysulphones	1 24	7		

not completely immune to acid attack, however, and may show surface pitting even when otherwise compatible. This tendency can be reduced by the addition of molybdenum, which also extends resistance to include acetic and phosphoric acids.

(iii) **Ferritic-austenitic steels.** They are normally 25/5 chromium/nickel content, together with small additions of molybdenum and copper. Resistance to attack by acids is superior to the two other types with additional advantage of improved erosion resistance since the alloy is also hardenable. They are a particularly attractive choice for chemical pump components including pump shafts and wear rings, in which the relatively high cost of the material can be justified.

A further category of high alloy steels of interest to process pump manufacturers is the heat-resisting alloys. The corrosion-resistant alloys have generally higher maximum service temperatures than plain steels or cast iron, but tend to scale at temperatures in excess of about 650°C. The heat-resistant alloys extend the useful temperature range from 700 to about 1 200°C, depending on the chromium content. The proportion is almost direct between 15 and 30% chromium. Such steels are also particularly resistant to sulphurous ambients. High nickel-chrome alloys are even more resistant to oxidizing and reducing atmospheres but not to sulphur fumes.

A further series of high alloy steels have balanced compositions permitting precipitation hardening a substantial increase in tensile strength (eg up to 120 h bar). In general, however, higher mechanical properties are only achieved at the expense of a reduction in resistance to corrosion. The extreme is reached with maraging steels which may have tensile strengths in excess of 150 h bar but very poor corrosion resistance.

Stainless Steels

This is a general term covering a wide range of chromium alloy steels containing 10% or more chromium. They comprise three basic types:-

(i) **Martensitic.** Heat-treatable to enhance mechanical properties but having the lowest resistance to corrosion. 1-3% nickel content.

(ii) **Ferritic.** Containing 17-20% chromium, non-hardenable by heat treatment and having moderate corrosion resistance. Nought or 1-5% nickel content.

(iii) **Austenitic.** Again non-hardenable by heat treatment but the most important type because of good resistance to corrosion and good mechanical properties.of the nickel alloys make them materials which must always be considered. Some brief data on the chemical resistance properties of typical nickel-based alloys are summarized in Table V.

Superalloys

Superalloys are special compositions developed for stressed high temperature working (ie with high creep resistance) and/or satisfactory scaling and corrosion resistance at elevated temperatures. They are thus primarily intended for high temperature applications. The range of such alloys is vast, but their application to pump components is strictly limited because of their high cost and difficulty of fabrication, which latter further increases component cost. They may, however, need to be used in specialized high temperature process services. (Table VI).

Titanium

Titanium is a useful alternative material for high temperature working because of its high inherent strength and good creep properties. It also has excellent resistance to attack under oxidizing conditions due to the stability of the surface oxide film. General corrosion resistance can also be improved by small additions of other metallic elements, eg traces of palladium to improve resistance to attack by sulphuric and hydrochloric acids.

The corrosion resistant properties of titanium are at least equal to those of the austenitic stainless steels. Moreover the material is stronger and has a considerably better strength-to-weight ratio over a higher temperature range. It does have two disadvantages: it is expensive and is difficult to cast or even melt, so that titanium components have to be machined and consequently cost more to produce.

Bronzes

Bronze or bronze-fitted pumps are standard for slightly acidic liquids (pH 4-6), gunmetal being the most widely used alloy. It contains various proportions of tin, zinc and lead as the alloying elements with copper, and up to 5% nickel. Lead-free gunmetals (actually containing up to 1.5% lead) were originally widely used, but leaded gunmetals are now generally preferred as they produce superior castings with excellent corrosion resistance, high strength and improved wear resistance. The main alloy used is BS1400 - LG2C, or the later BS1400 - LG4C where optimum mechanical properties and corrosion resistance are required. The latter has a higher tin and nickel content (at the expense of lead) and is thus more costly.

Aluminium bronze offers the highest strength of all the bronzes and has good corrosion

TABLE 5 – Corrosion resistant properties of high-nickel alloys

Alloy	Remarks
Nickel	Excellent resistance to hot concentrated alkalis. Good resistance to dilute acids and salt solutions
Nickel-copper	Excellent resistance to hot concentrated alkalis. Excellent resistance to brines. Good resistance to acids and reducing solutions (not sulphurous). Excellent resistance to hydrofluoric acid
Nickel-molybdenum	Exceptional resistance to hydrochloric acid. Resistant to pure phosphoric acid. Good resistance to sulphuric acid (but not sulphurous)
Nickel-molybdenum- -chromium	Improved resistance to oxidizing solutions. Resistant to dry and wet chlorine
Nickel-chromium- iron	Good resistance to oxidizing acids. Excellent resistance to hot concentrated alkalis

TABLE 6 – Proprietary high-nickel alloys

Name	Corrosion Resistance
Hastelloy B	Good resistance to sulphuric and phosphoric acids. Excellent resistance to boiling hydrochloric acid and wet HCL gas.
Hastelloy C	High resistance to sulphuric, phosphoric, acetic, formic acids, etc. Resistant to oxidizing agents.
Hastelloy G	Extreme resistance to acids (but less used than B and C for pump components)
Langalloy 4R	Good resistance to acids.
Langalloy 5R	Good resistance to acids, oxidizing agents, wet chlorine and hypochlorites
Langalloy 6R	Particularly resistant to sulphuric acid
Langalloy 7R	Highly resistant to sulphuric acid and acid mixtures

resistance especially to saline solutions. But its main application is for sea water duties which do not normally come within the range of chemical or process pumping. Corrosion resistance of straight aluminium-copper alloys is improved by the addition of manganese.

The use of phosphor bronze is mainly limited to bearing sleeves and wear-resistant components in bronze-fitted pumps. It does warrant consideration on its own for other abrasion resistant pump components such as impellers and casings where the fluid is moderately acidic.

Brasses

Only the specialized brasses need be considered under this heading, and even these have strictly limited application. High tensile brasses have a certain application for impellers, being cheaper than bronzes and having better corrosion resistance than ordinary brasses, although inferior to bronzes. Aluminium-brass also has good resistance to erosion, corrosion and impingement attack when handling high liquid velocities. It thus has a certain application as an impeller material, but more so far tubes. A cheaper alternative, with inferior corrosion resistance, is Admiralty brass.

Aluminium

Aluminium has become an inexpensive, and readily castable and machinable material since the rapid development of high- strength aluminium alloys, starting with the Y alloy (copper 4%, nickel 2%, magnesium 1.5%).

Aluminium and aluminium alloys have good resistance to corrosion although not necessarily immunity from attack. Thus an initial corrosion product is formed as a surface layer which inhibits further attack unless the initial attack is strong enough to dissolve or penetrate the surface layer, as occurs with strong alkalis, sulphuric acid, hydrochloric acid, carbonates and fluorides. The corrosion products formed by surface attack are colourless, odourless, tasteless and non-toxic, so that aluminium is widely used in the processing, handling and storage of a broad range of chemicals such as acetic acid, hydrogen peroxide, concentrated nitric acid, ammonium nitrate, numerous petrochemicals, and in particular food products.

Alluminium is rather less widely used as a pump material, however, although it is often available as an alternative material for process pumps, or is specified for certain food processing duties. Thus an all-aluminium pump is commonly specified for handling milk.

Certain limitations apply to aluminium. Apart from mechanical properties the most serious is the readiness with which galvanic corrosion can be initiated when in contact with other metals that are higher in the galvanic scale. The coefficient of expansion of alluminium is higher than those of other metals (eg twice that of steel), which can produce problems in 'mixed' constructions operating over a temperature range.

Lead

Lead is one of the traditional materials imp pumps designed for handling acids. It is used in the form of hard lead or lead-antimony alloy containing 6 to 12% antimony. The hardness of the alloy is usually of the order 8-10 Brinell and may be doubled or trebled after casting, by annealing, quenching and ageing. Heat treatment also raises the tensile strength by 8 h bar. The corrosion resistance of lead is generally excellent, lead coatings being widely used on other metals, notably steel, to provide good chemical resistance. The

Figure B.1
Volute pump with split casing.

degree of resistance achieved is dependent on the alloying additions employed (Type B lead), if chemical lead of 99.99% purity (Type A lead) is not satisfactory. Thus, linings or coatings can be chosen from Type A or Type B lead, as appropriate. Castings for acid pumps, on the other hand, are invariably made from lead- antimony alloys because of the greater strength and hardness of this material.

Ceramics

Traditionally, ceramic materials such as stoneware, porcelain and glass have always had a place as constructional materials for chemical and process pumps. since these materials are hard and brittle they are employed on as linings set or cemented in a cast iron housing for the construction of pump casings and inlet covers. This technique can be applied to the largest sizes of pump required, although difficulties may be experienced with differential thermal expansion when the pump is subject to thermal cycling. Thus, whilst the use of such materials persists, the tendency is to employ more flexible or semi- rigid moulded linings in rubber, PVC or other plastics which can be assembled by clamping in place, or coatings may be bonded to the inner surface of the casing. Epoxy resin is particularly attractive on account of its good chemical resistance and excellent bonding properties, although the cost is high. The point should not be overlooked that ceramic linings can be run at higher service temperatures than other non-metallic materials (except carbon), provided that the material is not unduly stressed by thermal cycling.

References

29. WARRING, R. H., 'Materials for Chemical and Process Pumps', Pumps-Pompes-Pumpen, Jan 1969.
68. TRUSCOTT, G.F., 'A Literature Survey on Abrasive Wear in Hydraulic Machinery' Brit Hydromechanics Res Assn. TN1079. 1970.

APPENDIX C

EXAMPLE OF CENTRIFUGAL PUMP DESIGN - DOUBLE ENTRY PUMP
PRACTICAL APPLICATION OF DESIGN FACTORS

Duty

150 l/s 55 metres head, 5 m suction lift, 50 Hz supply. The chart of duties on Fig 17.1 indicates that this pump would be single stage type.

Fig 10.1 shows that for 55 m head a suitable shape number is 1 000 from which the operating speed would be 1 650 rev/min. The next lower 50 Hz slip speed is 1 470 rev/min giving a shape number of 890.

$$N_s = 1\ 470 \ x \ \sqrt{150/55}\ ^{3/4} = 890$$

Suction Performance

The product of the flow and the square of the speed is

$$QN^2/10^9 = 2 \ x \ 75 \ x \ 1\ 470^2/10^9 = 0.163 \ x \ 2$$

Fig 18 shows that 5 m suction lift can be obtained using an operating inlet shape number

$$N_{ss} = 1\ 470 \ x \ \sqrt{2 \ x \ 75}/(10 - 5)^{3/4} = 3\ 900 \ x \ \sqrt{2}$$

A higher suction lift of 6 m head could be obtained with slightly larger eye.

The multiplication by 2 and by $\sqrt{2}$ is to indicate that the impeller is double entry, each side having

$$QN^2/10^9 = 0.163 \text{ and } N_{ss} = 3\ 800$$

Efficiency

From Fig 7.1, the efficiency would be, say, 80% making the losses 100 − 80 = 20%. For a volute pump the losses would be initially estimated as follows:-

(Final values may differ slightly)

1/20th part leakage (1% in this example)	5%	100%
1/5th part mechanical (4% in this example)		95%
3/4th part hydraulic friction (15% in this example)	15%	
Useful work	80%	80%
Total	100%	

The above totals 20% loss of efficiency. The hydraulic efficiency used in head calculation is, therefore, 80/95, leakage and disc friction being assessed separately from hydraulic passage friction.

These values of losses differ from the 3% constant loss of the efficiency formula since this example has a lower specific speed than optimum.

Area Ratio

From Fig 10.1, a suitable area ratio for N_s 890 is 1.5 giving a head coefficient of 1.20 and a quantity coefficient of 0.41. The head coefficient will give the Newton head which is generated by the dynamics of the pump. The test head must allow for friction losses and is equal to the Newton head multiplied by the hydraulic efficiency. The test head coefficient is, therefore

$$1.20 \times \frac{80}{95} \text{ per cent} = 1.01$$

This is equivalent to test head/peripheral velocity head and equals $\dfrac{55 \text{ m}}{u^2/2g}$ from which

u = 32.8 m/s and impeller diameter equals 425 mm.

The Newton head is 55 m x $\dfrac{95}{80}$ equals 65 m

The quantity coefficient is 0.41 equals

$\dfrac{\text{Throat velocity}}{\text{Peripheral velocity}}$ from which throat velocity is 0.41 x 32.8 = 13.6 m/s

For 150 l/s 13.6 m/s gives a throat area of 11 140 mm² (119 mm diameter).

Impeller

The impeller outlet area between vanes is the product of the area ratio and the throat, ie 1.5 x 11 140 = 16 700 mm2. A suitable outlet angle is 24°, making an outlet width of 32.4 mm when 5% is allowed for vanes. (Outlet area between vanes equals diameter x width x sine 24° x 0.95 x π).

Outlet Angle

Designing experience on the principle of the area ratio shows that the impeller outlet angle is by no means a critical feature; for example, satisfactory pumps have been designed with impeller outlets as low as 15° and as high as 40° without undue loss of efficiency. In each case, however, the impeller width has been altered to suit the change of impeller angle and result in the same area between the outlet blades of the impeller, that is to say, a smaller angle would result in a wider impeller.

This is of advantage on very low specific speeds where difficulties are experienced in casting a very narrow impeller. The optimum angle from an efficiency point of view would appear to be in the neighbourhood of 22° or 24°.

Eye and Shaft Design

Eye diameter is given by the empirical formula

$$\frac{\text{Eye dia}}{\text{Outlet dia}} = \frac{\sqrt{N_S}}{73} \text{ from which dia of eye} = 173 \text{ mm}$$

Shaft diameter for split casing pumps is usually one-seventh to one-sixth of outlet diameter, that is to say 65 mm, for which a suitable hub is 90 mm. (There is a direct relationship between impeller diameter and shaft diameter which is independent of shape number at constant QN^2; increase of N_S involves reduction of actual speed, thus giving almost constant torque stress and rigidity).

Eye area for one side of the imeller is 17 140 mm² (eye 173 hub 90) giving a flow velocity of 4.5 m/s. Peripheral eye velocity is 13.3 m/s. Inlet angle at periphery has tangent equal to

$$\frac{4.5}{13.3} \text{ ie} = 18.30°$$

with larger angles towards the hub since peripheral velocity falls with diameter.

Inlet Angle

Inlet angles are made slightly larger than the above calculations to allow for vanes and for a margin of safety on quantity. This margin is necessary on pumps for general sale to the market as otherwise quantities slightly above the best efficiency quantity would show a severe drop in head and efficiency due to inlet choking. (See Ref 38, example of varying inlet angles).

Number of Impeller Vanes

In general, for specific speeds between 400 and 2 000, eight impeller vanes can be used. A smaller number of vanes is sometimes employed to give a steeper falling curve or a smaller output from the same casing, and, of course, on unchokable duties, 2, 3 or 4 vanes are used.

Casing

Volute area increases uniformly around the circle from zero to a throat of 11 140 mm². The discharge cone will taper at $5^1/_2°$ initial angle, with a trumpet shape, to a delivery branch of 200 mm, where the velocity will be 4.66 m/s. A suitable suction velocity will be given by a 250 mm branch, namely 3 m/s. (Suction pipes should be at least 300 mm diameter to keep friction losses down).

The purpose of the trumpet shape of discharge passage is to ensure minimum losses at the initial high velocity. A $5^1/_2°$ cone for the whole of the passage would result in excessive length (Ref 17).

The inlet passage should be of generous cross section (as shown dotted in Fig 27.3) with a rib to prevent prerotation at impeller inlet. The volute will be wider than the imeller; eg, double the impeller passage width as in the section of Fig 27.3. Volute width is not critical: for example, Fig 51.6 shows Tracey and Grand Coulee pumps having volute widths respectively three times and equal to impeller widths.

Variations

The following describes the various alternatives, itemizes the losses and considers the mechanical stresses.

Characteristics

At closed valve, experience shows that the head coefficient is generally about 1.2, giving

$$1.2 \times \frac{32.82}{2g} = 65.5 \text{ m}$$

Fig 7.1, shows the general shape of the characteristics of head, efficiency and power. Power absorbed at duty with 80% efficiency is 101.5 kW for which a 110 kW mtor will be suitable, being slightly above the peak of power curve.

Alternatives

A positive head could permit a higher operating speed and efficiency, and a smaller pump. For example, 2 m inlet pressure would permit 2 940 rev/min operation and a shape number of 1 780 where $QN^2/10^9$ would become 0.652 x 2. Efficiency would be slightly higher - 83% - due to higher specific speed.

Operation inlet shape number would be

$$N_{ss} = 2\,940 \sqrt{2 \times 75}/(10 + 2)^{3/4} = 3\,940 \times \sqrt{2}$$

For a steeper head curve and a power curve which is non-overloading at best efficiency, a lower value of the area ratio may be used. This would give a larger impeller diameter and throat, and a higher closed valve head. Efficiency would be slightly reduced by increase of disc friction. Conversely, a smaller and slightly more efficient pump may be made by increasing the area ratio, giving a humped head curve and a steep horsepower curve. This requires a larger motor only in cases where the head is liable to fall. (See Fig 146).

The area ratio curves permit selection of several characteristics for a given shape number, an advantage over curves where head and quantity coefficients are plotted against N_s only. Suction performance may be improved with slight sacrifice of efficiency by making a larger eye. (See Fig 9.3).

Choice of Shape Number for Best Efficiency

Efficiency may in some cases be improved by altering the operating speed so as to bring the shape number bearer to the optimum. This would involve increasing the running speed of high head pumps (which can only be done where adequate inlet pressure is available) or by reducing the running speed of low head pumps.

Efficiency Losses in this Example

Disc friction	3	kW	=	3 % approx	
Glands	0.5	kW	=	0.5% approx	
Bearings	0.5	kW	=	0.5% approx	$5\frac{1}{2}$
Neckring leakage	23	l/s	=	$1\frac{1}{2}$ % approx	

0.25 mm clearance diametrically.

Hydraulic losses by difference 10 m	$14\frac{1}{2}$%
Liquid power as useful work	80%
Total	100%

Radial Loading at Zero Flow

0.3 x 0.425 m x 32.4 mm x 5.4 bar = 22.3 hectonewtons.

Shaft Stress

Bending stress due to 22.3 hectonewtons radial central load on 610 mm span with 65 mm shaft is 1.28 hectobar at zero flow. Shaft deflection is 0.057 mm, slightly less than the radial clearance.

Torque Stress

Based on motor power of 110 kW on 57 mm shaft at coupling is 1.92 hectobar at full flow when radial load is absent.

These stresses are low since the duty is a modest one for 1470 rev/min. Higher speed duties are therefore practicable from this pump.

In general, pumps up to 600 mm branches are designed as a range to give the greatest flexibility in meeting the various duties that the market demands (see Chapter 13). For example, if this pump were offered in the American market for 60 Hz, 1760 rev/min operation, the duty could be 180 l/s (2880 US gal/min) 79 m, 81% efficiency (increased because of higher flow) absorbing 174 kW, 195 kW to be provided.

Torque stress will increase to 2.88 hectobar which is still reasonable. Suction lift would be reduced to 3.7 m since $QN^2/10^9$ is now 0.282 x 2 for the two sides of the impeller and inlet shape number

$$N_{ss} = 1760 \sqrt{2} \times 90/(10 - 3.7)^{3/4} = 4200 \sqrt{2}$$

NPSH and Atmospheric Pressure

The figure of 10 m, used in these examples for determination of NPSH, is the difference between the 10.15 m absolute atmospheric pressure and the water vapour tension of approximately 0.15 m at 15°C. This is equivalent to saying that with zero pressure drop between inlet branch and impeller eye a suction lift of 10 m should be attained at sea level. (Corrections for temperature and for altitude may, of course, be required). In practice, vapour pressure of air in water prevents a greater suction lift than about 8 m being attained, 4.5 to 6 m being normal.

Geometric designing, mentioned above, is repeated again for convenience.

Change of Speed for the Same Size of Pump

For a given operating point on the characteristic of the pump, flow varies as rev/min. Head varies as rev/min^3. Since power is the product of flow and head divided by efficiency, it follows that power is proportional to rev/min^3. A small change of impeller diameter gives similar variations.

Geometric Change of Size for the same rev/min

Flow varies as diameter^3; Head varies as diameter^2; power varies as diameter^5. (Power may, however, be slightly affected by a change of efficiency.)

Geometric Change of Size with Inverse Change of Speed

This is the most important relationship and is the bases of model tests. It is a combination of the above relationships and maintains constant head, peripheral and flow speeds and constant stresses in the structure. A model test of moderate size can in this way be used to forecast performance of a full-scale prototype. Head is constant; flow varies as diameter^2; power varies as diameter^2; rev/min varies as inverse diameter. The above assumes no change in efficiency. In practice, efficiency increases slowly with increase of size and speed (see above), thus slightly reducing the power on larger machines.

CALCULATIONS

The Use of a Slide Rule or Calculator without Fractional Powers

A slide rule can often be quicker than a calculator for certain frequently used formulas.

Determination of Shape number

Shape number can be quickly calculated, thus avoiding the need for tables of H$^{3/4}$. The procedure is as follows:

$$N_S = \text{rev/min} \times \sqrt{\text{flow}} \div \sqrt{\text{Head}} \div \sqrt{(\sqrt{\text{Head}})}.$$

When determining $\sqrt{\text{Head}}$, note the value of $\sqrt{\text{Head}}$, then, for the next step, take the square root of $\sqrt{\text{Head}}$. Provided care is taken to keep the position of the decimal points of $\sqrt{\text{flow}}$ and $\sqrt{\text{H}}$, no difficulty should be experienced.

Example: 1 480 rev/min 4 000 l/s 250 metres

$$N_s = 1\ 480 \times \sqrt{4\ 000} \div \sqrt{250} \div \sqrt{(\sqrt{250})} = 1\ 490$$

The root of 250 is 15.8 which is noted and memorized for the next step. For this reason the two division steps are carried out consecutively instead of alternatively. Similar procedure with calculators which have no y^x key.

Fractional Power

The normal engineer's slide rule contains A B C D scales for first power and squares, a single scale for cubes, a scale for reciprocals and the log scales. If the reciprocal scale be replaced by a second cubic scale, so that a cubic scale appears on the stock, with another on the slide, a considerable advantage is found for hydraulic calculations of pumps and water turbines.

For example, an increase of speed results in flow varying as the first power, head as the square and power as the cube. An increase of size at the same speed results in the head varying as the square and the flow varying as the cube of the size change. These changes can be carried out very rapidly with their slide rule having two cubic scales.

Using this cubic slide rule and setting the kilowatts on one cubic scale against the rev/min on another cubic scale will give a direct reading on the first power scale of the shaft diameter in millimetres for a torque stress of approximately 4.7 hectobar.

The additional cubic scale has saved its cost many times over in calculations similar to those above for pumps and water turbines.

References

17. ANDERSON, H.H., 'Modern Developments in the Use of Large Single Entry Centrifugal Pumps'. Proc IME 1955, Vol 169 No 6.
38. BLOM, C., 'Development of the Hydraulic Design for the Grand Coulee Pumps'. ASME, 49 – SA 8.

APPENDIX D

QUALITY ASSURANCE (Ref 21)

This section describes the various test programmes and procedures established to ensure that the pumps operate as intended, and equally important that future production maintains the highest standard quality that can be obtained.

Pressure Vessels - Cycling Tests

Here two problems arise. Firstly, to determine the ultimate bursting pressure so that a suitable margin of safety can be established and secondly to assess the life of the unit under normal pressure so that premature failure can be eliminated. It will be appreciated, of course, that boiler feed pumps are cycled during their lives about 15 000 times and therefore the problem of low cycling fatigue must be investigated. A quarter size model of the two stage pump casing was cycled to the hydrostatic test pressure 25 000 times, after which it was sectioned for examination and found to be free from cracks or deformation.

A 250 mm model of the self-seal joint was cycled from zero to 50 MPa 20 times, to 43 MPa 100 times and to 32 MPa 20 000 times and to 2 000 000 cycles at ±1 MPa. The absence of cracks was proved by die penetrant and ultrasonic testing and by sectioning for examination.

Photoelastic Tests on a 120 mm Araldite Vessel (epoxy resin)

Tests were made by the National Engineering Laboratory at 2.6 times elastic stress concentration for the geometry corresponding to the 250 mm test barrel. It was stated that any yielding would redistribute the stresses. This statement was confirmed by the actual performance of the barrel as described above.

Bursting Tests

A small replica of the self-seal joint was loaded mechanically by a plug to simulate hydraulic end load and determine the strength of the shoulder which takes the main thrust. This model failed at a load corresponding to 160 MPa hydrostatic pressure.

A further test to failure might be mentioned. This was a quarter size model of the stainless steel spun cast barrel for the 200 MW pump mentioned above and again this was found to fail at 8 times the maximum working pressure.

Stress Analysis

Finite element techniques were applied to various critical regions of the pump casing including the region around the main high pressure joint. Precise procedure was laid down for hydrostatic tests with a view to limiting the number of applications and maximum hydrostatic pressure to which the casing is subjected. During the manufacture of the pump special care was taken to ensure that areas of stress concentration were free from surface defects and that adequate radii were provided at all critical points.

Photoelastic tests were carried out on model impellers under thrust and torque loadings.

Materials for Rotating and Stationary Surfaces

Elaborate rubbing tests were taken in water on various materials with contact pressure up to 1 MPa at 60 m/s. The principal conclusion was that the differential hardness between the two materials was not the important factor but that the absolute hardness of the stationary material was the most important feature. Plated and coated materials, nonferrous alloys, austenitic cast iron etc were found to be totally unsatisfactory on high speeds and high loads. Fully hardened stainless steel of the 17% chrome, 4% nickel, precipitation hardening, and 13% chrome, 1% nickel were found to be acceptably compatible under these conditions.

High Speed Impeller and Diffuser Tests

Endurance tests were carried out on a 150 mm diameter impeller with associated diffuser in order to determine the most suitable geometry and to establish parameters of inlet conditions etc.

The pump was run at 22 700 rev/min to generate a head of 1 860 m in one stage, 17.3 MPa at 120°C. Two step-up gear boxes were required, the driving unit being a 2.3 MW motor.

Reference

21. ANDERSON, H.H., 'Improvement of Reliability in Thermal Station Pumps', Reliability Conference, Loughborough 1973. IMechE. 84. 'Fluid Machinery Failures' – Prediction Analysis and Prevention. Convention, I.Mech.E. 1980.

APPENDIX E

Some knowledge of reversed centrifugal pumps, known as water turbines, is essential for a complete understanding of centrifugal pumps and vice versa.

Historical

The forerunner of the modern water turbine was the water wheel or millers' wheel, some of which are still to be seen in operation.

One or two hundred years ago every river of any size had several of these wheels and to this day the old weirs, lades and ruined mills are a familiar sight in our countryside.

The weir wheel was generally arranged with a horizontal shaft and a rim carrying several buckets to which the water was admitted at the top (over shot wheels) half way up (breast wheels), or near the bottom (under shot wheels). Work is done on the wheel by the weight of the water falling from the head race to the rail race.

It will be seen therefore that only a part of the wheel circumference receives water and that the water enters the wheel at atmospheric pressure.

Increase of Power

When consideration was given to increasing the power output of the wheel such increase could occur in two ways:

(i) The water could be contained in a pressure pipe so as to take advantage of a head race above the wheel.

(ii) The water could be admitted to the whole circumference of the wheel.

In the above cases:

(i) The energy of the water due to pressure and quantity would be converted into velocity so that the water issuing from the penstock strikes the wheel at a very high speed.

(ii) The wheel would be enclosed in a casing under pressure.

The containing of the water and conversion of its mass times pressure energy to velocity energy so that it could strikes the blades at a greater force produced the Pelton Wheel or Impulse Turbines.

The increasing of the admission to the whole circumference and the fitting of a pressure casing around the wheel produced the Francis Turbine and as a later development the Kaplan Turbine for very low heads.

It will be seen, therefore, that the three major types of turbines - Pelton, Francis and Kaplan - have grown out of the original water wheel.

Development of Turbines

The original Francis turbine had a runner whose inlet diameter was greater than the outlet diameter and was suitable for medium heads. When the need for higher speeds to suit electric generators arose, particularly in connection with lower heads, it was necessary to reduce the inlet diameter of the turbine until it was less than the outlet diameter. This produced the high specific speed Francis turbine which was further developed for low heads until the axial flow portion only remained, which became a propeller turbine, and with moving blades became the Kaplan turbine.

Modern Turbines

Modern turbines range in head from a few cm to 2 000 m and in power from fractional kW on oil servo duties to the largest prime movers yet made. Several units of 700 MW one million HP have been made.

Specific Speed

The specific Speed N_s of a turbine is a convenient number to represent the shape of a turbine and indicate into which of the above three main groups (Pelton, Francis, Kaplan) the machine will fall.

The number is derived as follows:

$$\text{Specific speed } N_s = \text{rev/min} \times \frac{\text{Square root of Power}}{\text{Head } 5/4}$$

Operating Head
The operating heads for the various types of water turbines are as follows:_

Large Pelton or Impulse Wheels
N_s 1 to 10 - 300 to 2 000 m. For small sizes Pelton wheels are suitable for heads below 300 m.

Francis Turbines
N_s 20 to 100 - 30 to 300 m.

Kaplan Turbines
N_s 100 to 200 - up to 50 m

The aforementioned limits are very approximate and naturally are affected by progress in technique, particularly in metallurgical fields, where better metals permit operation at high heads and speeds without undue risk of damage by high water velocities.

Choice of Turbine

For a given duty the diameter of the runner depends upon the head and the speed; the width of the runner depends upon the quantity of water; and the combination of head and quantity of water determines the power generated by the turbine. If we reduce the quantity of water and the power for the same head, a narrower runner will be required. Obviously a narrow runner is more difficult to cast and beyond the limit of narrow casting, that is the lowest specific speed of a Francis wheel, further reduction of power and specific speed can only be obtained by admitting the water at one point instead of around the circumference of the wheel. It is here that the change takes place from Francis to Pelton turbine.

There is a considerable gap between the Francis (lowest $-N_s$ 20) and the Pelton (highest $-N_s$ 10) which is made up by the introduction of double jet and four jet Pelton, where a compromise is obtained between the single admission of the pelton and the full circumferential admission of the Francis. The jets are arranged in equal steps around the wheel (N_s for a four jet pelton is $\sqrt{4}$ or twice the N_s of a single jet) 3, 5 or 6 jets are also used but are less usual.

A special form of Pelton wheel with side entrance so as to permit a larger jet and a much greater quantity of water for a given wheel diameter is called the Turgo and competes with the two jet and four jet Pelton wheels.

Limiting Speed

The speed of water flow through the wheel and the speed of blades relative to the water imposes a limit on the revolutions per minute and consequently upon the head against which the wheel may operate, since excessive speeds give rise to cavitation with consequent damage to the wheel element.

In the case of a Francis wheel if the power is very small the runner may be of an awkward shape for casting and may require to run at a speed inconvenient to the generator. It is for this reason that on low powers the Pelton is used for heads as low as 30 m.

Application of Water Turbines

Water turbines provide power for all purposes. In the larger sizes the power produces is almost invariably used for generating electricity for the national grid or for industry.

There are certain pump schemes particularly in the chemical field for cooling or scrubbing duties, where the return main has a greater height or pressure than can be recovered by a syphon. In many cases the water, after having passed through the process, enters a turbine so as to recover as much power as possible. These turbines are very often arranged in tandem with the pump so that the motor power necessary is reduced by the contribution made by the turbine.

Problems in the Operation of Turbines

A steam turbine is governed by stop valves controlled by a speed governor and it is possible to close this stop valve quickly. In the case of a water turbine, however, there is in the penstock a large mass of water approaching at a relatively high speed.

Pipeline Inertia.

To close the turbine gates too quickly would give rise to a considerable increase of pressure since the long column of water would suddenly be brought to rest. In order to avoid a pipe fracture, therefore, it is necessary to bring this column of water slowly to rest.

Water turbo alternators differ from some other prime movers in that they load, in this case the alternator torque, can vanish instantly without any warning. If an electric storm strikes the grid, the circuit breakers may disconnect the generating station from its load within a fraction of a second. At this time, however, the turbine is being fed with water and generating full power, and until the water column can be brought to rest, power is still being fed to the turbine. As a result, its speed increases and in the limit may reach the runaway speed, which is nearly twice the normal running speed.

Rotor Inertia

The only way in which this rate of increase of speed can be regulated is by building a greater flywheel effect or inertia into the rotor of the electric generator. This is a particularly difficult problem for the electrical designers, since on the one hand they must build as much metal into their rotor as possible, and on the other hand this rotor should be capable of running at nearly twice normal speed without bursting (this is vital in case of governor failure). To this end the shortening of the penstock to the minimum possible length is essential and therefore on any power scheme it is usual to provide a horizontal canal or tunnel from the upper lake to a point as near the power house as possible, followed by a very short pipeline. At top of this pipeline is erected a surge tank with an open top, so that the water flowing in the horizontal tunnel or canal can escape and the turbine designer need only consider the amount of water in the length of the relatively short penstock approaching the turbine.

It will be seen that this problem resolves itself into the balancing of the water inertia in the pipe against the inertia of the rotating parts of the turbo alternator.

A further means of helping this problem is to provide a by-pass so that as soon as the governor starts to close the gates beyond a certain speed the by-pass is opened, permitting the penstock water to run to waste instead of through the turbine. This by-pass or relief valve is only opened if the speed of governor movement exceeds a certain valve. On the other hand, even with a pressure relief valve on certain turbines, the incidence of over speed causes in itself a reduction of quantity which in turn gives rise to a further pressure increase, the cumulative effect of which on lowest specific speed of Francis turbines is extremely difficult to forecast.

In the case of a Pelton wheel, by-passing can be carried out more conveniently with a deflector which removes the jet entirely from the wheel. The deflectors have the advantage that no pressure rise whatever occurs, since the water is not deflected until it is passed out of the nozzle into the atmosphere.

Construction of Turbines

The Francis turbine usually comprises a scroll casing which leads the water from a pipe into the periphery of the runner. Before entering the runner the water passes through a series of gates which are controlled by the governor, thus admitting a larger or smaller quantity of water into the turbine. A Kaplan turbine is similar, but has in addition the problem of moving the rotating blades by a servo motor in the shaft. The blades and the gates normally move in unison. With a Kaplan turbine there is, furthermore, a risk of exceptionally high overspeed should the correct phasing of blades and gates accidentally get out of order.

In addition, constructional features such as the shaft for supporting the runner bearings, gland, casing, gate mechanism, etc, are involved. The casing may be of concrete up to 90 ft head or of steel for higher heads.

The Pelton wheel is rather more simple, comprising only the wheel and the nozzle at the end of the penstock, which is controlled by a needle valve arranged so as to reduce the size of the jet without altering its speed or causing any undue frictional loss. The Pelton wheel is contained in a light casing so as to prevent undue splashing. The buckets are of double hemispherical shape. The wheel runs at half the jet speed and the jet is turned through nearly 180° in the bucket. The water speed when leaving the bucket is theoretically zero, all its energy being given to the wheel.

APPENDIX F

PERMITS TO WORK

The Health & Safety at Work etc. Act 1974 requires all employers to provide and maintain "safe systems of work". The Management of Health and Safety at Work Regulations 1992 require employers to provide "effective planning, organisation and control.... measures". The Offshore Installations (Operational Safety Health and Welfare) Regulations 1976 (SI 1976/1019), more usually referred to simply as "ten-nineteen", specifically requires the provision of "written instructions" and "work permits".

A permit to work is not merely a document giving authority to start work, since in addition it contains clear instructions as to how the work should proceed, and what precautions need to be taken. By requiring certain persons to sign indicating that specific actions have been taken, it also confirms responsibility, and hence accountability, for any failure which results in an accident.

The design of a permit system is a critical task since any fault in it is eventually certain to cause a serious accident, since an employee working under a permit to work tends to treat it as a guarantee of his safety.

Isolation

Isolation involves a mechanical separation of a supply of electricity, steam, compressed air, hydraulic fluid, process materials etc. from a place where persons have to work, by imposing a physical barrier such as the installation of a spade, blank flange or spectacle plate in a pipeline or the breaking of a supply path such as the removal of fuses, isolating links, spool pieces or sections of pipe.

Its purpose is to provide an additional safeguard so that in the event of leakage through a valve or accidental resupply of power, it is impossible for the supply to reach the work area involved.

Lock Off Systems

Lock off systems are used in conjunction with permits to work to provide positive mechanical confirmation of the administrative steps taken by the permit itself. These may vary from the simple "fuses in the hip pocket" or "stop cock handle in the toolbox" variety, to quite complex sequential trapped key systems and multiple personal padlocks. A separate "isolation certificate" may be used to confirm that isolation has been completed. The activities of "isolation" and "reconnection" are themselves hazardous tasks, since they involve work on "live" lines and reinstatement of the supply may also give rise to flange leaks and the like.

Vessel Entry

Where the work to be done requires personnel to enter into "confined spaces" very great care must be taken. The atmosphere inside the tank or vessel may be contaminated by flammable or toxic gases or it may contain either insufficient oxygen to maintain consciousness (or even to support life), or an excess of oxygen above that which is normally available, which gives rise to a very serious risk of fire.

There may be a sludge at the bottom of the tank which, if disturbed, may release gases into an otherwise hazard free atmosphere. Gas tests are essential before any attempt is made to enter the tank and they may need to be repeated either continuously or at intervals as the work progresses. Breathing apparatus may be required for the persons entering the enclosed space, as their only guarantee of a safe breathing atmosphere, and will certainly be needed, even where the atmosphere is proved to be safe, as emergency equipment for the stand by team which is an essential feature of all vessel entry work.

"Vessel entry" hazards are often thought to be exclusively restricted to tanks which have held chemicals of one kind or another. This is most certainly not so and many people have been killed by the lack of oxygen caused by the rusting of steel or by toxic gases arising from bacterial action or curing of paints in tanks or spaces which have been constantly open to atmosphere.

The permit to work system must be completely effective, not only to remove the plant to be worked on from its "operational" environment and control to the "maintenance" environment and control, and to prepare for and safeguard operations up to the completion of the work, but must also facilitate its return to its working state without undue risk.

Many otherwise well designed permit systems have serious weaknesses at the stage where the plant is to be returned from the maintenance environment.

On offshore installations, permit systems must cater, not only for routine maintenance work, but may also have to allow for simultaneous operations involving construction, "hook-up". drilling, stem testing, completion, production, well logging, workover, N.D.T., radiographic and high pressure mechanical and electrical testing and diving operations.

There are several clear steps in the life of a simple maintenance permit:-

(a) A request from "production" (the owners' of the plant, usually referred to as the "area authority") to "maintenance" (usually referred to as the "performing authority") to carry out a certain task or tasks. At this stage, "maintenance" may not know either of the need for the work or the type of work required, although, of course, their advice may be needed, and thus formal consultation may be required. Where planned preventive maintenance procedures are involved, the initial work request may come initially from the maintenance department.

(b) A request by maintenance to production, for the plant concerned (and others which may be directly or indirectly involved) to be taken out of use, and made safe to work on. This may involve depressurising, draining out chemicals, flushing out, purging, steam cleaning, disconnecting electrical, mechanical, hydraulic, steam or compressed air power supplies, disconnecting feedstock supply lines and product drainage lines and blocking, spragging or otherwise mechanically restraining parts from moving under the influence of weight, vibration, wind or tidal forces.

Where supplies are disconnected, it is normal practice to "isolate" the supply by removing fuse links or sections of pipe, by fitting blanking flanges or spades, or by ensuring that a section of pipe between two valves is vented to atmosphere. Leakage past the upstream valve, into the vented section will not cause leakage at the downstream valve, since there is no difference in pressure across it. This work may or may not be carried out by "maintenance" personnel, (whoever does it usually being referred to as the "isolating authority") and is frequently backed up both by mechanical "lock off" devices and a separate "isolation confirmation certificate". It is important that the person who does the isolation is not the same person who will carry out the work.

(c) A statement confirming "b" above, that the plant concerned is now in a state in which the specified work may be safely carried out, and what may or may not be done in connection with the work. If, once the job has started, additional work is found to be necessary, the existing permit must be suspended and a new permit raised for the now different scope of work.

(d) A request by the performing authority for certain preparatory work to be done in connection with the work, for example building scaffold access, stripping lagging off boilers or pipework or arranging the availability of special lifting gear.

These separate steps may all be combined where, for example, production do their own preparatory work, but it must be clearly understood as to what work is done by whom and as to what stage has been reached.

(e) At this stage there must be a clearly identifiable transfer of authority over, and hence responsibility for, the plant to be worked on.

(f) An instruction to the performing authority personnel to carry out the relevant task in a particular way and using certain tools, equipment and protective clothing. At this stage, in addition to safety requirements, there may also be the need to consider the "quality control" aspects of the job with material specifications, welding standards, limits, fits, tolerances, clearances and surface finish parameters as appropriate.

(g) A statement by the performing authority that the task is now complete, that men, materials, tools and equipment have been withdrawn and that it is safe for power supplies etc. to be reinstated. Where the task cannot be finished within the duration of the permit, or when unforeseen problems cause the task to be stopped, the statement will confirm that the work is incomplete but has been made safe and that the work site has been cleared and is safe for other unconnected work to be carried out. It is at this stage that many permit systems have their greatest weakness.

Modifications to the wiring of a three phase motor, requires a check its direction of rotation, but this cannot be done until power is restored. Power cannot, however, be restored until the motor and its drive train is totally enclosed and thus one cannot easily check rotation. Temporary reconnection certificates may be required at this stage, or even separate "work permits", in order to allow pressure testing, radiography etc to be performed.

Other problems which may be encountered at this stage are the presence of air (with oxygen) or water, which are critical in an oil production or petrochemical plant as are dirty footprints in a food, cosmetic or pharmaceutical plant. Nitrogen purging or solvent washing may introduce further hazards at this stage.

(h) A statement by the area authority, stating that the plant has been inspected to confirm that it is safe, accepting it back from the control of the performing authority and assuming responsibility for it.

It is an essential feature of all work permits that there is independant confirmation of each step in the permit process, so as to minimise the risk of a simple mistake escalating into a disaster.

Permit systems may be of a standard type for all kinds of activity carried out in an organisation, or may be divided into specialist types.

There may be, for example, "electrical permits", "naked flame hot work permits", "spark potential hot work permits", "cold work permits", "radiography permits" and "vessel entry permits". Permits may be issued within the relevant department or may all be routed via an "issuing authority", in order to prevent two or more incompatible operations being done simultaneously, such as scaffolding work over an area where divers are working.

The work to be covered by a permit may vary from the replacement of a light bulb or gauge glass to the upgrading of a pressurised system or the installation of a turbo-alternator.

The personnel covered may range from one maintenance fitter to several hundreds employed by different contractors and working different shift patterns.

A permit may cover a complete activity or only a part of it, and may last for a considerable period of time or may be cancelled and re-issued at the change of each work shift.

All permits become invalid as soon as a general alarm sounds.

The supervisor's responsibility under permits to work

If an employee suffers an eye injury whilst working under a work permit which states that eye protection must be worn, then, clearly some responsibility must lie with the injured person. The supervisor, however, has the greater responsibility, since his signature is written proof of his knowledge of the hazard and the injury proves that he has failed adequately to supervise his work force.

Another area of concern occurs when the supervisor, who fills out a work permit, requests that the work should comply with a standard operating procedure contained in a large and complex operating manual, without checking exactly what is required and without checking to ensure that the workforce is fully conversant with what that procedure entails.

In the event of an accident, the supervisor may find that, since the work was not being done in the manner required, (especially if it is impossible to do the work in that manner), he has placed himself in a completely indefensible position.

LEGAL REQUIREMENTS RELATING TO THE DESIGN, MANUFACTURE, IMPORTATION AND SUPPLY OF PUMPS AND ASSOCIATED TYPES OF MACHINERY

The legislation covers transport, installation/erection, commissioning, maintenance, adjustment, repair, dismantling and disposal.

The health and safety at work etc Act 1974 (Ch 37) referred to as HASAWA, forms the foundation of all safety legislation in the UK. It covers all types of work activities and mainly refers to the duties of employers towards their employees. Its general duties are supplemented by more detailed legislation in the form of Statutory Instruments, many of which are supported by Approved Codes of Practice. HASAWA applies wide ranging duties on designers, manufacturers, importers and suppliers of "articles for use at work" in relation to the manufacture, installation, erection, setup, adjustment, maintenance, cleaning, dismantling and disposal. More specific duties are contained in the Supply of Machinery (Safety) Regulations 1992 (SI 1992 No 3073), referred to as SMSR and in the Provision and Use of Work Equipment Regulations 1992 (SI 1992 No 2932), referred to as PUWER.

Many of these legal duties are qualified by the phrase "so far as is reasonably practicable". On the one hand, this means that if the cost and the inconvenience involved in preventing or minimising a hazard far outweigh the risk of that hazard causing an accident, then there is no legal duty to do anything. However, in the event of a prosecution, for any offence which is alleged to involve failure "to do something so far as is reasonably practicable", then the accused person is, in effect, considered to be guilty of that offence until he is able to prove to the court that he was entitled to allow the accident to occur i.e. he has to prove that there was nothing more that he needed to do in order to prevent the accident from happening. Thus any foreseeable accident which occurs can result in a successful prosecution.

It is important to recognise that safety legislation only relates to prevention of injury to personnel, and is not concerned with property damage, apart from (under offshore legislation only), the prevention of damage sufficient to prejudice the safety of the offshore installation itself. Thus, if a company simply seeks to comply with its legal duties, concentrating on the "Safety of its employees and others", it fails to recognise the importance of reducing, not only the risk of personal injury, but also the risks of property damage, product loss, plant downtime, non-availability of assets and damage to its reputation and public image, which, in the event of a serious accident, can result in catastrophic loss of industry, public and shareholder's confidence, even to the extent of threatening the survival of the company itself.

A Health and Safety Executive (HSE) Factory Inspector has the authority to issue an improvement or prohibition notice. An improvement notice states that the inspector is of the opinion that a certain activity is illegal and describes the action he believes is required to correct it and states the date by which the action must be taken. A prohibition notice states that the inspector is of the opinion that a certain defined activity involves "imminent risk of serious personal injury" and requires that the activity be stopped, either immediately or, in the case of a "deferred prohibition notice", at a later stated time.

Since these notices are a statement of the opinion of an inspector, his opinion may be challenged before an Industrial Tribunal. In the case of an improvement notice, the notice may be ignored pending appeal, once an appeal has been made. Entering an appeal does not, however, allow a prohibition notice to be ignored.

Failure to comply with either notice is punishable by a possibly unlimited fine. Failure to comply with a prohibition notice can also result in up to two years imprisonment.

The management of health and safety at work regulations 1992 (SI 1992 No 2051) require that every employer must assess, "suitably and sufficiently", the risks to the health and safety of his employees and anyone else who might be affected by his work activities, so as to identify what preventive and protective actions will be necessary. Where five or more employees are employed, the employer must record the "significant factors" of these assessments and must identify and record any group or groups of employees who are exposed to significant risk.

Once these assessments have been made, he must make arrangements for putting into practice various preventive and protective measures. This involves planning, organisation, control measures and activities, monitoring and review of the steps taken to manage health and safety. Where more than five employees are employed, the "significant factors" of these assessments must be in writing or held as electronic data.

He must arrange for health surveillance of his employees where it is appropriate to do so.

He must appoint competent persons from within or outside the workforce, in order to help to devise and apply the preventive and protective steps which the assessment shows to be necessary.

He must set up procedures to be taken in an emergency and give information on health and safety matters to his employees.

The Manual handling operations regulations 1992 (SI 1992 No 2793) require that every employer must avoid the need for his employees to undertake any manual handling operations at work which involve a risk of their being injured.

Where it is not reasonably practicable to avoid the need for his employees to undertake any manual handling operations at work which involve a risk of their being injured, a suitable and sufficient assessment must be made of all such manual handling operations so as identify ways to minimise the risk of injury. Wherever possible, mechanical handling aids must be used.

"Work equipment" is defined widely enough to apply to a complete refinery, to an individual item of plant such as an emergency fire pump or to a single item such as a screwdriver.

"Machinery" is similarly widely defined but at least one part of the machinery must move. Thus a ratchet screwdriver is machinery, but a hammer is not. Both, however, are work equipment.

The provision and use of work equipment regulations 1992 PUWER replace much existing legislation relating to machinery and equipment. The regulations bring UK legislation into line with other EC countries by implementing EC Directives in order that machinery and equipment used throughout the EC are designed and constructed to the same high standard. These PUWER Regulations apply duties on employers and self employed persons in relation to work equipment which they supply for their employees to use. They incorporate the Supply of machinery (safety) regulations 1992, SMSR, which apply EC directives to designers, manufacturers, importers and suppliers of work equipment.

Schedule 3 of the SMSR quotes Annex 1 of the EC Machinery Directive, which is a list of the "essential health and safety requirements relating to the design and construction of machinery". Where a hazard exists the appropriate "essential requirement" must be incorporated into the equipment. All machinery, regardless of the hazards which may or may not arise in connection with its use, must comply with "the principles of safety integration" which are a list of design and construction method statements. The machinery must have been submitted to an approvals authority, must have their approval certificate and must carry the "EC mark" and other appropriate markings fixed to it and must be accompanied by a comprehensive instruction manual.

Work equipment which is designed and manufactured by an employer for use in his own operations must also comply with these SMSR regulations.

The general standards of safety listed in the schedules attached to both the SMSR and PUWER regulations apply to all work equipment.

Higher standards are applied to especially dangerous machinery, ("Schedule 4 machinery" defined in Annex 4 to the EC Machinery Directive), such as power operated saws and mechanical power presses.

The SMSR regulations do not apply to hand powered machines such as torque spanners and breast drills. They do, however, apply to hand powered lifting machines, such as jacks, chain blocks, tirfors and pull-lifts. They do not apply to flammable and dangerous substance pipelines and storage vessels, means of transport, passenger lifts, seagoing vessels and mobile offshore installations. PUWER, however, apply to all such equipment and do apply to offshore mobile units.

Most modern equipment should already conform to these new requirements, but older, non conforming, equipment must be replaced or updated by 1st. January, 1997.

The SMSR regulations require that manufacturers and importers must submit a technical file to a national approvals authority before they may supply the machinery. The technical file consists of general arrangement and detailed part drawings, material and technical specifications, wiring diagrams, control circuit diagrams, calculation notes, test results and, for machinery classified as especially dangerous, copies of the instruction manual which will be supplied to the user of the machinery.

Extensively detailed specifications for different types of machinery are contained in the Schedules appended to the regulations.

When the approvals authority is satisfied that the machinery is in fact safe and that it conforms to the relevant EC specification, it will issue a "EC type examination certificate".

On receipt of the EC type examination certificate, the manufacturer must then issue an "EC declaration of conformity certificate" and mark on the machinery the "EC mark". The EC mark consists of the letters "CE" and the two final digits of the year of manufacture. Type approved machinery manufactured and approved in 1994 would thus be marked "CE 94".

In addition to the "EC mark" the machinery must carry the name and address of the manufacturer, a designation of series or type, information such as the maximum rotational speed, operating pressure etc, and its serial number. If the machine or its component parts have to be moved using lifting gear, the whole machine and each of its component parts must be fitted with suitable lifting gear attachments, (unless they are of a shape which allows easy slinging), and be clearly marked with its mass. If the machinery or its component parts have to be manhandled, they must be suitably shaped in order to allow this or must be fitted with suitable handles. The mass must be marked.

Work equipment must be supplied complete with technical specifications, relevant warnings and an instruction manual. The instruction manual must give detailed information on what the equipment is designed to do, what precautions need to be taken when being used, information regarding foundation requirements, assembly, setting up, proper use, adjustment, cleaning, maintenance, handling and dismantling. Where appropriate, the instruction manual should give advice on any training which should be given and should define ways in which the equipment should not be used. Information on hazards such as vibration levels produced, and noise levels emitted by the equipment must also be included.

The PUWER regulations require that employers supply only EC type examination certificate approved and EC marked equipment for use in their undertaking. These regulations are concerned with the health and safety of persons at work. Other EC based legislation requires equipment to comply with standards such as noise and vibration levels and structural strength.

"Use" is widely defined in PUWER and includes starting, stopping, installing, dismantling, programming, setting, transporting, maintaining, servicing, repairing and cleaning.

Where a maintenance logbook is provided with the equipment, it is a legal duty to keep it up to date.

Specific duties under PUWER apply to:

1. *Suitability of work equipment for the work to be done.*

    Every employer must consider the various working conditions and hazards at the workplace when choosing work equipment, so as to ensure that the equipment is suitable for what it is be used for, that it is not used in unsuitable circumstances and that it is properly maintained. The risk assessments which an employer must make under the Management of health and safety at work regulations 1992 will help to identify the type of work equipment which must be provided, who must use it, what precautions must be taken and the circumstances in which it should be used. Compressed air driven or hydraulic equipment may, for example, be more appropriate than electrically driven equipment for use in damp or flammable environments.

2. *Maintenance operations.*

Maintenance must be done with the machinery isolated from power supplies. Depending on the complexity of the equipment this may simply involve removing the plug from the socket or may involve complex sequential trapped key isolation facilities. Any lubrication or adjustment which must be done when the machinery is not isolated must be possible without exposing persons to any danger.

3. *Specific risks to health and safety.*

Where the use of particular equipment will involve specific risks to health and safety, the employer must restrict the use of work equipment to nominated persons who have been adequately trained. Maintenance work on such equipment must also be restricted to nominated trained personnel.

4. *Information, instruction and training.*

Specific information, instruction and suitable training must be given to those who are to use work equipment and to those who will supervise the use of work equipment.

5. *Conformity with EC directives on work equipment.*

Every employer and self-employed person who provides work equipment for use at work must ensure that the equipment complies with any UK legislation which implements the relevant EC Directive. Work equipment supplied in the UK must comply with the Supply of machinery (safety) regulations 1992, which implement various relevant EC Directives.

6. *Guarding of dangerous parts of machinery.*

Detailed information is given in the regulations as to how the guarding should be designed, materials which should be used in construction and how machinery not capable of being guarded may be protected by such equipment as infra-red barrier, automatic cut-outs and the like. The use of jigs, holders, push sticks etc. is required where appropriate. The guarding must be capable of keeping persons away from danger and must also retain ejected parts, dust and fumes.

7. *Protection against specific hazards.*

Measures must be taken to prevent the occurrence of, and reduce the severity resulting from, failure of work equipment and of articles falling from or being ejected from work equipment. Steps must also be taken to prevent or minimise overheating, fire, discharge of gas, fume, vapour or dust and the possibility of explosion within work equipment.

8. *Parts or materials at high or very low temperatures.*

Where work equipment contains materials or has external surfaces which are at high, or very low temperature, specific steps must be taken to protect any person from being burned, scalded or seared.

9. *Control systems and control devices.*

Work equipment must be provided with control systems which must be as far as possible designed to be "fail safe". Controls for starting up, changing operating conditions and stopping the equipment must be clearly marked, must operate in a logical manner and must require positive action to operate. Separate emergency stop controls must be provided where it is appropriate to do so.

10. *Isolation of equipment from sources of energy.*
Appropriate facilities must be available to allow work equipment to be isolated from all sources of energy. Although this is usually assumed to mean isolation of the power supplies, the duty extends to include lubricant supply, hot or chemically reactive products, electrical heat tracing lines and the like.

11. *Stability of equipment.*
This duty requires work equipment to be bolted down to suitable foundations, lashed or tied back to structural members or fitted with outriggers or ballast in order to prevent risk of collapse or overturning.

12. *Lighting.*
Adequate lighting must be provided, both in the general work area where work equipment is used and at suitable places on and inside work equipment where it is appropriate to do so to allow safe operation, adjustment, cleaning, lubrication etc. The provision of higher lighting levels may be appropriate where eye protection is worn continuously in order to take account of the reduction in light level at the eye caused by the eye protector's lenses.

13. *Markings.*
Work equipment must be clearly marked with all the information necessary for it to be used safely. This includes confirmation of EC conformity, manufacturer's name and address, identification and marking of controls, directions of operation, maximum operating pressures, speeds and safe working loads, fluid levels, hazardous substances etc.

14. *Warning notices, markings and devices.*
In addition to the markings mentioned above, warning notices may be required. These may indicate a permanent hazard such as high temperature, voltage or noise levels. They may be temporary, eg indicating that the equipment is broken or is undergoing maintenance. Flashing lights, horns or sirens may be required to indicate malfunction or overload situations. Where such warnings are necessary they must be visible and available at all times. The information should be in a form approved by the Safety signs regulations 1980.

A pump imported into the UK from a manufacturer whose factory lies within the European Union, can be assumed to comply with all of the quoted legislation. It can also be assumed that a pump bought from the UK subsidiary of a non-EU foreign manufacturer will fully comply with the legislation. In each case this can easily be confirmed by the presence of the "EC mark". Where a pump is imported directly from a non-EU country, the responsibility for compliance with this legislation lies with the importer, who must process the necessary application to the approvals authority. In the case where a UK company buys a pump directly from a foreign manufacturer, the UK company is legally the "importer" and thus becomes responsible for all the legal duties.

Installation, setup and commissioning on a site for the company who controls the activities on that site (legally the "occupier"), by the pump manufacturer or supplying agent is usually considered to relieve the occupier of any responsibility for the work. This assumption is only correct, however, if the work is carried out by the visitors on a separate part of the site, with no access to the worksite by the occupier's own staff. If, however, the occupier's staff retain access to the workplace, so as to monitor the work, then legal duties are shared by both parties, with the major responsibility for the way the work is carried out falling on the occupier. The exact wording of the relevant contract document is of critical importance and will decide who will be prosecuted in the event of an accident.

Maintenance work carried out by employees and by visiting subcontract labour on site is always considered to be the responsibility of the site occupier.

Further problems arise when a pumpset is sent back to its manufacturer or agent for repair or refurbishment.

Often forgotten is the need to fasten the pump properly to the lorry which carries it. Reference should be made to the Department of Transport publication Safety of Loads on Vehicles published by HMSO (ISBN 0 11 550666 7).

It is normal practice to issue a "plant clearance certificate" before any plant item is allowed to leave a site, which confirms that all toxic and hazardous residues have been cleaned out and that the pump is now safe to be worked on without risk from contaminating substances. The pump must not leave the site with hazardous residues still inside it as they are liable to leak out during the journey.

The risks associated with the process of cleaning and washing out a pump before shipping off site for repair or disposal need to be assessed and appropriate procedures need to be instituted and followed. Compliance with a Permit to Work procedure may be appropriate. The residues from the cleaning process must then be discarded as hazardous waste. Contaminated water which has been used to clean up a sewage pump can easily be processed on site. Pumps, valves and pipework on petro-chemical installations, are likely to be contaminated with pyrophoric scale which must be kept wet due to its ability to catch fire on contact with air. Equipment at the upstream end of a refinery may also be significantly contaminated with LSA "low specifict (radio) active" scales. Removal of these substances may require the assistance of speciallist contractors and disposal of the residues will require compliance with waste control legislation.

If flammable solvents are used for cleaning residues from pump casings and impellers, and greases and oils from transmission systems, there is an obvious risk of fire and explosion. Non-flammable solvents avoid these risks, but impose a burden on the atmosphere and particularly on the ozone layer. 1.1.1 trichloroethane is now withdrawn because of its ozone depletion potential (ODP). Trichloroethylene, which is now being used, has a lesser ODP but is more toxic. If either of these substances is heated, for example in the burning end of a cigarette, they will break down into highly toxic phosgene gas.

The control of substances hazardous to health regulations 1987 (SI 1987/1657) usually referred to as COSHH, require employers to assess the risks posed by such substances to their employees and other persons. Once the assessment has identified the problem areas, exposure to hazardous substances must be prevented or controlled to a minimum level.

Work on a pump set must also involve the provision of safe access with handrails, ladders, scaffolding, safety harness attachment points etc.

SMSR regulations require proper sling attachment points to allow safe slinging of complete pumpsets and individual components, if they are too heavy to be manhandled. Welded or cast-in-place lugs and pad-eyes are often removed from component parts before the pump leaves the factory, thus causing problems for the maintenance staff. If they are needed during manufacture and assembly they should be left in place. They must be positioned so as to allow them to accept angular loadings during any upending and overturning operations during maintenance or dismantling. The holes in them must be large enough to be able to accept the appropriate sized shackle, despite the presence of several coats of paint. Drilled and tapped holes provided for eyebolts should be protected by blanking plugs and should be marked as eyebolt holes with the thread size clearly identified. Since dynamo type eyebolts can only be used with the load applied directly in line with the screwed pin, the marking should include reference to whether or not dynamo type eyebolts should be used. HSE publication PM 28 Eyebolts gives information on the use of eyebolts and warns of the many different thread types which are available. In the case of a complete pumpset which has its centre of gravity displaced from the centre of the unit, the manufacturer should identify the position of the C/G and the size, length and capacity of the slings which should be used for lifting so that the pump hangs level in its slings. In the case of pumps which are to be used for pumping acids, a warning should be given against the use of "Grade T" (grade 80) chain slings which are seriously affected by acid embrittlement. Special swivel lifting brackets may be necessary for eg down hole pumps so as to permit safe upending of the pump before it is dropped into the borehole. The length of a complete submersible motor and pumpset may impose excessive bending loads on the flanges during the upending process. It may be preferable to upend each component first and mate them in the vertical position. Suitable support brackets may need to be included in the installation equipment to hold the vertical motor securely while the crane is lifting the pump.

Pump houses are often provided with an overhead travelling crane or at the very least with an overhead runway beam on which may be mounted a beam trolley and suitable lifting appliance. Where these are not available, the building architect should give consent before chain blocks and other lifting appliances are used on the roof trusses. A "Blondin" is often used in these circumstances. This is a horizontal wire rope stretched tightly across the width of the building with a chain block hung from it so as to allow the block and its suspended load to be slid along the wire. Provided the loads on the anchorages are properly calculated and the architect permits, the use of a Blondin is acceptable. Buildings have, however, been destroyed by the excessive loads applied to the rope anchorages. Pumps which are installed outside may be accessible to a suitable mobile crane, but limited ground loadbearing capacity or the presence of underground culverts and pipelines may require the use of mattresses or extensive timbering to accept the loads imposed by the crane outriggers.

Special lifting equipment designed and manufactured in house must have a suitable factor of safety (usually 5:1) and must be proof tested to twice the working load limit for limits up to 50 tonnes. While this testing can legally be done in-house, the insurance company who will pay out in the event of failure of the lifting gear may insist on independant testing by a member of the Chain Testers Association of Great Britain Ltd.

Submersible and shaft-driven borehole pumps lifting salt water for cooling, firefighting and freshwater making purposes have to be removed for maintenance. Pumps and rising mains become heavily encrusted with silt, shellfish and weed, especially when they are installed inside a caisson. This will make them very much heavier than when first installed. If the pump is pulled out by a crane, which gives the operator no direct feel on the controls as to the weight he is lifting, there is a risk of breaking either the lifting gear or the rising main connections if the pump is cemented solidly into the caisson. The insulating flange fitted between a submersible pump and its motor is a common site of failure in these circumstances. Heat exchanger headers can easily be pulled off the tube bundle in this way. It may be safer to lift the pump out by hand, using chain blocks or lever ratchet chainpullers. The operator will be able to feel the presence of shellfish or silt by the excessive effort required to operate the equipment. The use of a load cell in the sling setup will show the actual load being applied to the lifting gear and, of course to the equipment and its slinging points.

EDITORIAL INDEX